# Basics of Hydraulic Systems

*Second Edition*

# Basics of Hydraulic Systems

*Second Edition*

By

Qin Zhang

CRC Press
Taylor & Francis Group
Boca Raton London New York

CRC Press is an imprint of the
Taylor & Francis Group, an **informa** business

CRC Press
Taylor & Francis Group
6000 Broken Sound Parkway NW, Suite 300
Boca Raton, FL 33487-2742



Printed and bound by CPI Group (UK) Ltd, Croydon, CR0 4YY

International Standard Book Number-13 978-1-1384-8466-5 (Hardback)



**Library of Congress Cataloging-in-Publication Data**
A catalog record has been requested for this book

**Visit the Taylor & Francis Web site at**
http://www.taylorandfrancis.com

**and the CRC Press Web site at**
http://www.crcpress.com

# Table of Contents

# *Preface*

Power transmission is the delivery of power from its point of generation to where it is deployed to do work. Hydraulic power transmission is one of the premier methods of power transmission and is applied in many machinery systems. This book is written as an instruction material suitable for both engineering and technical management students, as well as professionals in relevant industries, without prior knowledge or training in hydraulic power systems, allowing them to easily understand its contents. Therefore, this book is suitable as a primary textbook for those students with a college sophomore/junior academic standing, as well as continuing education material suitable for engineers and service representatives in industry.

This textbook covers the fundamentals of operating principles, configuration features, functionalities, and applications of core composing elements in typical hydraulic systems, and presents these materials in a systematic way. Energy transmission within a hydraulic system, ranging from power generation to distribution to deployment, is explained in the first seven-chapters. Chapter 1 introduces the basic concepts and theoretical basis of hydraulic power transmission. Chapter 2 focuses on describing the basic principles and configuration features of hydraulic pumps and explains how to control the power generation process on typical hydraulic pumps. Chapter 3 provides a thorough explanation of how different types of hydraulic control valves, separately or collectively, regulate the power distribution process in hydraulic systems. Hydraulic manifolds and the application of different types of conductors are also introduced in this chapter as part of an introduction of major components for power distribution. Chapter 4 discusses power deployment using either linear or rotary actuators. As a common application, hydrostatic transmissions are also introduced in this chapter. Chapter 5 deals with power storage and regeneration components and applications. Chapter 6 gives an introduction to auxiliary components, such as the reservoirs, filters, seals, and heat exchangers. Commonly used hydraulic fluids are also introduced in this chapter. Chapter 7 uses examples to describe how individual components can be connected differently to form hydraulic circuits of different functionalities.

To make it a commercial-free learning medium, this textbook explains operating principles, configuration features, functionalities, and applications of core hydraulic elements without relating to any specific products from particular manufacturers. All the graphical illustrations created in this book are intended solely for the purpose of explaining operating principles, configuration features, or functionalities of the elements or systems, and it is very important to remember that all these graphical illustrations *cannot be used as a design exemplar in engineering practice.*

Finally, I would like to express my special thanks to Mary Schultze, Ryan Kingdon, and Dr. Bo Jin for their proofreading prior to making this book available to students.

**Qin Zhang**
*University of Illinois at Urbana-Champaign*

# Preface to the Second Edition

I would like to acknowledge the Executive Editor for his warm encouragement in my efforts to update this book. The second edition addresses one of many valuable suggestions from users of the first edition, namely, the addition of some essential information on controls to help the reader better understand the technologies of electrohydraulic systems and controls. Two chapters, Chapters 8 and 9, covering hydraulic systems modeling and controls, were added for this purpose. It should be emphasized, however, that this second edition remains a textbook on hydraulic systems, not on automatic controls, as it merely provides the reader with some fundamental knowledge needed to understand automatic controls. It also gives the reader the capability of communicating with control system engineers in a professional manner. For those who wish to obtain the knowledge needed to design control systems for electrohydraulic systems, I would strongly suggest taking courses on control systems.

I would like to take this opportunity to express my deep thanks to many colleagues, who teach hydraulic system courses at different universities, for providing me with extremely valuable suggestions for making this textbook more comprehensive to meet the needs of students taking this course. I especially want to point out that it was my students who taught me how to present the materials in an easy-to-understand flow so that they could best follow and learn the subject matter. Finally, I would like to express my special thanks to Linda Root who helped me proofread the entire book of this second edition and thereby, in this second edition, present students with the best quality possible.

**Qin Zhang**
*Washington State University*

# About the Author

**Dr. Qin Zhang** is the Director of the Center for Precision and Automated Agricultural Systems (CPAAS) of Washington State University (WSU) and a Professor of Agricultural Automation in the Department of Biological Systems Engineering, WSU. His research interests are in the areas of agricultural automation, agricultural robotics, and off-road equipment mechatronics. Prior to his current position, he was a faculty member at the University of Illinois at Urbana-Champaign (UIUC), worked at Caterpillar Inc., and taught at Zhejiang Agricultural University in China, involved in hydraulic systems teaching, research and development in all those positions. Based on his research outcomes, he has authored/co-authored two textbooks, edited/co-edited three technical books, written nine separate book chapters, edited three conference proceedings, published over 160 peer-reviewed journal articles, and been awarded eleven U.S. patents. He is currently serving as the Editor-in-Chief for *Computers and Electronics in Agriculture*. Dr. Qin Zhang received his B.S. degree in engineering from Zhejiang Agricultural University (ZJU), China; M.S. degree from the University of Idaho (UI); and Ph.D. degree from the University of Illinois at Urbana-Champaign (UIUC), both in Agricultural Engineering. Dr. Qin Zhang is an American Society of Agricultural and Biological Engineers (ASABE) Fellow and is serving as a guest or an adjunct professor for nine other universities.

# 1

# *Introduction to Hydraulic Power*

## 1.1 Power Transmission and Hydraulic Systems

### 1.1.1 Power Transmission in Machinery Systems

A machinery system, in general, consists of a prime mover, a power transmission, and an implement end-effector. The prime mover is used to convert energy of various potential forms into the mechanical form to provide the needed power to drive the machinery; the implement end-effector is created to perform the designated work; and the power transmission is designed to deliver the regulated power from the prime mover to the end-effector. For example, an automobile uses an internal combustion engine to convert potential chemical energy carried in the fuel into mechanical power. A mechanical power transmission mechanism then delivers and regulates the power to the wheels to drive the vehicle at a desired speed. An electrical ceiling fan uses an electric motor to convert electrical energy, delivered using an electrical power line, into mechanical power to drive the fan. A hydraulic jack converts biological energy, carried by a person, into mechanical power via a hand pump to lift a heavy automobile. Common among these three examples is that all three devices convey a certain amount of power, using mechanical parts, electricity or pressurized fluids, from the prime mover to drive the end-effector to perform designated functions. These examples illustrate three basic methods of power transmission: mechanical, electrical, and fluid power transmissions.

A ***mechanical power transmission*** normally employs gears, belts, chains, and couplings to deliver energy and motion to perform designated tasks. One basic principle of mechanical power transmissions is to use less force, but a longer traveling distance, to perform the task. In other words, by using a mechanical power transmitting device, it is possible to use less force but not less work to perform a certain task. Depending on the type of mechanism applied, mechanical power transmission systems can offer the most diversified designs and allow us to use mechanical advantages in many different formats. However, this feature often turns the construction of such a mechanical system into a highly complicated task. Another noteworthy feature of mechanical systems is that the major form of energy loss during power transmission is the friction between contacted components. Because of the small amount of energy required to overcome the friction between mechanical components, mechanical systems normally offer high efficiency in power transmission.

An **electrical power transmission** applies electrical motors, relays, and wiring circuits to deliver and regulate the potential energy, and is one of the most effective and economical power transmission methods. However, the electrical power, normally generated at centralized power stations, often employs high voltage power grids to deliver electricity to remote users. Even with the technology available for storing certain amounts of electricity in batteries for mobile uses, the limited capacity of energy storage in today's battery technology has greatly restricted the applications of electrical power transmission on mobile equipment. Moreover, one should keep in mind that this obstacle may be removed by technological advancement in the near future.

A **fluid power transmission** uses pressurized fluids to deliver energy and often employs hydraulic/pneumatic cylinders or motors to convert the delivered energy into a mechanical form to perform useful work. The unique feature of fluid power transmission is its power-carrying media. Pressurized fluids can be shaped into any geometric shape, depending on the container, and delivers the power to all directions as needed, making it possible to place a fluid power system in a contained space and make it act like the muscle in a human body to move loads in required patterns. One example is power steering: pressurized hydraulic fluid is used to help the driver turn the wheels to steer the vehicle, thereby reducing the effort needed by the driver.

In terms of the type of fluids being used, a fluid power system can be further classified into a **hydraulic power system** if a liquid fluid is used or a **pneumatic power system** if a gas fluid is used. While these two types of fluid power systems have many features in common, the most distinguishing feature is the compressibility of liquids and gases. Liquids, due to their incompressible nature, can carry very high pressure with little change in their volume, which makes a hydraulic power system capable of transmitting a large amount of power using a small volume of liquid. In comparison, a pneumatic power system of similar size carries much less power, due mainly to its low operating pressure. Because the size and weight of a power transmission system are important design concerns for mobile equipment, the high-energy carrying capability of hydraulic power transmission systems over pneumatic ones makes hydraulic systems widely used for power transmission on mobile equipment.

### 1.1.2 Hydraulic Power Systems on Mobile Equipment

As defined by its name, a hydraulic power transmission system is designed to transmit energy from its source to its place of deployment using pressurized hydraulic fluids. This energy transfer is accomplished by a three-step conversion: mechanical kinetic energy is converted to pressure potential energy. This potential energy is delivered to where it is converted to work and then converted back to kinetic energy. Figure 1.1 illustrates the energy conversion process in a typical hydraulic power system.

As shown in Figure 1.1, a hydraulic power system uses a hydraulic pump to convert the mechanical power supplied by a prime mover (often an internal combustion engine on mobile equipment), which raises the energy level carried by the pressurized hydraulic fluid from zero to the maximum. The potential energy is first delivered to a control valve through a hose. This control valve then regulates the direction and amount of the pressurized fluid to different ports, with more hoses being used to deliver the pressurized fluid to an actuator to drive the load performing the desired work. It should be noted that some hydraulic energy will be lost during the transmission process due to the friction of the fluid flowing from the pump to the actuator, as well as leaking of the pressurized fluid.

Based on their functions, a hydraulic power system can be divided into the subsystems of power generation, distribution, deployment, and regulation. The **power generation** subsystem consists of mainly the hydraulic pump, with the main function of converting the kinetic energy supplied by the prime mover to the potential hydraulic energy in the pressurized fluids. The **power distribution** subsystem often employs various control valves to control and/or regulate the fluid pressure, flow rate, and directions to deliver the pressurized fluids to appropriate hydraulic actuators. The **power deployment** subsystem uses

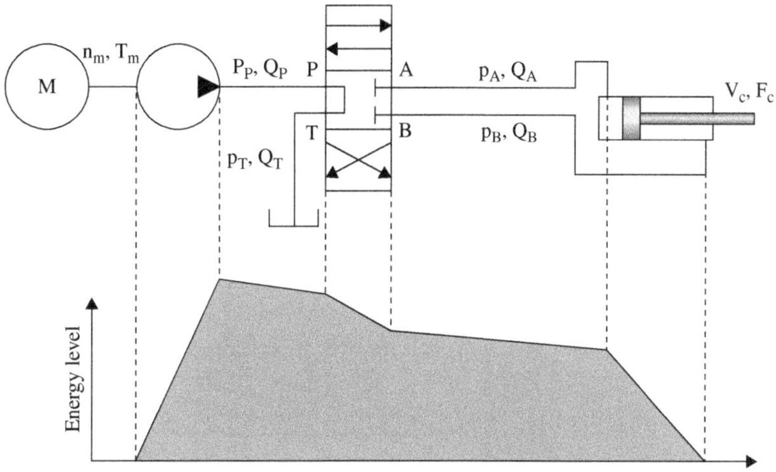


**FIGURE 1.1**
Illustration of the principle of energy transmission process in a hydraulic power system.

hydraulic actuators, commonly hydraulic cylinders or motors on mobile equipment, to utilize the potential energy carried by the pressurized fluids to move the loads. The **power regulation** subsystem normally includes the reservoir (often called the tank), hydraulic hoses, filters, and accumulators. All these components provide a sequence of supporting functions, including but not limited to energy storage and power regeneration.

Mobile equipment is designed mainly to perform various tasks in motion or needing frequent relocation, and is widely used in agriculture, construction, mining, and other field operations. Agricultural tractors, hydraulic excavators, bulldozers, backhoe loaders, tunnel drills, and feller bunchers are a few examples of mobile equipment. Different types of mobile equipment are designed to perform different tasks, and therefore their appearances and structures may be very different. For example, an agricultural tractor and a hydraulic excavator are different in both appearance and structure because the tractor is designed mainly to pull various implements to perform crop production tasks in motion and the hydraulic excavator is designed to perform earthmoving tasks of frequent relocation. However, these two types of machinery have one thing in common: both consist of a prime mover, a power transmission, and an implement end-effector. The most commonly used prime movers on mobile equipment are diesel engines, which convert the chemical potential energy of diesel fuel into mechanical kinetic energy to drive the load via power transmission. The power transmissions commonly use either mechanical or hydraulic systems to transmit the mechanical energy converted by the prime mover to the end-effector. Thus, the end-effectors are basically the tools that perform the designated tasks. This textbook introduces the fundamental principles and the basic configurations of mobile hydraulic systems and their major components.

As stated earlier, a mobile hydraulic system uses a pump, at least a control valve, a hydraulic actuator, a fluids reservoir, and connecting hoses to transmit power. A typical application on mobile equipment is a hydraulic steering system, as depicted in Figure 1.2. The control valve in a typical hydraulic steering system is often a hand pump. As shown in the figure, while steering, the steering wheel turns the hand pump via the shaft to direct the pressured fluid supplied from the hydraulic pump to different chambers of the cylinder actuator, which in turn drives the steering linkage to steer the wheels to complete the turn. This hydraulic system also uses a filter in the fluid return line.

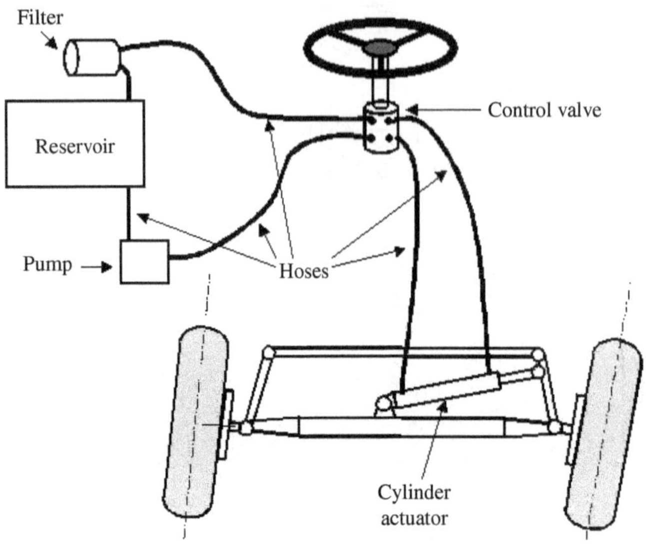


**FIGURE 1.2**
Configuration illustration of a typical hydraulic steering system on mobile equipment.

Heavily influenced by physical properties of the pressurized fluids, hydraulic power transmission systems can offer many attractive features for mobile applications and, therefore, have been widely applied on mobile equipment as the most common power delivery method for end-effector actuating. One of the most noteworthy features is the **unrestricted geometric shape formation capability,** which allows the hydraulic potential energy to be delivered to all directions with the same capacity. Another attractive feature is the **high power-to-weight ratio,** making it possible to build a compact power transmission system able to deliver sufficient power to drive heavy loads. Other features important for operation control are the fast response to control inputs; the capability of instantly starting, stopping or reversing the motion on the actuator; the capability of obtaining a constant force or torque at infinitively variable speeds in either direction with smooth reverses; and the ability of being stalled without damaging the system which provides very reliable overload protection. The major shortcomings of hydraulic systems are the difficulty to achieve an accurate speed conversion ratio, caused mainly by the hard-to-eliminate fluid leaking and compressibility and the generally lower efficiency of power transmitting than other means of power transmission.

### 1.1.3 Hydraulic Components and System Schematics

As shown in Figure 1.1, the major components in a typical hydraulic system include a hydraulic pump, a control valve, and a hydraulic cylinder. Because many hydraulic components are complicated in structure, the industry often uses some standardized graphical symbols to represent different hydraulic components in a system circuit. Such a system circuit, drawn using standard graphical symbols, is often called the **system schematic.** A simplified system schematic for a hydraulic implement system commonly seen on mobile equipment is shown in Figure 1.3. In this circuit schematic, one sees that all the major components are represented using graphical symbols. For example, item 4 is the symbol for a fixed-displacement pump, item 15 is the symbol for a directional control hydraulic

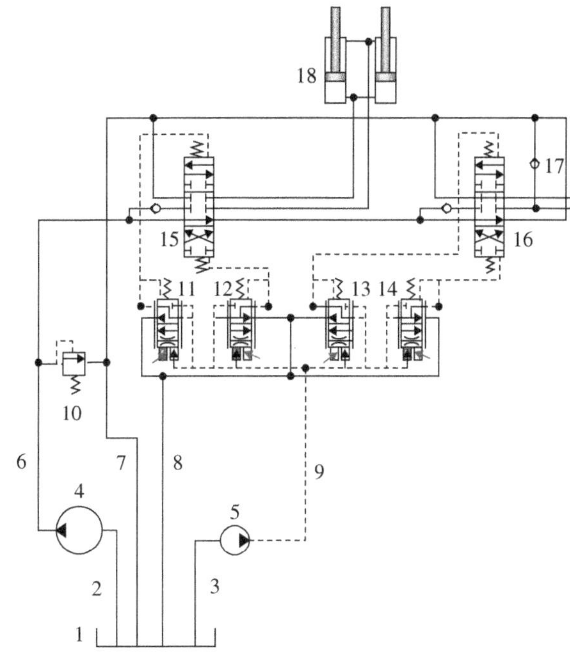


**FIGURE 1.3**
A simplified system schematic of a typical hydraulic implement system in mobile equipment.

valve, and item 18 is the symbol for a single-rod double-action hydraulic cylinder. The solid lines of 2, 6, 7, and 8 represent hydraulic hoses. From this system schematic, the readers may also find that those symbols, along with the associated lines, not only indicate how these components are connected, but more importantly show the basic function of those elements.

To ensure that these symbols can be widely understood by professionals all over the world, the industry, the government, and professional organizations have worked together to create a family of standard graphic symbols to represent fluid power components for fluid power system schematic drawings. In the United States, the National Fluid Power Association (NFPA) has coordinated the creation and modification graphical symbols for all the fluid power components through an industry-wide effort. The American National Standards Institute (ANSI) is responsible for coordinating the creation and change of these symbols and proofing the standard symbols as needed. The International Standards Organization (ISO) has the same responsibility for symbols used internationally and has issued an international standard on the graphical symbols for fluid power systems and components. Based on those standards, the symbol for a hydraulic component is used to present the function and connections of the component. It does not show the actual structure or parameters of the component. The connections indicated in a system schematic only show how two components are connected. It will not indicate actual installation locations of these components on equipment, but the symbols in a system schematic do normally show their neutral or initial positions of the represented components in the system. For example, all control valves in Figure 1.3 are presented in their neutral positions. A separate section later in this chapter (Section 1.4) will briefly introduce some of most commonly used standard graphical symbols to represent different hydraulic components.

## 1.2 Fundamentals of Hydraulic Power Transmission

### 1.2.1 Multiplication of Force

The reason fluids can transmit energy when contained is best stated by the French physicist, Blaise Pascal (1623–1662). Pascal made the following declaration, which is now often referred to as **Pascal's law**: Pressure in a confined body of fluid acts equally in all directions and at right angles to the containing surfaces. Pascal's law is one of the basic laws for fluid power and provides the foundation for formulating the principle of fluid static pressure transmission. This basic principle of fluid power systems can be explained using a hydraulic jack as an illustration. As shown in Figure 1.4, assume we are using the hydraulic jack to lift a heavy load, such as a car. The return valve (7) provides a designated pass to bleed the liquid from the large cylinder to release the load. Consequently, it must be turned off before operation of the jack can be accomplished. During jacking, one pushes the lever (5) down to create a force $F_1$ on the pump piston (3). The pump piston in turn pushes the liquid in the hand pump (4) out through a check valve (6) to the actuator cylinder (8). Due to the incompressibility of liquid, the additional liquid pumped into the large cylinder will then push the actuator piston (9) up and therefore lift the car.

During the car-lifting process, the hydraulic jack allows us to apply a small force on the lever to raise a heavy car. How can this jack help us accomplish the work we are normally unable to do? Pascal's law tells us: *when a force is applied on a confined container of liquid at rest, it will induce the same pressure on throughout the liquid in this container, and this pressure carried in a fluid at rest will transmit this force equally in all directions.* So the total force applied to a specific surface of a contained space is proportional to the area of this surface. Pascal's law defines the basic relationship between **force** ($F$), **pressure** ($p$), and **area** ($A$) in a hydraulic system: the force exerted on a surface equals the pressure exerted on the surface times the area of the surface. Mathematically, we can express the relationship as follows:

$$F = pA \tag{1.1}$$

Instead of using hydraulic cylinders, some hydraulic systems use hydraulic motors as actuators. In this case, the hydraulic force applied on the motor is a rotary force, often called **torque** ($T$). Similar to the force applied on a hydraulic cylinder, the torque on a

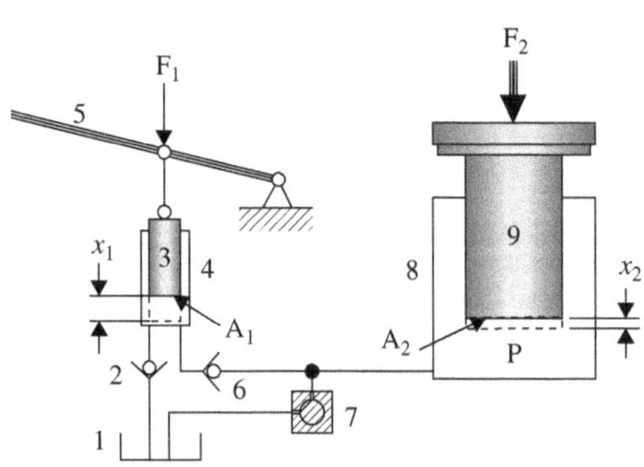


**FIGURE 1.4**
Concept illustration on the operating principle of a simple hydraulic jack.

motor can also be determined in terms of fluid pressure and motor **displacement** ($D_v$) using Pascal's law.

$$T = \frac{pD_v}{2\pi} \tag{1.2}$$

Using the same hydraulic jack example, based on Pascal's law, when a small force is applied on the small piston of the hand pump (3), the result will be a larger force on the actuator piston (9) because of the difference in piston areas. Another fact we can learn from this example is that when the pump piston (3, piston area $A_1$) moves downward for $x_1$, the amount of liquid volume ($V_1$) being pumped out from the hand pump (4) is:

$$V_1 = x_1 A_1 \tag{1.3}$$

Because this hydraulic jack is a confined system, all liquid pumped out of the pump cylinder can only be sent to the actuator cylinder. Due to its incompressibility, the liquid pumped into the actuator cylinder has to push the actuator piston up to get more space for itself. If the leakage is ignored, the volume of liquid entering the actuator cylinder is exactly the same as the volume of liquid pumped out of the pump cylinder. Because the piston area of the actuator cylinder ($A_2$) is greater than that of the pump piston ($A_1$), the inlet liquid will push the actuator cylinder for a smaller distance ($x_2$) inversely proportional to the ratio of piston areas. Their relationship can be defined using the following equation:

$$x_1 A_1 = x_2 A_2 \tag{1.4}$$

Dividing the travel distances on both sides of Eq. (1.4) by the time interval $t$ results in the traveling speeds of both pistons ($v_1$ and $v_2$). Readers may also find that the piston speed is proportional to the flow rate entering or leaving the cylinder and is proportional to the inverse of the piston areas:

$$v_1 A_1 = v_2 A_2 \tag{1.5}$$

One should pay special attention to an important phenomenon exhibited in the simple hydraulic system illustrated in Figure 1.4. The pump only delivers a certain volume of liquid to the cylinder from each action of pumping, and the back pressure acting on the check valve (6) determines how high the pump discharges pressure in order to pump liquid into the cylinder. This back pressure on the check valve equals the pressure in the cylinder chamber, which is solely determined by the external load acting on the piston (9) if the weight of the piston is ignored. This phenomenon reveals a fundamental fact of hydraulic power systems. The *operating pressure in a hydraulic system is determined only by the load,* and the pump produces only the flow not the pressure.

**Example 1.1: Application of Pascal's Law**

Assume the hydraulic jack depicted in Figure 1.4 has a small piston of 20 mm in diameter, and a large piston of 100 mm in diameter. If there is a 2500 N load acting on the large piston, what is the pressure of the confined fluid in the fluid chamber of the jack, and how much force is needed on the small piston to lift this load (assume friction and fluid leakage are negligible)?

a. Based on Pascal's law, the pressure of the confined fluid in the jack can be determined using the represented equation (1.1) as follows:

$$\begin{aligned} p &= \frac{F}{A} \\ &= \frac{2500}{\frac{\pi}{4}\times\left(\frac{100}{1000}\right)^2} \\ &= 318,471(Pa) \end{aligned}$$

b. Based on the calculated pressure, we can find the driving force on the small piston by applying Eq. (1.1) directly.

$$\begin{aligned} F &= pA \\ &= 318,471\times\frac{\pi}{4}\times\left(\frac{20}{1000}\right)^2 \\ &= 100(N) \end{aligned}$$

DISCUSSION 1.1: The results indicate that when friction and leakage are not considered, the pressure in the confined fluid is the same everywhere, and the forces acting on the two pistons are proportional to their effective areas. Does this mean that a hydraulic jack (in general speaking, a hydraulic system) can gain more force out of nothing? The next section will give an explanation.

### 1.2.2 Conservation of Energy

Like any power transmission system, a hydraulic power system is merely a method of energy transmission. From physics, we have learned that energy can neither be created nor be destroyed, although it can be transferred from one form to another. The energy conversion in a hydraulic power transmission system (excluding the prime mover) is twofold. Kinetic mechanical power is first converted to hydraulic potential energy via a hydraulic pump and then from hydraulic potential energy to kinetic mechanical power via a hydraulic actuator. In engineering, we often use a concept of **power** to quantitatively determine how much **energy** has been converted from one form to another in a certain period of time to do the **work.**

In a hydraulic power system, the energy is kept as potential energy due to both the elevation and pressure of the fluid, and as kinetic energy due to the flowing of the fluid. Theoretically, a hydraulic power system can be modeled as a contained pipeline with various tunings and different sizes (Figure 1.5). Because the pressurized liquid within the pipeline is contained by the pipe wall, the energy transmission can occur only at cross sections 1 and 2, namely, the inlet and outlet ports of the pipeline as shown in Figure 1.5. Therefore, an energy balance equation can be created to describe the energy conversion within this contained fluid power system as follows:

$$gz_1+\frac{v_1^2}{2}+\frac{p_1}{\rho}=gz_2+\frac{v_2^2}{2}+\frac{p_2}{\rho} \tag{1.6}$$

where $z_1$ and $z_2$ are the elevation of fluid surfaces; $v_1$ and $v_2$ are the velocity and $p_1$ and $p_2$ are the pressure of the fluid at cross sections 1 and 2; $g$ is the acceleration of gravity; and $\rho$ is the fluid density.

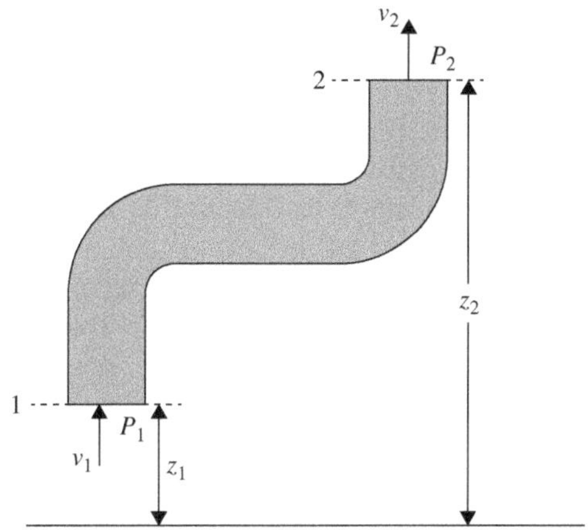


**FIGURE 1.5**
Illustration of the principle of three forms energy within a control volume of fluid.

Equation (1.6) is often called **Bernoulli's equation**, one of the fundamental equations for describing energy conversion within hydraulic power transmission systems. In general, Bernoulli's equation states the energy conservation in a perfect, frictionless fluid under steady-state conditions. In a typical mobile hydraulic system, the elevation difference between any two points is limited, resulting in a very small portion of its potential energy being affected by the elevation difference. Therefore, it is often acceptable in energy conservation analysis for mobile hydraulic systems to ignore the contribution of elevation. Bernoulli's equation can then be simplified as follows:

$$\frac{2}{\rho}(p_1 - p_2) = v_2^2 - v_1^2 \tag{1.7}$$

This equation reveals that during operation, the potential energy (in the form of pressurized fluid) in a hydraulic system is commonly transferred to kinetic energy (in the form of fluid velocity), where the decrease in pressure will result in an increase in velocity, and vice versa. An illustrative example of the simplified Bernoulli's equation is demonstrated in the case of a hydraulic orifice, normally a small fluid passage area, in a fluid transmitting line. As shown in Figure 1.6, when the fluid flows through an orifice, the velocity of the fluid will be increased due to the reduction of the fluid passage area. Such a flow velocity increase will result in a pressure drop across the orifice. After the fluid flows through this orifice

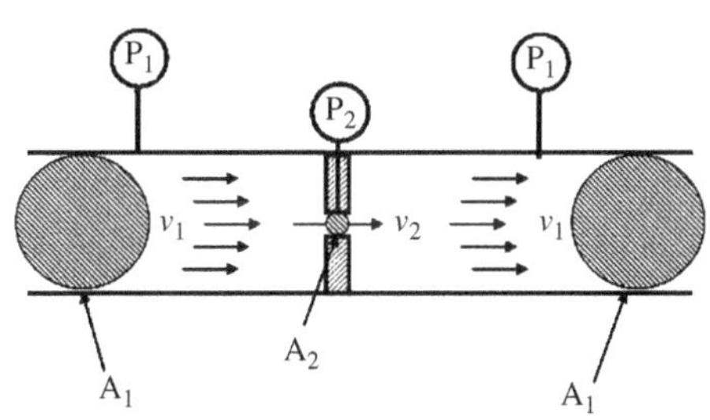


**FIGURE 1.6**
Concept illustration of pressure and flow velocity variation through a hydraulic orifice.

and the fluid passage area resumes its original size, the flow velocity will also reduce to its original value, along with the pressure. This phenomenon is very useful because it provides the theoretical basis for performing hydraulic power transmission control.

An **orifice equation**, derived from Bernoulli's equation, is commonly used to estimate the flow rate passing through the orifice in terms of the measurable pressure drop across the orifice. Some basic assumptions made to derive this orifice equation include: (1) the orifice is a small round hole on a thin wall; (2) the orifice area is much smaller than the upstream and downstream flow passage areas; and (3) the upstream flow velocity is negligible because it is much lower than the flow velocity in the orifice. Based on the above assumptions, the simplified Bernoulli's equation can be represented as:

$$v_2 = \sqrt{\frac{2}{\rho}(p_1 - p_2)} \tag{1.8}$$

The flow rate passing through the orifice can, therefore, be determined using the following orifice equation:

$$Q = C_d A \sqrt{\frac{2}{\rho}(p_1 - p_2)} \tag{1.9}$$

where $Q$ is the flow rate, $\rho$ is the fluid density, $A$ is the orifice area, and $C_d$ is the orifice coefficient, used to determine the effective flow passage area of the orifice due to the flow contraction. In engineering practice, $C_d$ is often selected between 0.6 and 0.8, depending on the shape of the orifice.

**Example 1.2: Application of Orifice Equation**

The pressure drop between the upstream flow and the in-orifice flow (as shown in Figure 1.6) is 1.0 MPa. Compare the flow rate passing through the orifice when a sharp-edged orifice of 10 mm in diameter or a squared-edged orifice of 12 mm in diameter is used. (Assume the specific fluid density is 900 $kg \cdot m^{-3}$.)

a. When a sharp-edged orifice is used, an orifice coefficient of 0.8 is commonly used, and the flow rate can be determined as follows:

$$\begin{aligned} Q &= C_d A_2 \sqrt{\frac{2}{\rho}\Delta p} \\ &= 0.8 \times \frac{\pi}{4} \times 0.010^2 \times \sqrt{\frac{2}{900} \times 1000000} \\ &= 0.0030\left(m^3 \cdot s^{-1}\right) = 180\left(L \cdot \text{min}^{-1}\right) \end{aligned}$$

b. When a square-edged orifice is used, an orifice coefficient of 0.6 is often used, and the flow rate can be determined as follows:

$$\begin{aligned} Q &= C_d A_2 \sqrt{\frac{2}{\rho}\Delta p} \\ &= 0.6 \times \frac{\pi}{4} \times 0.012^2 \times \sqrt{\frac{2}{900} \times 1000000} \\ &= 0.0032\left(m^3 \cdot s^{-1}\right) = 192\left(L \cdot \text{min}^{-1}\right) \end{aligned}$$

DISCUSSION 1.2: As a flow control device, the shape of an orifice, other than the size of the orifice and the pressure drop across the orifice, is a nonignorable factor in affecting the amount of fluid flowing through the orifice.

### 1.2.3 Continuity of Fluids

One fundamantal principle of hydraulic power transmission is that the fluid flows continuously within the system. This continuous flow principle plays an important role in hydraulic system analysis. A **flow continuity equation** is often used to state the situation that for steady flow in a control volume, the total inlet flow rate is the same as the total outlet flow rate. A control volume in a hydraulic system is a boundary of the section of interest, which represents where and how the mass flow balance is performed. Again, let us use the hydraulic jack depicted in Figure 1.4 as an example to explain the fluid continuity principle. Define the volume of fluid being pumped out by the hand pump at every stroke as the control volume being transmitted in the hydraulic jack. As defined by Eq. (1.5), the outlet flow from the hand pump ($Q_1$) can be determined using the following equation:

$$Q_1 = v_1 A_1 \tag{1.10}$$

Because the hand pump has only one outlet, according to the fluid continuity principle, the inlet volumetric flow rate of the actuator cylinder ($Q_2$) should be the same as the outlet flow rate from the hand pump under the ideal condition (ignore the leakage).

$$Q_1 = Q_2 = v_2 A_2 \tag{1.11}$$

Equation (1.11) is the fluid continuity equation for hydraulic jack applications. It shows that, under a certain flow rate, a smaller piston area results in a higher stroking velocity and vice versa. The fluid continuity equation is one of the fundamental equations for hydraulic system analysis. To apply the fluid continuity equation in fluid distribution analysis without loss of generality, the control volume of fluid in a "T" connector, commonly seen in practical fluid power systems, is shown in Figure 1.7. The fluid continuity equation for this "T" connector can be written as follows:

$$q_1 = v_1 A_1 = v_2 A_2 + v_3 A_3 = q_2 + q_3 \tag{1.12}$$

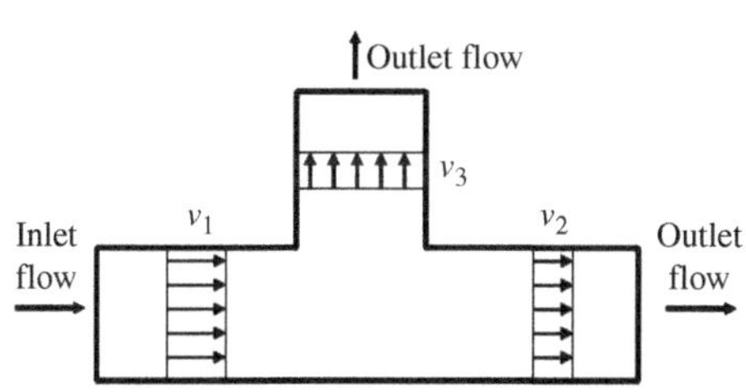


**FIGURE 1.7**
Concept illustration of continuity of fluid in a typical "T" type connector.

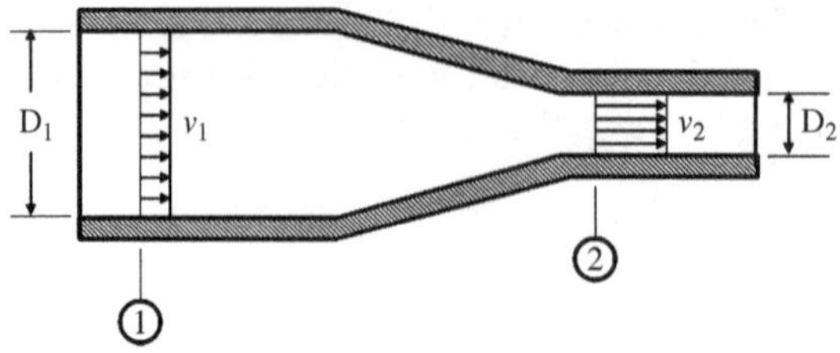


**FIGURE 1.8**
Concept illustration of continuity of flow in a changing cross section pipeline.

**Example 1.3: Application of Fluid Continuity Equation**

For a pipeline of $D_1 = 25$ mm, $D_2 = 15$ mm, and $v_1 = 2.0\ \text{m}\cdot\text{s}^{-1}$ (as shown in Figure 1.8), find (a) the volumetric flow rate Q and (b) the fluid velocity at cross section 2.

According to the fluid continuity principle, the volumetric flow rate should be the same regardless of the size of cross section if the fluid between two cross sections is confined.

a. The volumetric flow rate at cross section 1:

$$\begin{aligned} Q_1 &= v_1 \frac{\pi}{4} D_1^2 \\ &= 2.0 \times \frac{\pi}{4} \times 0.025^2 \\ &= 0.00098\left(m^3 \cdot s^{-1}\right) = 58.9\left(L \cdot \text{min}^{-1}\right) \end{aligned}$$

b. The fluid velocity at cross section 2:

$$\begin{aligned} v_2 &= v_1 \times \frac{\frac{\pi}{4} D_1^2}{\frac{\pi}{4} D_2^2} \\ &= 2.0 \times \frac{0.025^2}{0.015^2} \\ &= 5.56\left(m \cdot s^{-1}\right) \end{aligned}$$

DISCUSSION 1.3: The fluid continuity equation shows that the smaller the pipe size, the higher the fluid velocity, and vice versa.

## 1.3 Energy and Power in Hydraulic Systems

### 1.3.1 Energy Conversion in Hydraulic Systems

As illustrated in Figure 1.1, the total energy carried by the pressurized fluid reaches the highest level at the pump-discharge port. This energy level decreases as the fluid flows away from the port due to all kinds of energy losses. Such losses indicate that energy conversions occur during the delivery process. Figure 1.9 illustrates the energy flow in a typical hydraulic system during the power delivery process. To deliver the potential energy

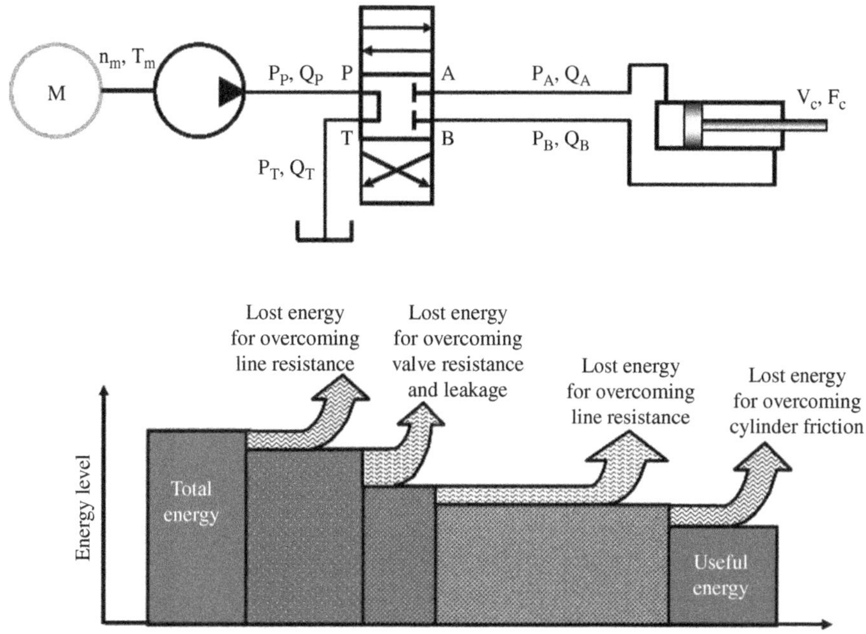


**FIGURE 1.9**
Concept illustration of energy conversions during a typical hydraulic power delivery process.

carried by the pressurized fluid to the actuator, there are various resistances and losses to be overcome, such as the line resistance, the valve resistance, and the friction on the actuator. To overcome those resistances, a certain amount of energy will be consumed. Because this amount of energy is not used to drive the load—in other words is not converted into kinetic energy to do useful work—it is always converted into thermal energy and results in a temperature increase in hydraulic fluids.

If there is no load to be driven by the pressurized hydraulic fluid, such fluid must be released from the system, often through a line-relief valve or other flow control means, to avoid excess pressure build-up within the system. Similar to the energy used to overcome resistance, the released hydraulic fluid does not perform any useful work. Instead, it consumes all the potential energy to overcome the resistance for releasing, which converts the potential energy into a thermal form. Therefore, it is necessary to design a hydraulic system with a minimal release of pressurized fluid to achieve high-energy efficiency. More discussions on energy efficiency enhancement methods and approaches will be provided in later chapters.

### 1.3.2 Hydraulic Power and Efficiency

From the discussion of energy conservation laws in the previous sections, we have learned that the power transmission in a typical hydraulic system includes the conversion from mechanical kinetic energy to hydraulic potential energy and reverse. One very important fact about this power transmission is that the output mechanical power is always less than the input level. The ratio of the output mechanical power to the input value is defined as the efficiency of the hydraulic power transmission system. To analyze the efficiency, it is important to know how the mechanical and hydraulic powers are determined in a hydraulic system.

The **mechanical power** in a typical hydraulic system is presented in the form of input power to drive the hydraulic pump and output power from the hydraulic actuator to drive the load. The input power to a hydraulic pump and the output power from a hydraulic motor are always determined by the torque and angular velocity using the following equation.

$$P_m = T\omega \tag{1.13}$$

where $P_m$ is the mechanical power, $T$ is the external torque applied on the shaft of either pump or motor, and $\omega$ is the angular velocity of the shaft. In Eq. (1.13), the torque is measured by $N \cdot m$, the angular velocity is in $s^{-1}$, and the unit of power is $W$ in SI units.

Different from the mechanical power, the **hydraulic power** is always determined by the system pressure, and the volumetric flow rate is as follows:

$$P_h = pQ \tag{1.14}$$

where $P_h$ is the hydraulic power, $p$ is the system pressure, and $Q$ is the volumetric flow rate. The unit of the pressure is Pa, the flow rate is $m^3 \cdot s^{-1}$, and the unit of power is $W$ in SI units.

In a case where a hydraulic cylinder is used as the actuator in a hydraulic system, the output mechanical power can be determined using a different equation in terms of the pressure difference in the cap-end chamber and the rod-end chamber of the cylinder, along with the bore and rod sizes of the cylinder $(A_1$ and $A_2)$, as defined in the following equation:

$$P_m = (p_1A_1 - p_2A_2)v \tag{1.15}$$

where $p_1$ and $p_2$ are the cylinder cap-end and rod-end chamber pressures, and $v$ is the piston moving velocity. As in previously defined power equations, the unit of the pressure is Pa, the velocity is $m \cdot s^{-1}$, and the power is $W$ in this equation when SI units are used.

Up to this point, we have ignored all the losses when analyzing the power transmission in a hydraulic system. In actual system analysis, we will have to take those losses into consideration. While the sources of energy loss during a typical power transmission are diverse in form, one needs to remember only the fundamental rule in energy balancing that the summation of total energy losses within a device and the remaining useful energy output from this device is always equal to the input energy to the device. Based on this rule, we can easily figure out that to compute how much power is needed to drive a device, it is necessary to request more power than the theoretical power the device can deliver, and when calculating the power from a device to drive a load, the available power is always less than the theoretical output power from the device. For example, when we compute the mechanical power required to drive a pump, we need to take the pump efficiency into consideration by adding **pump efficiency** $(\eta_p)$ to Eq. (1.14) to form a new equation, as follows:

$$P = \frac{pQ}{\eta_p} \tag{1.16}$$

When we compute the mechanical power available to drive a load, we need to add the **motor efficiency** $(\eta_m)$ to the original equation in a different way as follows:

$$P = pQ\eta_m \tag{1.17}$$

**Example 1.4: Calculating Hydraulic Power**

A hydraulic pump delivers 50 L/min flow at 9000 kPa. How much hydraulic power does this pump deliver? If the overall efficiency of this pump is 90%, how much mechanical power is needed to drive the pump? If all the pressurized fluid is delivered to a motor to drive a load, what is the maximum useful power the motor can deliver if the motor has the same overall efficiency as the pump?

a. The hydraulic power the pump delivers:

$$\begin{aligned} P &= pQ \\ &= 9000\times 10^3 \times \frac{50\times 10^{-3}}{60} \\ &= 7500(W) = 7.50(kW) \end{aligned}$$

b. The mechanical power needed to drive the pump:

$$\begin{aligned} P &= \frac{pQ}{\eta_p} \\ &= \frac{7.50}{0.9} \\ &= 8.33(kW) \end{aligned}$$

c. The mechanical power the motor can provide to drive the load:

$$\begin{aligned} P &= pQ\eta_m \\ &= 7.50\times 0.9 \\ &= 6.75(kW) \end{aligned}$$

DISCUSSION 1.4: Because a hydraulic system delivers energy by means of the pressurized fluid, the power transmitted by a hydraulic system is therefore determined by the flow rate and the pressure. It also needs to be remembered that the pump determines only how much flow it can deliver, and the system pressure is determined by the load.

## 1.4 Standard Graphical Symbols for Hydraulic System Schematics

A typical hydraulic system is constructed using many components. According to their functionalities in the system, these components can normally be classified as power generation, power distribution, power deployment, power regulation, and fluid conditioning components. Interconnecting these components in a certain manner will make a hydraulic system suitable to perform specially designed functions. However, if the interconnection logic of a specific system is conveyed using a traditional cutaway engineering drawing, the preparation of such a drawing would always be extremely laborious. To solve this problem, a Joint Industry Council (JIC) of the fluid power industry agreed to develop a set of JIC symbols to represent different components to simplify the logic drawing of a hydraulic system circuit. Those symbols were later standardized by NFPA and adopted by both ANSI and ISO as an international standard.

**TABLE 1.1**

ISO/ANSI Standard Symbols of Commonly used Power Generation Components.

| | | | |
|---|---|---|---|
|  | Fixed-displacement hydraulic pump |  | Variable-displacement hydraulic pump |
|  | Bi-directional fixed-displacement hydraulic pump |  | Bi-directional Variable-displacement hydraulic pump |
| | |  | Pressure-compensated Variable-displacement hydraulic pump |

It should be noted that the ISO symbols are currently *representing only the functionality and the connection* of hydraulic components. The structural parameters of these components, or their actual locations, are not provided in system schematics represented using those symbols. However, it does indicate the initial or neutral position of a component before the system is in operation.

Table 1.1 lists the ISO/ANSI symbols of some of the most commonly used power generation components. The two symbols listed in the left column represent **fixed-displacement pumps**. The outward triangle represents the fluid being pumped from the components. The double triangles indicate that the pump can be operated in both directions. An additional arrow on the symbols in the left column indicates that the displacement of these pumps is adjustable. This means that all those symbols represent **variable-displacement pumps**, either uni-directional, bi-directional, or pressure-compensated. The operation principles of different types of pressure-compensated variable-displacement pumps are described in detail in Chapter 2.

Another category of basic symbols is those representing power distribution components. Table 1.2 lists a few of the most commonly used symbols in this category. The symbols in the left column represent hydraulic lines. One distinguishing feature between symbols representing **working lines** (also called main lines or conduct lines) and **pilot lines** is the use of solid lines for the former and dashed lines for the latter. When two lines are connected at a specific location, a solid node is used to represent such a connection. Otherwise, a simple crossing of lines indicates that the lines are unconnected. The top four symbols in the middle column of Table 1.2 are used to represent **pressure control valves**. One common feature of this category of valves is that they are all operated in terms of pressure. (A detailed explanation of these valves is given in Chapter 3.) The bottom two symbols in the middle column are simple **flow control valves**. The most versatile power distribution components are probably **directional control valves**. The right column gives a few examples of symbols commonly used to represent this category of valves. A typical valve symbol can provide the basic functional information, such as the number of ports and normal operational positions of a valve.

The basic rules for directional control valve interpretation are that the number of closed envelopes represent the number of normal operational positions, and the number of intersections indicates the number of ports or connections. For example, the first

**TABLE 1.2**

ISO/ANSI Standard Symbols of Commonly used Power Distribution Components.

| | | | | | |
|---|---|---|---|---|---|
| | Working line |  | Check valve | 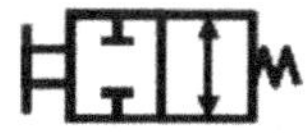 | Manual operated two-position two-way valve |
|  | Lines crossing |  | Shuttle valve |  | Button-operated two-position three-way valve |
|  | Lines joining |  | Pressure relief valve |  | Mechanically operated two-position four-way valve |
|  | Line with fixed restriction |  | Pressure reducing valve |  | Lever-operated three-position four-way valve |
|  | Flexible line |  | Adjustable flow control valve |  | Solenoid-operated tandem-center proportional valve |
| | Pilot line |  | On-off valve |  | Pilot-operated three-position six-way valve |

symbol in the right column consists of two closed envelopes, with two intersections on two opposite edges of each envelope. Based on the representation rules, this symbol represents a two-position, two-way directional control valve. When two parallel lines are added to the envelope block, as shown in the fifth symbol from the top in the right column, it indicates that this valve has an infinite "normal" operating position from its neutral position to the maximum opening, with its flow passage area in proportion to the opening of the valve, and therefore is often called a **proportional control valve**. In addition, the actuating method of such a valve is often depicted in a typical symbol for directional control valves.

The third group of hydraulic components is **power deploying components**. Table 1.3 lists a few symbols of the most commonly used power deployment components, namely,

**TABLE 1.3**

ISO/ANSI Standard Symbols of Commonly used Power Deployment Components.

| | | | |
|---|---|---|---|
| 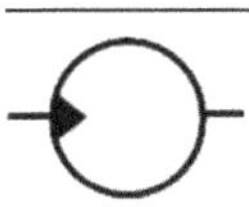 | Fixed-displacement hydraulic motor | 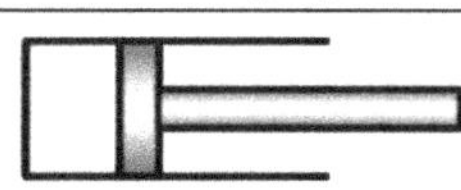 | Single-acting single-rod cylinder |
|  | Bi-directional fixed-displacement hydraulic motor | 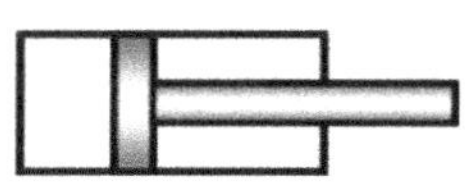 | Double-acting single-rod cylinder |
|  | Variable-displacement hydraulic motor | 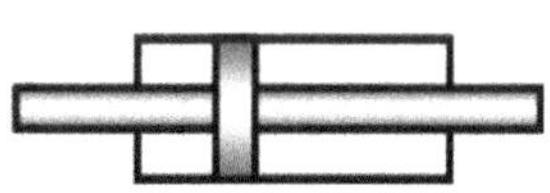 | Double-acting double-rod cylinder |
| 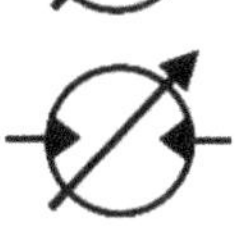 | Bi-directional Variable-displacement hydraulic motor | 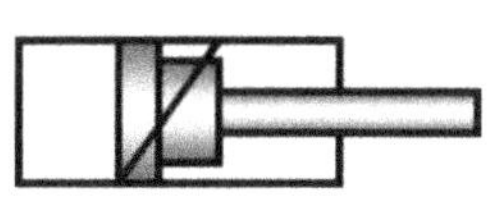 | Double-acting single-rod adjustable cushion cylinder |

**TABLE 1.4**

ISO/ANSI Standard Symbols of Commonly used Power Storage components.

| | | | |
|---|---|---|---|
| 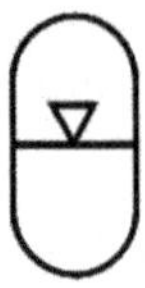 | Gas-charged accumulator | 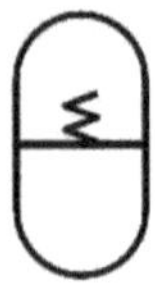 | Spring-loaded accumulator |

hydraulic motors and cylinders. The four symbols in the left column are four types of **hydraulic motors**. Similar in form to those used to represent pumps, but with the triangle pointed in an opposite direction, those motor symbols also provide the basic information of the characteristics of represented hydraulic motors, including whether it is a fixed- or a variable-displacement motor and whether or not the motor can input the fluid in both directions. The right column shows the symbols of most commonly used **hydraulic cylinders**, with their configuration characteristics.

The most commonly used **power storage components** are hydraulic accumulators. Table 1.4 contains the two most popular types of hydraulic accumulators: namely, the gas-charged and the spring-loaded accumulators.

Other than the main components, a complete hydraulic system needs to use many auxiliary components, such as hydraulic reservoirs, hydraulic filters or strainers, and heaters or coolers. Table 1.5 provides a list of fluid conditioning components, examples of auxiliary components.

More standard symbols of hydraulic components can be found in two ISO standards: *ISO 4391: Hydraulic fluid power—Pumps, motors and integral transmissions—Parameter definitions and letter symbols* and *ISO 5859: Aerospace—Graphic symbols for schematic drawings of hydraulic and pneumatic systems and components.*

## 1.5 Units and Unit Conversion in Hydraulic Systems

As with any other physical systems, a hydraulic power system involves many physical parameters, including pressure, force, velocity, density, temperature, and time. Traditionally, the **standard unit system,** which is based on the old ***British unit system*** (or often called **English unit system**), is widely used in the fluid power industry. With increasing globalization, the **SI unit system**, an international system of units approved by ISO, is replacing the old British unit system as the sole internationally

**TABLE 1.5**

ISO/ANSI Standard Symbols of Commonly used Fluid Conditioning Components.

| | | | | | |
|---|---|---|---|---|---|
| 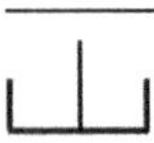 | Reservoir vented | 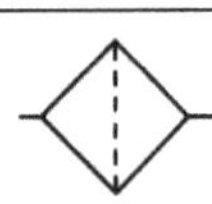 | Hydraulic filter or strainer | 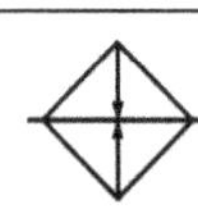 | Heater |
| 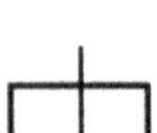 | Reservoir pressurized | | | 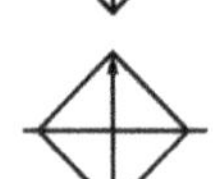 | Cooler |

accepted unit system. The transition from the traditional British unit system to the SI unit system will take considerable time and is a most challenging problem for many people.

One of the most commonly calculated system parameters in analyzing hydraulic power transmission is the power, both in mechanical form and in hydraulic form. One noteworthy difference between scientific computation and engineering calculation is that in engineering practice, people often use rotational speed ($n$, revolutions per minute or rpm) instead of the angular velocity ($\omega$, $s^{-1}$) to calculate the mechanical power. Therefore, Eq. (1.13) can be represented as one of the following two forms in engineering design practices, depending on the unit system being used:

In SI units:

$$P_m = \frac{Tn}{9550} \tag{1.18}$$

In British units:

$$P_m = \frac{Tn}{5252} \tag{1.19}$$

where the mechanical power is measured by $kW$ or $hp$ and the torque is $N \cdot m$ or $ft \cdot lbf$ in SI or the British unit system, respectively, with the rotating speed $rpm$ in both unit systems.

Similar to mechanical power calculation, one can also use one of the following two forms to replace Eq. (1.14) for calculating the hydraulic power delivered in a hydraulic system, depending on which unit system is being used:

In SI units:

$$P_h = \frac{pQ}{60000} \tag{1.20}$$

In British units:

$$P_h = \frac{pQ}{1714} \tag{1.21}$$

where the hydraulic power is measured by $kW$ or $hp$, the pressure is $kPa$ or $psi$, and the flow rate is $L \cdot \min^{-1}$ or $gpm$ in SI or British unit system, respectively.

The above equations indicate that the mechanical and hydraulic powers in a hydraulic system can be calculated using different equations, in terms of the unit system been used. But the conversion factors of those powers between SI and old British units are the same and can be determined based on the definition of power—the rate of performing work. The conversion factor of power is 1.34 $hp \cdot kW^{-1}$ if converting from SI to British unit, or 0.746 $kW \cdot hp^{-1}$ if converting from British to SI unit. Similar to power, most other system parameters used to describe the operation state of a hydraulic system can be measured using either SI or British units and can be converted back and forth using a conversion factor. Table 1.6 collects the conversion factors of common parameters used in analyzing hydraulic systems.

**TABLE 1.6**

Conversion Factors of Common Parameters of Hydraulic Systems.

| Parameter | SI Unit | Conversion Factor | British Unit | Conversion Factor | SI Unit |
|---|---|---|---|---|---|
| Length | meter (m) | 3.28 | foot (ft) | 0.305 | meter (m) |
| Volume | liter (L) | 0.264 | gallon (Gal) | 3.785 | liter (L) |
| Mass | kilogram (kg) | 2.2 | Pound (lbm) | 0.454 | kilogram (kg) |
| Force | newton (N) | 0.225 | pound force (lbf) | 4.45 | newton (N) |
| Torque | newton-meter (N·m) | 0.74 | pound-foot (lb-ft) | 1.36 | newton-meter (N·m) |
| Pressure | kilopascal (kPa) | 0.145 | pound per inch$^2$ (psi) | 6.89 | kilopascal (kPa) |
| | bar | 14.5 | psi | 0.069 | bar |
| Power | kilowatt (kW) | 1.34 | horsepower (hp) | 0.746 | kilowatt (kW) |
| Energy | kilojoule (kJ) | 0.948 | British-thermal-unit (BTU) | 1.055 | kilojoule (kJ) |

**Example 1.5: Unit Conversion**

Assume a hydraulic cylinder has a diameter of 2.5 in; compute the area of the piston. If the cylinder lifts an 8836 lbf load, what pressure will be developed in the system? What is the pressure in Pascal units?

a. The piston area can be calculated in terms of the diameter of the cylinder:

$$\begin{aligned} A &= \frac{\pi \cdot D^2}{4} \\ &= \frac{3.14 \times 2.5^2}{4} \\ &= 4.91\left(in^2\right) \end{aligned}$$

b. The system pressure can be determined using Eq. (1.1):

$$\begin{aligned} p &= \frac{F}{A} \\ &= \frac{8836}{4.91} \\ &= 1800(psi) \end{aligned}$$

c. A pressure unit conversion can be done by using the conversion factor (CF) from British unit (BU) to SI unit listed in Table 1.6:

$$\begin{aligned} p_{SI} &= p_{BU} \times CF \\ &= 1800 \times 6.89 \\ &= 12400(kPa) = 12.4(MPa) \end{aligned}$$

**DISCUSSION 1.5:** Unit conversion can easily be done by multiplying a conversion factor to the parameter to be converted, as shown in the example.

## References


1. Akers, A., Gassman, M., Smith, R. *Hydraulic Power System Analysis.* CRC Press, Boca Raton, FL (2006).
2. Backé, W. The present and future of fluid power. *Proc Instn Mech Engrs: J. Systems and Control Engineering,* 207: 193–212 (1993).
3. Batchelor, G.K. *An Introduction to Fluid Dynamics* .Cambridge University Press, Cambridge, UK (2000).
4. Burrows, C.R. Fluid power—progress in a key technology. *JSME Int. J., Series B,* 37: 691–701 (1994).
5. Cundiff, J.S. *Fluid Power Circuits and Controls: Fundamentals and Applications.* CRC Press, Boca Raton, FL (2002).
6. Esposito, A. *Fluid Power with Applications (6th Ed.).* Prentice Hall, Upper Saddle River, NJ (2003).
7. Goering, C.E., Stone, M.L., Smith, D.W., Turnquist, P.K. *Off-road Vehicle Engineering Principles.* ASAE, St. Joseph, MI (2003).
8. Guan, Z. *Hydraulic Power Transmission Systems* (in Chinese). Mechanical Industry Press, Beijing, China (1997).
9. Hedges, C.S. *Industrial Fluid Power (3rd Ed.).* Womack Educational Publications, Dallas, TX (1988)
10. Hydraulics & Pneumatics. *Fluid Power Basics.* http://www.hydraulicspneumatics.com/200/FPE/IndexPage.aspx. Accessed on November 20 (2006).
11. International Organization for Standardization. *ISO 4391: Hydraulic fluid power—Pumps, motors and integral transmissions—Parameter definitions and letter symbols.* ISO, Geneva, Switzerland (1983).
12. International Organization for Standardization. *ISO 5859: Aerospace—Graphic symbols for schematic drawings of hydraulic and pneumatic systems and components.* ISO, Geneva, Switzerland (1991).
13. Keller, G.R. *Hydraulic System Analysis.* Penton Media, Cleveland, OH (1985).
14. Lu, Y. Historical progress and prospects of fluid power transmission and control. *Chinese J. Mechanical Engineering* (in Chinese), 37: 9 (2001).
15. McClay, D., Martin, H.R. *The Control of Fluid Power.* John Wiley & Sons, New York, (1973).
16. Manring, N.D. *Hydraulic Control Systems.* J. Wiley & Sons, New York (2005).
17. Merrit, H.E. *Hydraulic Control Systems.* J. Wiley & Sons, New York (1967).
18. NFPA. *Fluid Power Training: Basic Hydraulics.* NFPA, Milwaukee, WI (2000).
19. Pease, D.A. *Basic Fluid Power.* Prentice Hall, Englewood Cliffs, NJ (1967).
20. Scott, T.E. *Power Transmission: Mechanical, Hydraulic, Pneumatic, and Electrical.* Prentice Hall, Upper Saddle River, NJ (2000).
21. Stringer, J. *Hydraulic Systems Analysis: An Introduction.* John Wiley & Sons, New York (1976).
22. Tanaka, H. Fluid power control technology—present and near future. *JSME Int. J., Series C,* 37: 629–637 (1994).
23. Thoma, J.A. *Hydrostatic Power Transmission.* Trade and Technical Press, Morden, Surrey, UK (1964).
24. Vickers, Inc. *Vickers Mobile Hydraulics Manual (2nd Ed.).* Vickers, Inc., Rochester Hills, MI (1998).
25. Watton, J. *Fluid Power Systems, Modeling, Simulation, Analog and Microcomputer Control.* Prentice Hall, New York (1989).
26. Welty, J.R., Wicks, C.E., Wilson, R.E. *Fundamentals of Momentum, Heat, and Mass Transfer (3rd Ed.).* John Wiley & Sons, New York (1984).
27. Yeaple, F.D. *Fluid Power Design Handbook.* CRC Press, Boca Raton, FL (1996).
28. Zhang, Q., Goering, C.E. Fluid power system. In: Bishop, R. (ed.), *The Mechatronics Handbook.* CRC Press, Boca Raton, FL, pp. 10–11 ~ 10–14 (2001).

## Exercises

**1.1** Use a layperson's language to define hydraulic power.

**1.2** Why is hydraulic power especially useful when performing heavy work?

**1.3** Why is hydraulic power transmission especially useful on off-road vehicles for driving heavy loads?

**1.4** Compare hydraulic, mechanical, and electrical power transmissions by listing the advantages and disadvantages of each.

**1.5** Hydraulic power transmission has many unique advantages. Try to find three applications in which different advantages of hydraulic power transmission are utilized.

**1.6** What hydraulic device creates a force available for pushing or pulling a load?

**1.7** Name five basic components required to construct a functional hydraulic system.

**1.8** List three applications of hydraulic power transmission on a hydraulic excavator.

**1.9** In Figure 1.4, assume the small piston area $A_1$ is 5 $cm^2$ and the large piston area $A_2$ is 50 $cm^2$. If the load applied on the large piston is 50 kN, calculate (a) the fluid pressure $p$ in the device; (b) the force $F_1$ required to apply on the small piston; and (c) how much the large piston will lift if one pushes the small piston down for 10 cm (assuming all the energy losses are negligible).

**1.10** In Figure 1.4, if the small piston area $A_1$ is 5 $cm^2$ the large piston area $A_2$ is 100 $cm^2$, and the load applied on the large piston is 50 kN, calculate (a) the fluid pressure $p$ in the device; (b) the force $F_1$ required to apply on the small piston; and (c) how many 10 cm strokes of the small piston are needed to lift the large piston for 1 cm (assuming all the energy losses are negligible).

**1.11** In Figure 1.4, assume the large piston diameter $D$ is 50 mm and the small piston diameter $d$ is 10 mm. If the total load applied on the large cylinder is 62.5 kN, calculate (a) the required force $F_1$ to apply on the small piston; and (b) the velocity of a large piston lifting the load if the actuating velocity of the small piston is 50 $mm \cdot s^{-1}$ (assuming all the energy losses are negligible).

**1.12** In a simple hydraulic system, as depicted in Figure 1.1, the cylinder bore diameter is 40 mm, and the rod diameter is 16 mm. If the pressures in the cylinder cap-end and rod-end chambers are 1000 and 800 kPa, respectively, during a no-load extension, what is the friction force of the cylinder?

**1.13** In the same simple hydraulic system, the cylinder bore diameter is 50 mm, and the rod diameter is 25 mm. If the back pressure in the cylinder rod-end chamber is 800 kPa and the friction of cylinder extension is 500 N, what will be the system pressure at the cylinder cap-end port during a no-load extension?

**1.14** An automotive lift raises a 6000 kg car 2 m above the floor level. If the hydraulic cylinder contains a 20 cm diameter piston and a 10 cm diameter rod, determine (a) the work needed to lift the car; (b) the required pressure; (c) the consumed power if the lift raises the car in 10 s; and (d) the flow rate for the automobile to descend in 10 s.

**1.15** A 2.0 mm diameter orifice is used to throttle hydraulic fluid that is flowing from a 12 MPa pressurized line to a reservoir at atmospheric pressure. Using the orifice equation (with a discharge coefficient of 0.65 and a fluid density of 850 $kg \cdot m^{-3}$), calculate the volumetric flow rate through this orifice.

**1.16** Estimate an appropriate valve opening for throttling a hydraulic flow of 30 $L \cdot min^{-1}$ from a 15 MPa line to a 7 MPa line. (Assume the discharge coefficient of 0.60 for this valve and a fluid density of 850 $kg \cdot m^{-3}$.)

**1.17** If a hydraulic pump discharges 30 $L \cdot min^{-1}$ of fluid to a system with an operation pressure of 10 MPa, what is the hydraulic power the pump is delivering? If the pump has 100% efficiency, what input power is required to drive this pump? What if the overall pump efficiency is 85%?

**1.18** If a hydraulic motor receives 90 $L \cdot min^{-1}$ of fluid to a system with an operation pressure of 10 MPa, what is the hydraulic power the motor received? If the motor is 85% efficient, what output power can this motor deliver to drive a load?

**1.19** A hydraulic motor receives 40 $L \cdot min^{-1}$ of fluid to a system with an operation pressure of 12 MPa. How much torque can this motor deliver when it operates at 800 rpm (assume a 90% motor efficiency)?

**1.20** A hydraulic pump delivers a certain flow to drive a hydraulic cylinder to do work. If this cylinder has a 40 mm bore and it takes 2 s to push this cylinder extending 20 cm, what is the discharge flow rate of the pump (assume a 100% volumetric efficiency)?

# 2

# *Hydraulic Power Generation*

## 2.1 Hydraulic Pumps

### 2.1.1 Overview of Hydraulic Pumps

Hydraulic pumps are hydraulic power-generating components in hydraulic systems. Their main function is to convert mechanical kinetic energy into hydraulic potential energy. Driven by a prime mover, often an internal combustion engine on mobile equipment, a hydraulic pump intakes fluid at atmospheric pressure to fill an expanding volume of space inside the pump through an inlet port and discharges pressurized fluids by reducing the volume of space at an output chamber of the pump. This method of pumping fluid is often called **positive displacement pumping.**

It is essential to notice that a hydraulic pump produces only the flow, not the pressure. The pressure of the pump-discharging flow, or the operating pressure of a system, is determined solely by the load of the system, which is combined by the resistance of the fluid flowing in the pipeline and the resistance to move an external load. In the meantime, pressure is also an important parameter representing the performance of a pump. While the operating pressure is determined by the load, pump manufacturers often use four pressure specifications, namely, the rated discharge pressure, the maximum discharge pressure, the minimum discharge pressure, and the maximum inlet pressure, to describe the performance of a hydraulic pump. The **rated discharge pressure** is defined as the maximum continuous operating pressure a pump can support under normal operating conditions. The **maximum discharge pressure** is the pressure under which a pump is allowed to operate for only a short period of time in special circumstances. The **minimum discharge pressure** is also called the **margin pressure** and is a threshold pressure to ensure the pump is operating properly, especially for variable-displacement pumps. The **maximum inlet pressure** is the required inlet fluid pressure for a pump to fully suck in the inlet fluid when it is running at its maximum allowed operating speed.

Another important parameter of a hydraulic pump is its operating speed. Pump manufacturers often provide three speed specifications—the rated speed, the maximum speed, and the minimum speed—to specify the speed performance of their products. The **rated speed** is defined as the speed at which the pump can continuously discharge flow at the rated pressure of the pump. While the **maximum speed** limits the highest speed a pump is allowed to operate temporarily, the **minimum speed** restricts the lowest speed a pump can normally operate in order to supply sufficient flow under rated pressure.

The capacity of a hydraulic pump is specified by its displacement and operating speed. The pump displacement is defined as the total theoretical volume of the fluid that can be delivered in one complete revolution of the pump shaft. It can be mathematically described as follows:

$$Q_T = D_v n \tag{2.1}$$

where $Q_T$ is the theoretical flow rate, $D_v$ is the volumetric displacement, and $n$ is the operating speed of a pump.

Based on their capability to change displacement, hydraulic pumps can be categorized into fixed-displacement pumps and variable-displacement pumps. Based on their configurations, hydraulic pumps can be categorized into a gear pump, a vane pump, or a piston pump. Normally, gear pumps are fixed-displacement pumps, while vane pumps and piston pumps have both fixed or variable-displacement designs. Different types of industry have a preference in choosing the design of the pumps. For example, machine tool manufacturers often select vane pumps because of their low noise and ability to deliver a variable flow at a constant pressure. Mobile equipment manufacturers commonly like to use piston pumps due to their high power-to-weight ratio, and agricultural equipment manufacturers prefer gear pumps for their low cost and robustness.

### 2.1.2 Principle of Positive Displacement Pumping

Hydraulic power systems universally use positive displacement pumps as their power generation components. As implied by its name, a positive displacement pump repeatedly discharges a certain amount of pressurized fluid in every rotation of pump shaft. Its operational sequence can be well illustrated using a reciprocating-type pump. As illustrated in Figure 2.1, the piston extends during the sucking stroke, which creates a partial vacuum as the pump chamber expands. This vacuum then forms a pressure difference between the reservoir and the pump chamber, and pushes the inlet check valve open and the outlet check valve closed. The fluid is then drawn into pump chamber by the ambient pressure (Figure 2.1(a)). After the piston is fully extended, the pump chamber draws in the maximum amount of hydraulic fluid (Figure 2.1(b)). The pumping process then continues to retract the piston to compress the fluid. Because of the incompressible nature of the hydraulic fluid, the pump instantly increases the pressure of the fluid in the chamber, which in turn closes the inlet check valve and pushes the outlet check valve open to eject the pressurized fluid into the hydraulic system (Figure 2.1(c)).

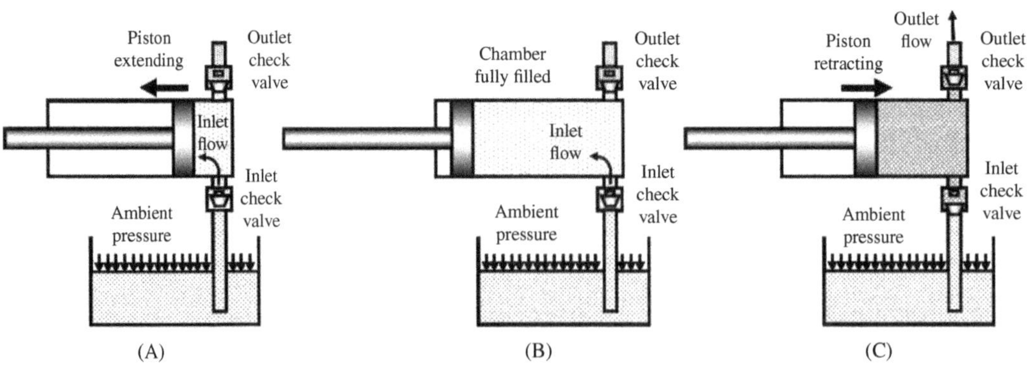


**FIGURE 2.1**
Illustration of the principle of the pumping process in a typical reciprocating-type positive displacement pump.

The pressure needed to push the outlet check valve open is determined by the system pressure applied on the back of the valve.

Because a positive displacement pump converts mechanical energy to hydraulic energy by means of high pressure with a comparatively small quantity and velocity of fluid, it is also termed a **hydrostatic pump**. If the length of piston stroke is kept the same in every cycle, this pump will discharge the same amount of hydraulic fluid each cycle. By altering the length of piston stroke, a change in displacement chamber volume will be achieved. Consequently, the pump can alter the amount of discharge flow to become a variable-displacement pump.

While the principle of positive displacement pumps is illustrated using a **reciprocating-type pump**, the same principle is also applicable to **rotary-type pumps**.

### 2.1.3 Gear Pumps

**The gear pump** is a type of positive displacement pump that forms two continuous-change chambers (also called **suction chambers**) between the teeth of two meshing gears to carry out the pumping process. In general, one of the gears is driven by the drive shaft, and the other is an idler gear turned by the drive gear. Based on the ways the gears mesh, gear-type pumps can be classified as **external-gear** and **internal-gear pumps** (Figure 2.2). Straight spur, helical, or herringbone gears are commonly used in gear pumps. Straight spur gears are easiest to make and are the most widely used. Helical and herringbone gears run more quietly than straight spur gears but usually cost more. Because of their simple structure, compact size, high reliability, and wide adaptability in applications, gear pumps are widely used in both mobile and stationary hydraulic systems.

As a continuous pumping device, a gear pump forms two separate chambers of suction and compression in between the meshed gear teeth, their adjacent teeth, and the pump housing. The volumetric displacement of a gear pump is calculated by the empty space between the gears and the pump housing. An approximate method to estimate gear pump displacement is to calculate the volume of a ring cylinder enclosed by the outside and inside diameters of the gear teeth (often called the diameters of the **addendum** and **dedendum circles** of the gear) and the gear face width using the following equation.

$$D_v = \frac{\pi}{4}\left(d_a^2 - d_d^2\right)L \quad (2.2)$$

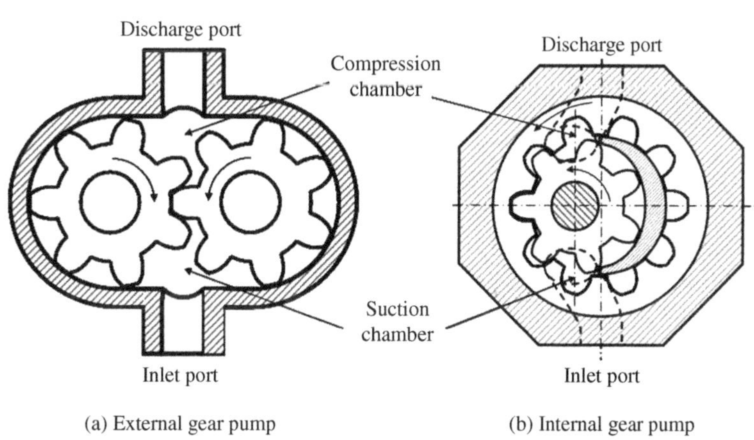


**FIGURE 2.2**
Illustration of the configuration feature and operating principle of typical gear pumps.

where $D_v$ is the volumetric displacement of the pump, $d_a$ and $d_d$ are the diameters of gear addendum and dedendum circles, and $L$ is the face width of the gear.

In actual situations, when a gear turns one circle, the fluid-carrying space between the gear and the pump housing can carry an amount of fluid very close to half of the ring cylinder space, enclosed by addendum and dedendum circles of the gear. The other half of the space is taken up by the gear teeth. For an external gear pump, both the drive and driven gears are often the same size; therefore, the two gears can carry a fluid volume that is almost the same as the volume of one complete ring cylinder. In the case of an internal gear pump, the drive gear is often the external gear. For one complete revolution the external gear turns, the internal gear will turn the same number of teeth as the external gear, which can carry almost the same amount of fluid to the compression chamber of the pump as the external gear. Therefore, the volumetric displacement of an internal gear can also be calculated using Eq. (2.2) in terms of the size of the external gear. In both cases, the calculation accuracy is satisfactory for most engineering applications.

Regardless of the specific design of a gear pump, a partial vacuum is created in its charge chamber as the gear teeth unmesh, which brings the fluid flowing in to fill the chamber. As the gear continues turning, the adjacent teeth and the pump housing form contained chambers that carry the inlet fluid to the compression chamber. As the teeth mesh again at the compression chamber, the fluid is discharged from the pump. Because the space of both chambers is determined by the contact point between two meshed gears, the changing of the contact point will result in a variation in the volume of both chambers. Such a variation will in turn induce a periodic pulsation on the discharged flow. For an external spur gear pump, such instant discharge flow pulsation can be mathematically calculated using the following equation:

$$Q = \omega L\left(R_a^2 - R_p^2 - f^2\right) \tag{2.3}$$

where $Q$ is the instant flow rate discharge from the pump, $\omega$ is the angular velocity of the pump shaft, $L$ is the face width, $R_a$ and $R_p$ are the radius of gear addendum and pitch circles, and $f$ is the distance between the contacting point to the pitch point of the meshed gears.

Because the $f$ value decreases as the number of teeth increases for spur gears, we know that the instant discharge flow pulsation can be reduced by increasing the number of teeth. A **flow variation index** is commonly used to evaluate the scale of flow pulsation caused by the structural impact of hydraulic pumps:

$$\delta_q = \frac{q_{\max} - q_{\min}}{\bar{q}} \tag{2.4}$$

where $q_{\max}$, $q_{\min}$, and $\bar{q}$ are, respectively, the maximum, minimum, and average instant flow rate discharge from a hydraulic pump.

Calculations performed by using Eqs. (2.2) and (2.3) to evaluate the effect of teeth number on the constancy of the discharge flow from a spur-type external gear pump showed that if a pair of 10 teeth spur gears was used, the flow variation index was over 0.21. By doubling the number of teeth, it could reduce the variation index to less than 0.11. The flow variation index provides a measure of the constancy of flow supply from the pump to a hydraulic system and therefore is an important indication of the pump performance. It is desirable to have a pump that can provide a constant flow, and one way to achieve that goal is to increase the number of gear teeth to a reasonable level.

Conversely, when the gears have too many teeth, the result is more than two teeth meshing simultaneously. A **mesh overlap index**, defined as the average number of meshed pairs of teeth during the meshing process, can be used to quantitatively assess the degree of mesh overlap. With a mesh overlap index value greater than, but very close to 1.0, the seal between the suction and compression chambers improves and in turn helps to improve the volumetric efficiency of the pump. However, the mesh overlap will form an enclosed chamber between the two pairs of meshed teeth, and the volumetric space change of this chamber due to the continuous turning will cause a pressure surge on the liquid carried in this chamber. Such a quick pressure rise is one of the major sources for hydraulic noise and vibration in a gear pump, which in turn will cause the pump efficiency to decline. One practical approach to solve this problem is to create a pressure relief groove connected to the pump suction chamber to release this pressure surge for smoother operation. Due to its structural advantages, internal gear pumps will not have this enclosed chamber induced pressure surge problem.

Because it is necessary to have a structural clearance between the gears and the pump housing to allow gears to turn, there unavoidably exists some internal leakage of the pressurized fluid from the compression chamber to the suction chamber. Such internal leakage will form a pressure gradient in the fluid carrying chambers formed by the adjacent teeth. Figure 2.3 depicts how the pressure gradient is formed among those chambers in an external gear pump. These pressure profiles form a total hydraulic force ($F_P$) acting in a tangent direction of gear turning at the meshing point. This hydraulic force, together with gear contact force ($F_T$) that acted on the meshing point, will form a radial force ($F_A$ or $F_B$) on each gear, which makes the external gear pump operate as an unbalanced load on both gear bearings. This unbalanced load is the major contributor to the uneven wear of the pump. To solve this problem, pump manufacturers often apply a pressurized lubricating technology to form hydrodynamic films to support the uneven load on bearings. Pressure-loaded side plates are also used on many gear pumps to provide hydrodynamic films that will form an optimal clearance between gear faces and the pump housing, preventing metal-to-metal contact that results in excessive wear, and to minimize fluid leakage from those clearances.

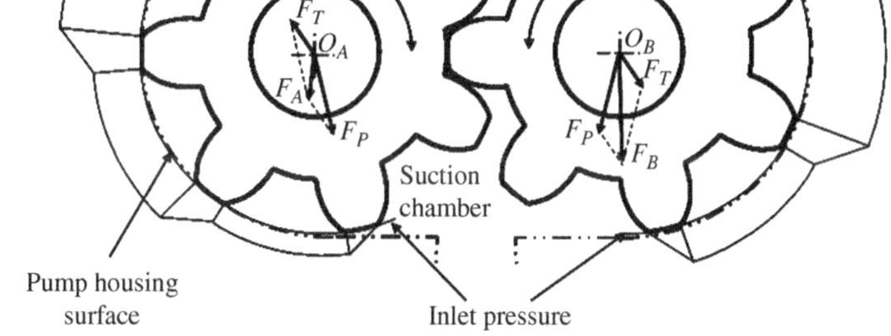


**FIGURE 2.3**
Illustration of the principle pressure and force profiles in a typical external-gear pump.

The **hydrodynamic sealing technology** requires both a minimum pump operating speed and a minimum fluid viscosity to form an adequate thickness of the oil film. This means that the hydrodynamic films cannot be adequately formed at low speeds or when fluid temperature is too high, and consequently, the running clearances between gear faces, gear tooth crests, and the housing will be noticeably increased. Such features imply that gear-type pumps should always be run close to their maximum rated speeds for high efficiency and low wearing rate.

This sealing technology can improve the volumetric efficiency but cannot eliminate the internal leakage problem because there must exist a small clearance between the gears and the housing to allow gears to turn. This means that the actual flow rate $(Q_A)$ a pump can deliver, in terms of its displacement and operating speed, is always less than the theoretical flow $(Q_T)$. The difference between the actual and the theoretical flow rate is defined as the **internal leakage** of the pump (some people also called it the **pump slippage**) and is a function of pump operating pressure (Figure 2.4). This internal leakage can be measured quantitatively in terms of the volumetric efficiency using the following equation:

$$\eta_v = \frac{Q_A}{Q_T} \times 100\% \tag{2.5}$$

where $\eta_v$ is the volumetric efficiency of the pump, and $Q_A$ and $Q_T$ are the actual and theoretical flow rates discharged from the pump.

In general, the volumetric efficiency of a positive displacement pump is around 90% when operating at its designed pressure range. Supported by adequate hydrodynamic sealing technologies, the volumetric efficiencies of gear pumps can run as high as 93% under optimum conditions. However, when the operating pressure of a pump is higher than its designed level, the volumetric efficiency will be decreased because of the excessive internal leakage caused by the higher pressure.

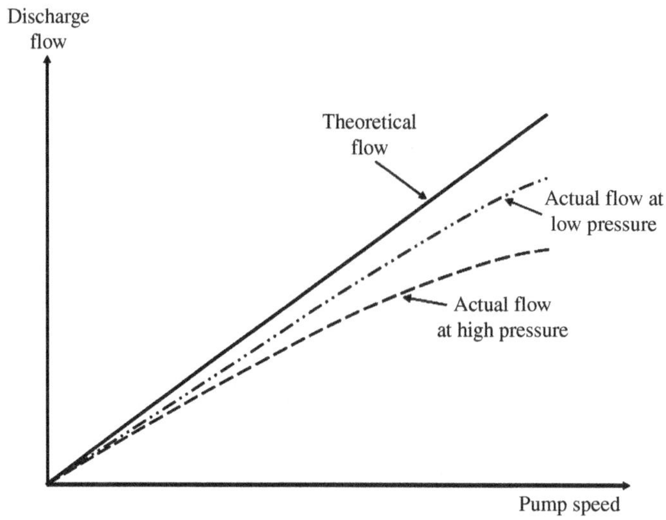


**FIGURE 2.4**
The relationship of between theoretical and actual flow rates of a positive-displacement pump.

**Example 2.1: Gear Pump Volumetric Efficiency**

An external gear pump has two gears with an addendum circle diameter of 80 mm, a dedendum circle diameter of 60 mm, and a face width of 20 mm. If the actual flow rate discharged from the pump is 78 L min$^{-1}$ when the pump is operating at 2000 rpm under its rated pressure, what is the volumetric efficiency of the pump?

a. The volumetric displacement of the pump can be calculated using Eq. (2.2):

$$\begin{aligned} D_v &= \frac{\pi}{4}\left(d_a^2 - d_d^2\right)L \\ &= \frac{\pi}{4}\times\left(0.080^2 - 0.060^2\right)\times 0.020 \\ &= 4.396\times 10^{-5}\left(m^3\right) = 4.396\times 10^{-2}(L) \end{aligned}$$

b. Using Eq. (2.1) to determine the theoretical flow rate of the pump:

$$\begin{aligned} Q_T &= Dn \\ &= 4.396\times 10^{-2}\times 2000 \\ &= 87.92\left(L\cdot \text{min}^{-1}\right) \end{aligned}$$

c. Based on Eq. (2.5), the volumetric efficiency of the pump is:

$$\begin{aligned} \eta_v &= \frac{Q_A}{Q_T}\times 100\% \\ &= \frac{78.0}{87.92}\times 100\% \\ &= 88.7\% \end{aligned}$$

DISCUSSION 2.1: Pump manufacturers often specify volumetric efficiency as the pump rated pressure. While the actual discharge flow from a pump will decrease as the operating pressure increases, the theoretical flow is a constant for a pump regardless of the operating pressure. Consequently, the pump volumetric efficiency will also decrease as the pressure increases.

There are many different types of gear pumps other than the spur gear pump. A few commonly seen examples are lobe, gerotor, and screw pumps. A **lobe pump** is a rotary, external-gear pump and operates in a similar way to a conventional external gear pump. The major difference is that both lobes are driven externally so that they do not actually contact each other and are therefore much quieter than a conventional gear pump in operation. Another noticeable difference is that a lobe pump generally produces greater pulsation on the discharge flow due to the smaller number of teeth than a conventional gear pump. A **gerotor pump** is a rotary, internal-lobe pump and operates similarly to a conventional internal gear pump. In a typical gerotor pump, the inner rotor does mesh with the outer rotor to drive the pumping operation. Normally, the outer rotor has one more tooth than the inner one, and the pump displacement is determined by the space formed by the extra tooth in the outer rotor. A **screw pump** is an axial-flow gear pump. A typical screw pump consists of three screws, with a central-drive rotor meshing with two idler rotors inside a closed-fitting housing with no metal-to-metal contact. In the pumping process, the inlet flow is pushed uniformly through a screw pump axially in the direction of the drive rotor. Because the fluid delivered by the screw pumps does not rotate, and the rotors work like endless pistons that continuously move forward, it results in no pulsations at any speed of motor operation. These features make a screw pump operate very quietly and efficiently.

### 2.1.4 Vane Pumps

Vane-type pumps use a number of vanes sliding in slots in a rotor that rotates in a **cam ring** (often called the *housing*) to pump pressurized fluid without flow pulsation. The housing may be eccentric with the center of the rotor (Figure 2.5(a)), or it may be shaped in an oval to form two suction and two compression chambers (Figure. 2.5(b)). In some designs, centrifugal force holds the vanes in contact with the housing, while the vanes are forced in and out of the slots by the eccentricity of the rotor. In other designs, either light springs or pressurized fluids are used to push the vanes against the housing.

As illustrated in Figure 2.5, the volume of suction chambers enclosed by vanes, rotor, and housing will increase, and a vacuum will be created in those chambers as the rotor turns. The atmospheric pressure will force hydraulic fluids to fill this inlet space. Meanwhile, the volume of the compression chamber will be reduced to force the liquid discharged through the outlet ports during pumping. The pump illustrated in Figure 2.5(a) is an **unbalanced vane pump** because the pumping action occurs in compression chambers located on only one side of the rotor. This one-side pumping arrangement imposes a side load on the rotor and consequently on the drive shaft. Because of this, unbalanced vane pumps are often used in low-pressure applications to avoid large radial forces acting on pump bearings. The displacement of an unbalanced vane pump is heavily determined by the eccentricity of the rotor to the cam ring and the face width of the vanes. It is commonly calculated using the following equation:

$$D_v = \frac{\pi}{2}(d_c + d_r)eL \tag{2.6}$$

where $d_c$ and $d_r$ are diameters of the cam ring and the rotor, $e$ is the eccentricity of the rotor to the cam ring, and $L$ is the face width of vanes.

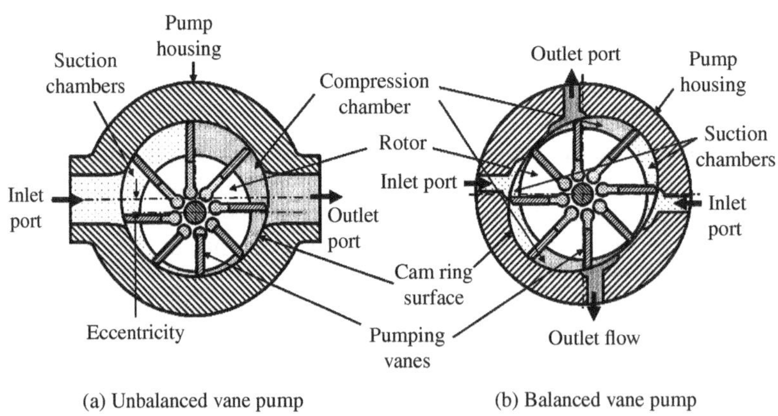


**FIGURE 2.5**
Illustrations of configuration features and operating principle of typical vane pumps. (a) Unbalanced and (b) balanced designs.

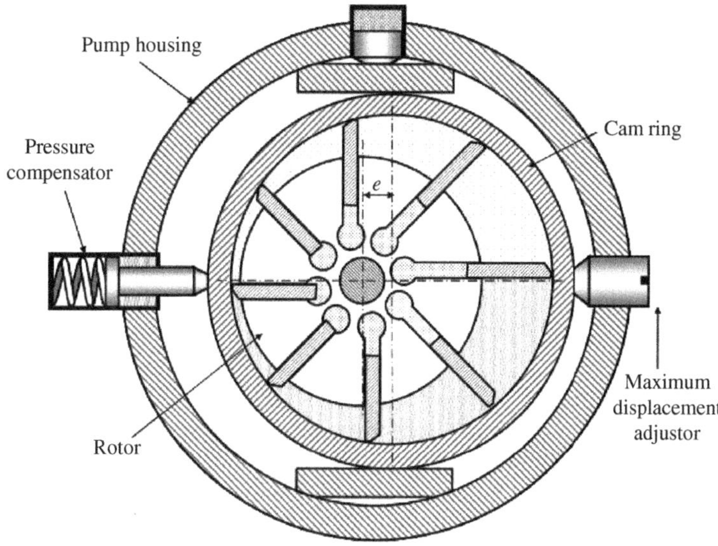


**FIGURE 2.6**
Illustration of the principle of typical pressure-compensated variable-displacement adjustment.

One important feature of the unbalanced design is its possibility of changing the eccentricity of the rotor to achieve the controlled variation on the pump displacement. As illustrated in Figure 2.6, the displacement of an unbalanced vane pump can be changed through an external control, such as a pressure compensator, to push the cam ring adjusting the eccentricity between the cam ring and the rotor. Expressed by Eq. (2.6), as the rotor turns, the displacement per revolution will be increased as the eccentricity value increases. In a pressure compensation process, it is very common to use a preloaded spring to balance the system pressure. By this means of adjustment, when the system pressure is high enough to overcome the compensator spring force, the cam ring shifts to decrease the eccentricity, and the displacement of the pump will be reduced.

When a vane pump uses an oval cam to form two separate pumping areas on the opposite sides of the rotor to cancel the side load, this pump is in a balanced construction as shown in Figure 2.5(b). While a **balanced vane pump** imposes very low radial force on its shaft and can be used in higher-pressure applications, it comes only in **fixed-displacement designs**. Determined by the geometric features, the displacement of a balanced vane pump can be calculated by the volume difference between the rotor and the oval-shaped cam ring. Based on an assumption that the short axle diameter of the cam ring is very close to the rotor diameter, the following equation can be used to calculate the volumetric displacement of this type pump:

$$D_v = \frac{\pi}{4}\left(d_{c1}d_{c2} - d_r^2\right)L \tag{2.7}$$

where $d_{c1}$ and $d_{c2}$ are diameters of the short and long axles of the oval cam ring, $d_r$ is the rotor diameter, and $L$ is the width of the vane face.

In many balanced vane pumps, the shape of the cam ring is specially designed to achieve optimal operation performance and maximum operation life. Different equations should be used for these vane pumps to more accurately calculate the pump displacement. Because centrifugal force is required in many vane pumps, either balanced or unbalanced, to hold the vanes against the cam ring to maintain a tight seal at those points, these pumps are often required to operate at a relatively high speed, normally above 600 rpm. For applications where a low-speed operation is required, it is recommended to select vane pumps whose vanes are held either by springs or by the pressurized fluids on the back of vanes.

Because of their capability of compensating for vane wear by pushing the vanes using either the centrifugal force or the external force, vane pumps can maintain high efficiency for a long period of time. While vane pumps are relatively good at tolerating contaminants in the fluids, they are sensitive to operating pressure, especially those with pressure-holding vanes. Long time use of a vane pump under high-pressure operations can noticeably reduce the expected life of the pump.

### 2.1.5 Piston Pumps

A piston pump works on the principle of reciprocal pumping using a piston–cylinder pair. Because reciprocal pumping is completed in two strokes of suction and pumping alternatively as illustrated by Figure 2.1, a single-cylinder piston pump can only supply pressurized fluid intermittently. To gain a continuous flow supply, typical hydraulic piston pumps use multiple piston–cylinder pairs, normally arranged in a circle on a rotating base, to deliver pressurized flow to the system in sequence.

Producing pressurized flow by forcing the fluid out of the cylinder using a well-sealed piston, a piston pump can generate very high pressure by pushing the fluid against heavy loads with high volumetric efficiency. It is reasonable to expect a piston pump to have a volumetric efficiency of over 97% and an overall efficiency of 90% or higher. Because of this feature, piston pumps are often used in heavy-duty applications with high operating pressures. It should also be noted that the high efficiency and high operating pressure are normally accompanied with high cost.

In terms of piston arrangement, piston pumps can be classified into two basic types, axial and radial pistons. Both are available in fixed-displacement and variable-displacement designs. Regardless of the structural arrangement of the pistons, an important part of the pump mechanism is to convert the rotational motion of the drive shaft into reciprocating motions on pistons to complete the pumping operations.

In an **axial-piston pump**, the pistons reciprocate in parallel to the centerline of the piston block. A unique feature of axial-piston pumps is their ability to convert drive shaft rotary motion into axial reciprocating motions of the pistons. One common design of such a motion conversion mechanism is the use of a **swash plate** in an **inline-piston pump**. As illustrated in Figure 2.7, the pistons are fitted to cylinders built in a rotating cylinder block with one end connected to a **shoe plate** (also called the **retracting ring**) via **piston shoes**. This shoe plate pushes the shoes against an angled swash plate. As the cylinder block turns, the piston shoes are forced to follow the angled surface of the swash plate, causing the pistons to reciprocate. A **valve plate**, consisting of an inlet slot and an outlet slot, is used to connect the cylinders to pump inlet and outlet ports to pull fluid into the cylinders as pistons extend and to discharge the fluid out of the cylinders as pistons retract.

Inline-axial pumps can be designed either as fixed- or variable-displacement models. The only difference between these two designs is in their swash plates. Compared to being fabricated as part of the pump housing in a fixed-displacement pump, the swash plate in a variable model is often mounted in a movable yoke (Figure 2.8). The angle of this adjustable swash plate ($\alpha$) can therefore be changed by pivoting the yoke on a pintle. Positioning of the yoke can be changed via manual adjustment, compensating control, or servo control. Despite the actuation methods for yoke repositioning, it follows the

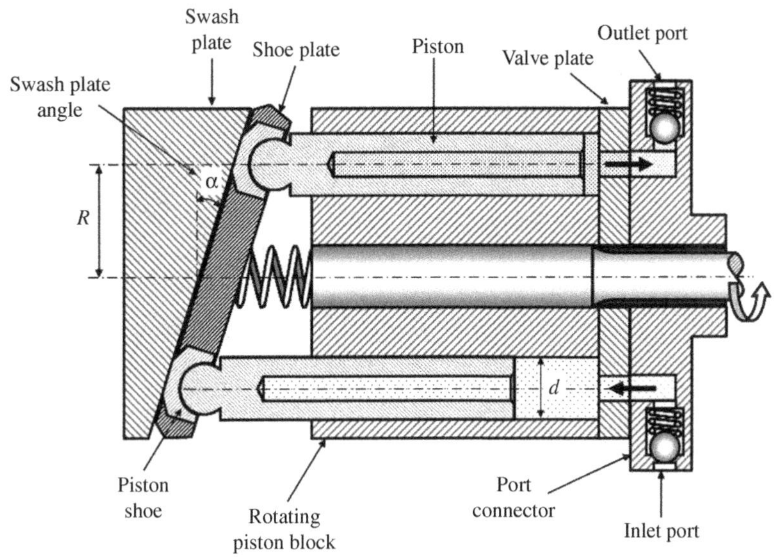


**FIGURE 2.7**
Illustration of the operating principle of typical inline-type fixed-displacement axial piston pumps.

same operation principle. Take a pressure-compensated variable-displacement actuation illustrated in Figure 2.8 as an example. The position of the adjustable swash plate is controlled by the balance between the spring force and the hydraulic force acting in the yoke pivot mechanism. In its normal condition, the spring force pushes the cylinder in the yoke pivot mechanism to its maximum extended position, as does the swash plate angle, to reach the maximum displacement. As the hydraulic pressure of the fluid supplied to the yoke pivot device increases, a larger hydraulic force will act on its cylinder and push

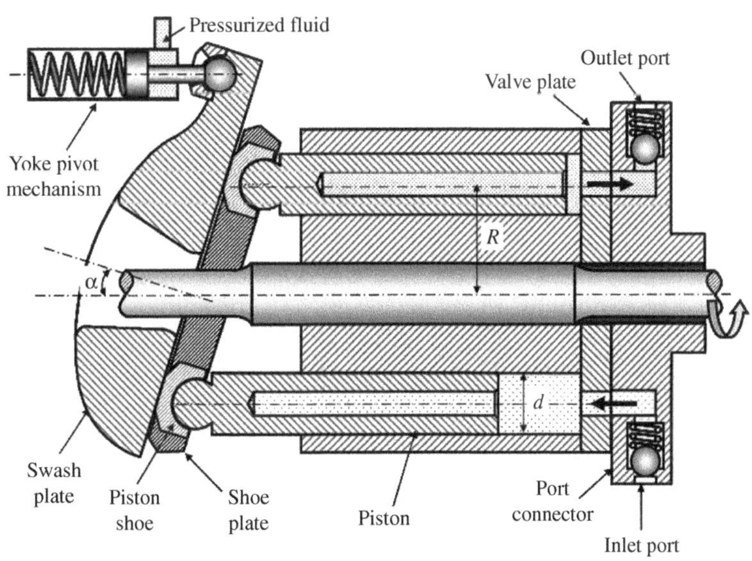


**FIGURE 2.8**
Illustration of the operating principle of typical inline-type variable-displacement axial piston pumps.

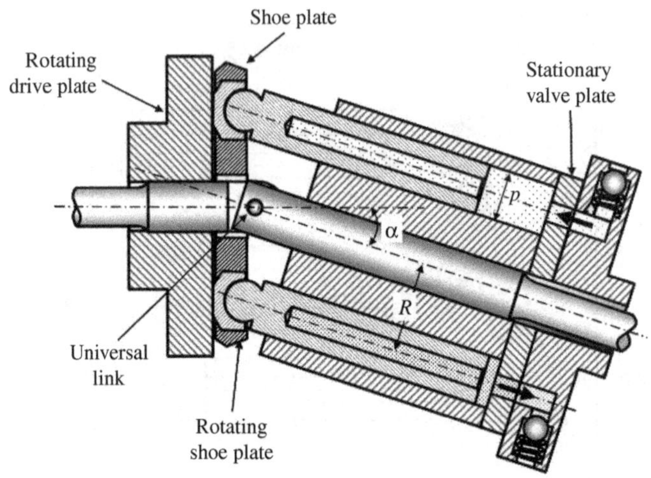


**FIGURE 2.9**
Illustration of the operating principle of typical bent-type axial piston pump.

it back against the preloaded spring force to turn the swash plate to a smaller angle to reduce the displacement until the spring force and the hydraulic force are re-balanced at a new level.

Another popular design of axial-piston pumps is the **bent-axis pump**. This type of pump consists of a rotating drive plate, a rotating cylinder block, a universal link, and a stationary valve plate (Figure 2.9). A notable structural feature of a typical bent-type axial-piston pump is that the axis of its cylinder block is set at an offset angle, often called the **bent angle**, relative to the axis of the drive plate. Rotation of the drive plate and the piston block is synchronized using the universal link connecting the drive shaft and the block shaft. As a result, all pistons will rotate with the drive plate to convert the rotating motion of the drive plate into the reciprocal motions of the pistons.

Both types of axial-piston pumps, with a few pistons extended and the rest retracted at the same time, can pump pressurized fluid out continuously. The volumetric displacement of an axial pump can be determined by the number and size of pistons as well as their stroke length, which is a function of the swash plate angle, using the following equation:

$$D_v = \frac{\pi}{2} d^2 m R \tan \alpha \tag{2.8}$$

where $d$ is the piston diameter; $m$ is the number of pistons; $R$ is the distance of cylinder centerlines to the centerline of the piston block; and $\alpha$ is the angle of the swash plate for an inline-type pump or the bent angle for a bent-type pump.

Equation (2.8) gives the theoretical volumetric displacement of axial piston pumps. The average discharged flow rate is defined as the product of pump displacement and rotating speed. By taking the volumetric efficiency into consideration, the following equation can be used to calculate the average flow rate discharged from an axial piston pump.

$$Q = \frac{\pi}{2} d^2 m n \eta_v R \tan \alpha \tag{2.9}$$

where $n$ is the rotating speed of the piston block and $\eta_v$ is the volumetric efficiency of the pump.

The above equation reveals that the average flow rate can be changed by adjusting either the rotating speed or the swash plate angle (or bent angle). Moreover, as the output flow is discharged from individual cylinders, the structural discontinuity positioning of those cylinders will result in a pulsation in the discharged flow. When using the flow variation index defined by Eq. (2.4) to analyze the flow pulsation, it can be found that such a pulsation could be reduced by increasing the number of cylinders. However, the total number of cylinders that can be placed in a piston block is limited. Another important fact is that the fluctuation in flow will certainly induce a fluctuation in system pressure, which in turn will affect the smoothness of the hydraulic system operation. To accurately analyze the flow fluctuation characteristics of an axial piston pump, it is necessary to study the instant flow discharge rate from the pump.

Figure 2.10 is the side view of Figure 2.9 from the valve plate view. In such a case, the piston in the cylinder located on the top (also called the start point), as shown in Figure 2.10, is at its fully extended position, namely, the full stroke of the piston. As the piston block turns clockwise to a certain degree, the piston stroke can be described using this equation:

$$x = R\tan\alpha(1-\cos\varphi) \tag{2.10}$$

where $R$ is the distance of the cylinder centerlines to the piston block centerline, $\alpha$ is the swash plate angle for an inline-type pump or the bent angle for a bent-type pump, and $\varphi$ is the positioning angle of the piston of interest on the valve plate in relation to the starting point.

Notice that the piston positioning angle is also a function of the rotating speed $\varphi = \omega t$. From Eq. (2.10), the piston reciprocal velocity can be derived as follows:

$$v = R\omega\tan\alpha\sin\varphi \tag{2.11}$$

where $\omega$ is the rotating speed of the piston block.

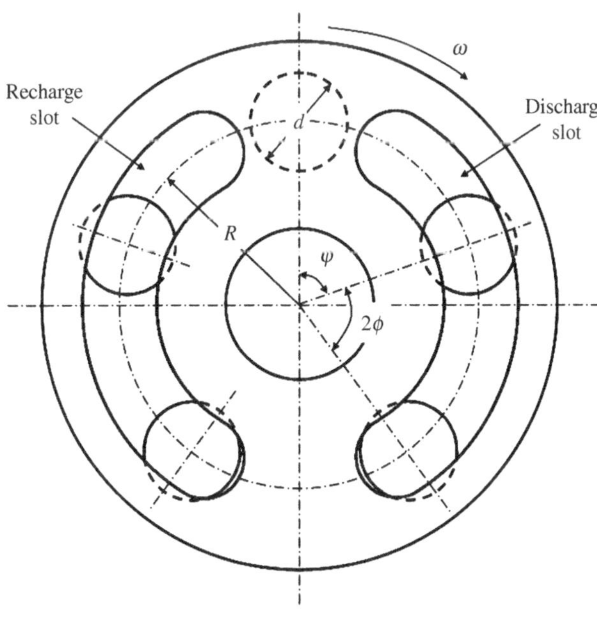


**FIGURE 2.10**
Configuration illustration of a typical valve plate in axial piston pumps.

Therefore, the instant flow rate discharged from a single cylinder can be determined using the following equation:

$$q = \frac{\pi}{4} d^2 R\omega \tan\alpha \sin\varphi \tag{2.12}$$

where $q$ is the instant flow rate discharged from one cylinder positioned at an angle $\varphi$ on the valve plate, and $d$ is the diameter of the piston.

The instant flow rate from the pump can now be determined by summating the instant flow from all cylinders connected to the discharge slot shown in Figure 2.10 using Eq. (2.13):

$$Q = \frac{\pi}{4} d^2 R\omega \tan\alpha \sum_{i=1}^{n} \sin\varphi_i \tag{2.13}$$

where $Q$ is the instant flow rate discharged from the pump.

Because the instant flow discharge is determined by the total number of cylinders, the summation value $\sum \sin\varphi$ is different for pumps with odd-number pistons from those with even numbers. When the piston block has an even number of cylinders, theoretically one-half of these cylinders are discharging fluid at any given time. If the central angle between two adjacent cylinders is $2\phi$, then the instant flow rate will vary in a $2\phi$ angular cycle, with the minimum instant flow occurring at $\varphi = 0$ or $2\phi$ and the maximum instant flow occurring at $\varphi = \phi$. When the piston block has an odd number of cylinders, for example, five, half of the time there are only two cylinders discharging fluid; the rest of the time, three are discharging. This results in a flow fluctuation at one $\phi$ angular cycle, namely, the minimum flow occurring at $\varphi = 0$ or $\phi$ and the maximum flow occurring at $\varphi = \phi/2$, as shown in Figure 2.11.

The flow variation index and variation frequency of an axial pump can be determined using the following sets of equations. For an even number of pistons:

$$\begin{cases} \delta_Q = 2\sin^2\left(\dfrac{\pi}{2m}\right) \\ \omega_Q = mn \end{cases} \tag{2.14}$$

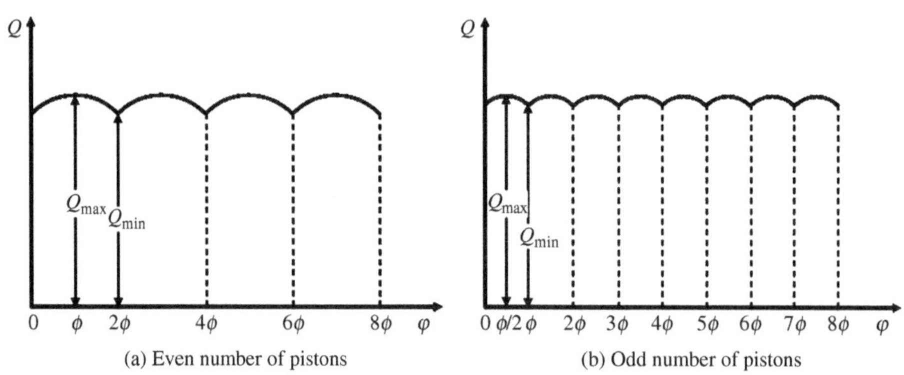


**FIGURE 2.11**
Typical variation of instant flow discharged from pumps with (a) even or (b) odd number of pistons.

And for an odd number of pistons:

$$\begin{cases} \delta_Q = 2\sin^2\left(\dfrac{\pi}{4m}\right) \\ \omega_Q = 2mn \end{cases} \tag{2.15}$$

where $\delta_Q$ is the flow variation index, $\omega_Q$ is the flow variation frequency, $m$ is the number of pistons, and $n$ is the rotating speed of the piston block of an axial pump.

The above equations reveal that an odd-number piston pump can generate flows with smaller and quicker pulsation, and therefore can achieve smoother and steadier operation than an even-number piston pump does.

A critical feature, minor to the structure but important to the performance in an axial piston pump, is the pressure-bleeding slots on the valve plate (Figure 2.12). As discussed earlier, an axial piston pump consists of multiple pistons installed in one piston block, with a few recharging fluids and the others discharging at the same time. A valve plate is used to connect all the cylinders operating under the same phase using either the recharge slot or the discharge slot as shown in Figure 2.10. To ensure high efficiency in hydraulic power transmission, the recharge and the discharge ports can never be connected under any circumstance. A common way to ensure this is to create a separation zone on the valve plate, longer than the cylinder diameter, thus preventing any cylinders from being connected to both recharging and discharging ports at the same time.

However, such a design will form a small, blocked chamber in the cylinder when its opening is completely covered by the separate zone. This often results in a deterioration of pump performance and a decrease in operation efficiency. As shown in Figure 2.12, assume that a complete cycle of the pumping process in a cylinder starts at location $A$,

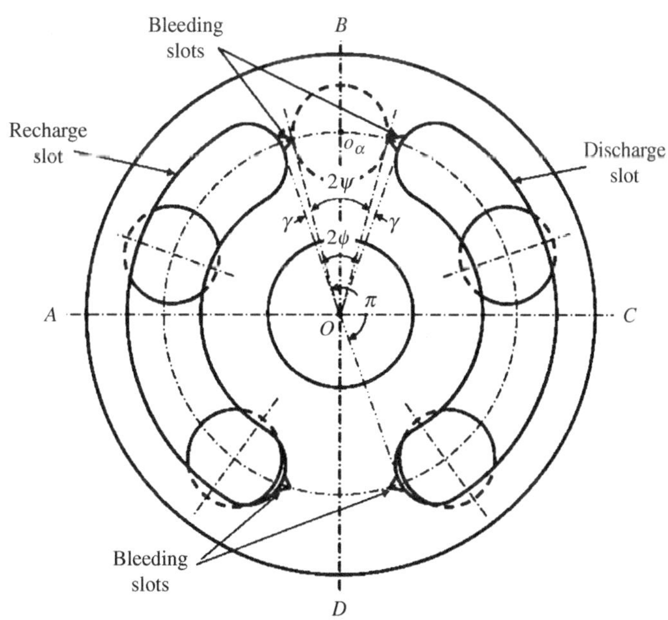


**FIGURE 2.12**
Illustration of the principle of pressure equilibrating using bleeding slots on a typical valve plate.

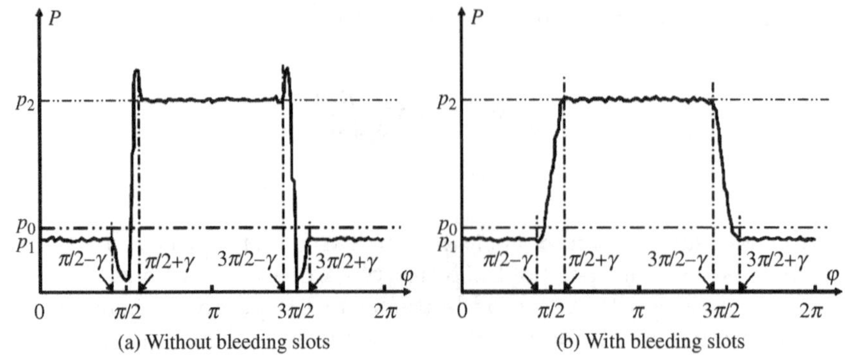


**FIGURE 2.13**
Typical pressure transition pattern within a cylinder chamber during one complete cycle of pumping process in an axial piston pump. (a) Without or (b) with bleeding slots.

the middle point of the fluid recharge process. As the piston block turns clockwise, the cylinder approaches the separation zone on the valve plate centered at location *B*. After the open cylinder becomes blocked (as the rotating angle of the cylinder centerline is at $\pi/2-\gamma$ point in Figure 2.12), the fluid supply to the cylinder is also blocked. With the piston continuing to extend, a vacuum is quickly formed in the cylinder, which brings the fluid pressure in the chamber to a very low level (Figure 2.13(a)). As the centerline passes point *B* on the valve plate at the $\pi/2$ point, the piston starts to retract and compresses the fluid to a very high pressure in a very short time due to the incompressibility of liquid fluid. After the open cylinder connects to the discharge slot again at $\pi/2+\gamma$ point, the fluid pressure in the cylinder chamber will then be equilibrated with the system pressure (Figure 2.13(a)). As the piston block continues to turn to the separation zone, from the discharge slot to the recharge slot, an opposite pressure change pattern will be formed during the transition period (Figure 2.13(a)). Such a phenomenon is often called **pressure overshoot** during the transition between pump recharge and discharge. An excessive pressure overshoot will not only weaken the performance, but also create a negative impact on the life of the pump.

The excessive pressure shoot problem can effectively be solved by creating a set of pressure-bleeding niches (often called slots) at both ends of the recharge and discharge ports, as shown in Figure 2.12. The basic function of these bleeding niches is to prevent either excessive high or low pressure from forming in the cylinder chamber during the blocked periods by offering a slow passage that will allow a very small amount of fluid bleeding in or out of the chamber to result in smoother in-cylinder pressure during the transition periods (Figure 2.13(b)).

Another type of piston pump is a **radial-type piston pump** in which the cylinders are arranged radially in a cylinder block and the pistons move perpendicularly to its shaft centerline. It generally consists of a cylinder block with pistons, a cam ring, and a porting plate. Based on the porting arrangement, a radial pump often uses either check valves or pintle valves to control fluid recharging and discharging. Mobile applications often choose the **pintle-type radial pump** for satisfying high-speed operation needs. Figure 2.14 illustrates the construction and operation principle of a pintle-type radial piston pump, in which the cylinder block rotates on a stationary pintle inside a circular cam ring. As the block rotates, the centrifugal force or charging pressure forces the pistons to be kept in contact with the inner surface of the ring. Since this ring is offset from the centerline of the

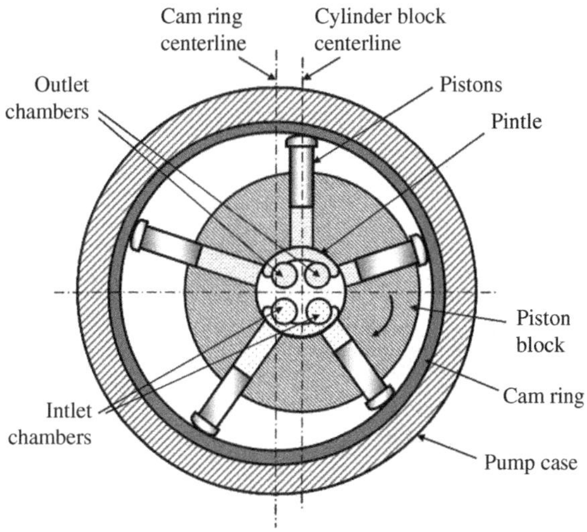


**FIGURE 2.14**
Illustration of configuration and operation principle of typical radial piston pumps.

cylinder block, it will cause the pistons to reciprocate in their bores as the cylinder block rotates. The pintle port permits the cylinders to take in fluid as the pistons move outward and discharge it as they move in. The size and number of pistons, as well as the length of their stroke, determine the displacement of a radial piston pump as follows.

$$D_v = \frac{\pi}{2} d^2 em \tag{2.16}$$

where $d$ is the diameter of piston, $m$ is the number of pistons, and $e$ is the eccentricity of the cylinder block to the cam ring.

Radial piston pumps are also available in both fixed- and variable-displacement designs. The adjustment of displacement is often accomplished by moving the cam ring to adjust the eccentricity of the cylinder block to the cam ring to increase or decrease piston strokes.

## 2.2 Control of Hydraulic Power Generation

### 2.2.1 Corner Power and Pump Efficiency

Hydraulic pumps are the energy conversion devices used to convert mechanical power into hydraulic potential energy, driving various hydraulic actuators to perform work. One fundamental requirement in designing a hydraulic system is to provide sufficient power to those actuators to do the designated work. In general, the amount of power delivered by a hydraulic pump is determined by the flow rate and the fluid pressure. As discussed in the previous section, the amount of flow a pump can deliver is determined by its displacement and operating speed, but the pressure is determined by the total load applied on the system.

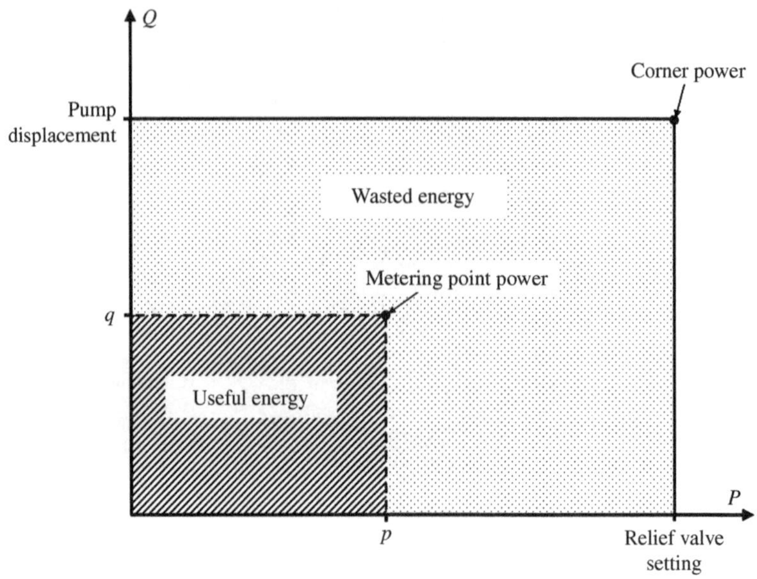


**FIGURE 2.15**
Graphical definitions of pump corner power and metering point power.

**Corner power**, defined as the product of the maximum operating pressure and the maximum flow supply capabilities of a pump, is often used to quantify the maximum power delivery capacity of a pump. Figure 2.15 provides a graphical illustration of corner power, which presents critical information on energy balance using a pressure-flow curve to show the relationship between the amount of energy being carried by a hydraulic system and the amount of energy being used to do useful work. Therefore, it is an effective tool for analyzing the efficiency of energy utilization in a hydraulic system.

When a fixed-displacement pump is running at a constant speed, this pump can supply a fixed amount of flow, namely, the theoretical flow determined by its displacement, to the system. The energy required to drive the pump discharging this amount of flow is dependent on the pressure of the flow until the pressure reaches the relief setting. This method of pressure control is known as **pressure limiting**. The hydraulic power delivered at this point is the corner power for the system, and it is a fixed value for a system constructed to use a fixed-displacement pump, operating at a constant speed with a fixed relief setting. In an actual operation, the actuator may not need all the flow to drive the load, and the system load often may not reach its maximum level. Consequently, only a portion of hydraulic power, often quantitatively represented using **metering point power** as defined graphically in Figure 2.15, is required to drive the load. As a result, the rest of the energy carried by the pressurized fluid is wasted, often as the excess flows through the relief valve and is converted into heat.

To improve the energy efficiency in a hydraulic power transmission, a commonly applied approach is to use variable-displacement pumps. By adjusting the pump displacement to deliver only the needed hydraulic flow to drive the load, a large portion of hydraulic power can effectively be saved by not supplying the excess flow (Figure 2.16(a)), especially when a low speed is required. However, the flow-pressure curve shown in Figure 2.16(a) reveals that this approach is unable to save the portion of energy carried by the excess pressure. To

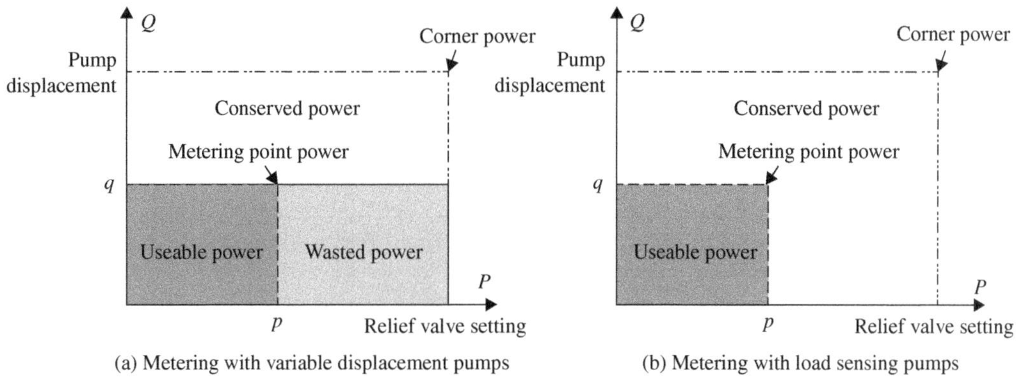


**FIGURE 2.16**
Illustration of the principle of energy-saving approaches using (a) a variable-displacement pump or (b) a load-sensing pump.

solve this problem, a load-sensing device can be used to adjust the pump-discharge pressure in terms of the actual need to further reduce the amount of wasted energy, as illustrated in Figure 2.16(b). The following sections will introduce the commonly applied pump control methods to improve the energy efficiency by means of enhancing the efficiency of pump operation. No matter which method is used, the fundamental principles are very similar, namely, to adjust the flow supply or the pressure setting, or both, to provide only the needed power to the system.

So far, all the discussion has been based on an assumption that the pump we used to generate the hydraulic power is an ideal pump, with its fluid delivery chamber always fully filled during the recharge process and with zero clearance between mating parts and zero friction between relatively moving parts in the pump. However, such an ideal condition never exist in real applications. The **power efficiency** (also called the **overall efficiency**) of a pump is commonly used to quantitatively assess the performance of a pump in comparison to its ideal case. This power efficiency can further be broken down into the two distinct components of **volumetric efficiency** and **mechanical efficiency** to track its main attributors. Their relationship can be presented using the following equation.

$$\eta_o = \eta_v \eta_m \tag{2.17}$$

where, $\eta_o$ is the overall efficiency, $\eta_v$ is the volumetric efficiency, and $\eta_m$ is the mechanical efficiency of a pump.

The volumetric efficiency of a pump indicates the energy loss during the pumping process in the form of flow loss, mainly caused by internal leakage between mating parts and less flow charged into the pump chamber during the recharge process, in comparison to its capability. Section 2.1.3 discussed how to determine the volumetric efficiency for a gear pump. Using the same approach, we can determine the volumetric efficiency for either a vane pump or a piston pump. Because the theoretical discharge flow is proportional to pump displacement and operating speed, a more general form for pump volumetric efficiency can be defined as follows:

$$\eta_v = \frac{Q_A}{D_v n} \tag{2.18}$$

where $\eta_v$ is the volumetric efficiency; $Q_A$ is the actual discharge flow from the pump; $D_v$ is the pump displacement; and $n$ is the driving shaft rotating speed of the pump.

Volumetric efficiencies typically run from 80 to 90% for gear pumps, 82 to 92% for vane pumps, and 90 to 98% for piston pumps.

The mechanical efficiency (also called the **torque efficiency**) indicates a form of energy loss used to overcome all resistance, including mechanical friction and fluid turbulence, during a pumping process. It is defined as the ratio of theoretical hydraulic power discharged from the pump to the mechanical power consumed to drive the pump:

$$\eta_m = \frac{pQ_T}{2\pi Tn} \tag{2.19}$$

where, $\eta_m$ is the mechanical efficiency; $Q_T$ is the theoretical discharge flow from the pump; $p$ is the pressure of discharged flow; $T$ is the input torque to drive the pump; and $n$ is the driving shaft rotating speed of the pump.

Using theoretical discharge flow to determine the mechanical efficiency can be justified by the fact that it is a measure of energy loss used to overcome *all resistance* other than the flow losses. Typically, the mechanical efficiency of a hydraulic pump runs from 90 to 95%.

**Example 2.2: Pump Efficiency**

A mobile hydraulic pump has a displacement of 70 mL. When operating at 1000 rpm, it can discharge 65 $L \cdot min^{-1}$ fluid to a 7 MPa system. If the engine inputs a 95 $N \cdot m$ torque to drive the pump, what is the overall efficiency of the pump, and what is the theoretical torque required for driving the pump if there is no mechanical loss?

a. To determine the overall efficiency, we need first to find the theoretical flow that can be delivered by the pump using Eq. (2.1):

$$\begin{aligned} Q_T &= D_v n \\ &= 70 \times 10^{-3} \times 1000 \\ &= 70\left(L \cdot \min^{-1}\right) \end{aligned}$$

The volumetric efficiency of the pump can be calculated by applying Eq. (2.5):

$$\begin{aligned} \eta_v &= \frac{Q_A}{Q_T} \times 100\% \\ &= \frac{65}{70} \times 100\% \\ &= 92.9\% \end{aligned}$$

The mechanical efficiency can be determined in terms of Eq. (2.19):

$$\begin{aligned} \eta_m &= \frac{pQ_T}{2\pi Tn} \\ &= \frac{7 \times 10^6 \times 70 \times 10^{-3}}{2 \times 3.14 \times 95 \times 1000} \times 100\% \\ &= 82.1\% \end{aligned}$$

Then, the overall efficiency can be computed using Eq. (2.17):

$$\begin{aligned}\eta_o &= \eta_v \eta_m \\ &= 92.9\% \times 82.1\% \\ &= 76.3\%\end{aligned}$$

b. The theoretical torque required to drive the pump is the portion of actual torque used to drive the pump; therefore, it can be determined as follows:

$$\begin{aligned}T_T &= T_A \eta_m \\ &= 95 \times 0.821 = 78.0(N \cdot m)\end{aligned}$$

**DISCUSSION 2.2:** Pump efficiencies, including the volumetric, mechanical, and overall efficiency, are important parameters for evaluating the performance of a pump.

### 2.2.2 Pressure Limiting

One of the most basic energy efficiency-enhancing methods for hydraulic systems is **pressure-limiting compensation**. By this approach, a pressure-limiting compensator is used to regulate the pump-discharge flow when the pressure at pump-discharge port reaches a preset limit. As illustrated in Figure 2.17, a typical pressure-limiting compensator consists of a pressure-limiting spool, a pressure-setting spring, and a pressure-setting adjustor, with a signal port connected to the pump-discharge port, an implementation port connected to the pump yoke control actuator, and a bleeding port connected to the reservoir. During operation, the pressure-limiting value is preset using a pressure adjustor preset by a preload spring. When the pump-discharge pressure is below the preset value, the yoke controller will keep the pump operating at its maximum displacement condition and supply the maximum flow capacity to the system. While the discharge pressure exceeds the preset value, the high pressure of the discharged flow will push the pressure-limiting spool downward, as illustrated in Figure 2.17, which will connect the ports of pump discharge and yoke control on the pressure-limiting compensator and allow the higher

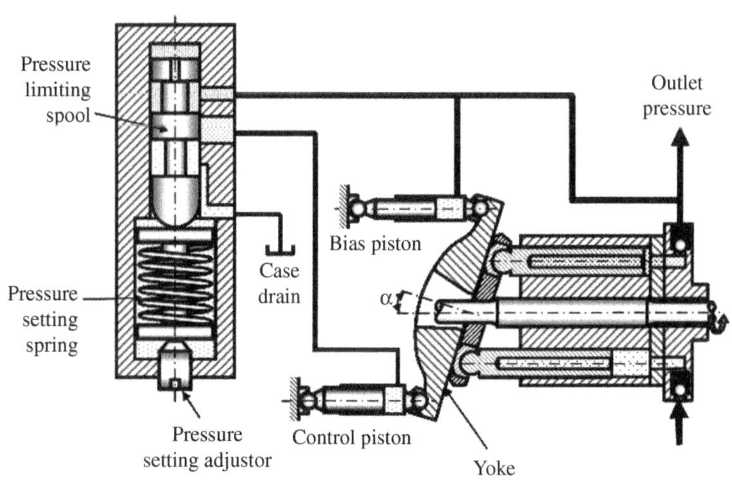


**FIGURE 2.17**
Illustration of the principle pressure-limiting compensation on a variable-displacement pump.

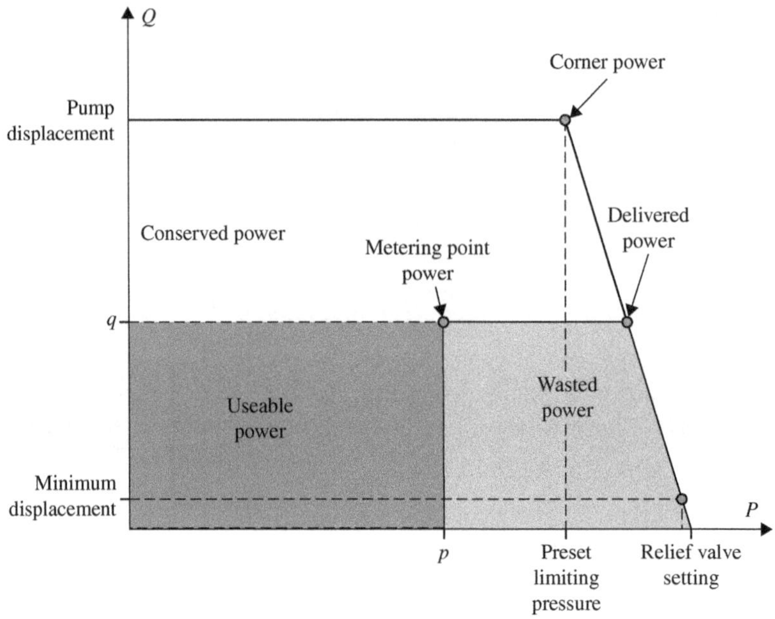


**FIGURE 2.18**
Change of delivered power with a pressure-limiting compensation on a variable-displacement pump.

pressure to push the yoke control actuator to reduce the pump displacement. The higher the discharge pressure, the wider the opening between the discharge port and the yoke control port will become, which in turn will reduce the pressure drop across those ports and push the yoke to further reduce pump displacement, and therefore the discharged flow, until it reaches the maximum allowed operating pressure.

Figure 2.18 shows the flow-pressure curve when a pump is controlled using a pressure-limiting compensator. Comparing this curve to the one presented in Figure 2.15, one sees that the maximum delivered power from this pump is not a single point but a function of pump displacement controlled by the pressure, in the flow-pressure relationship chart. This means that the pump can only supply the maximum deliverable flow under the set displacement limited by the preset pressure limit. Such a fact reveals the basic principle of a pressure-limiting approach to improve energy efficiency on a variable-displacement pump: to deliver only the needed flow for driving the load to improve the efficiency. Many of the load-moving operations are actually designed based on such a control philosophy, making the pressure-limiting approach very attractive.

However, this approach has a limitation: it can achieve a high-efficiency operation only when the system requires using the full pressure capacity to drive the load. When the system is operating a light load, namely, when the operating pressure is below the preset limiting pressure, this energy-saving device will not be able to provide full advantage. In other words, when the operating pressure is below the preset level and the demanding flow is low, that is, when the required metering point power is less than the corner power, the pressure-limiting function alone can save only a limited amount of energy, as illustrated in Figure 2.18. Therefore, it is insufficient to achieve high-energy efficiency in the entire operation range. The major reason for the insufficiency is the pump displacement being controlled solely by the discharge pressure, an indirect measure on actual discharged flow against the demanding flow and therefore is a **flow-sensing control**. One approach to solving this insufficiency is to use a more advanced pump control strategy of load sensing.

### 2.2.3 Load Sensing with Pressure Limiting

As a means of saving the amount of energy carried by the unnecessary high pressure, the **load-sensing control** is designed to supply the flow, with only sufficient pressure to drive the load, and therefore it is also called a **power-matching control**. To make the load sensing fully functional, it is often integrated with the pressure-limiting control. The basic principle of a load-sensing control for pump displacement is illustrated in Figure 2.19. Different from pressure-limiting control, an integrated load-sensing control applies a two-stage pressure compensator control: with one stage for limiting pressure and the other for load sensing. When in an unloaded operating condition, the bias spring in the yoke controller pushes the yoke to its maximum angle to set the pump to discharge the maximum flow. As the discharge pressure increases, the pump outlet pressure is transmitted to both pressure-limiting and load-sensing spools, and attempts to push both spools downward against the preset pressure control springs. The pressure-limiting spring is normally preset at a predetermined level at or below which the pump can supply the full flow, and it operates the same way as described in the previous subsection. The load-sensing spring is normally preset at a **margin pressure** (also called **standby pressure**) required for proper operation of the system. This margin pressure is usually set at the assembly line, in a typical range of 1.0 to 3.0 MPa. During operation, the load-sensing spool is controlled by a balance between the pump outlet pressure and the system operating pressure, plus the preset spring load. Normally, the system pressure is lower than the preset pump outlet pressure. Under the push of the pump-discharge pressure, the load-sensing spool will move downward to open a second passage between the pump-discharge port and the

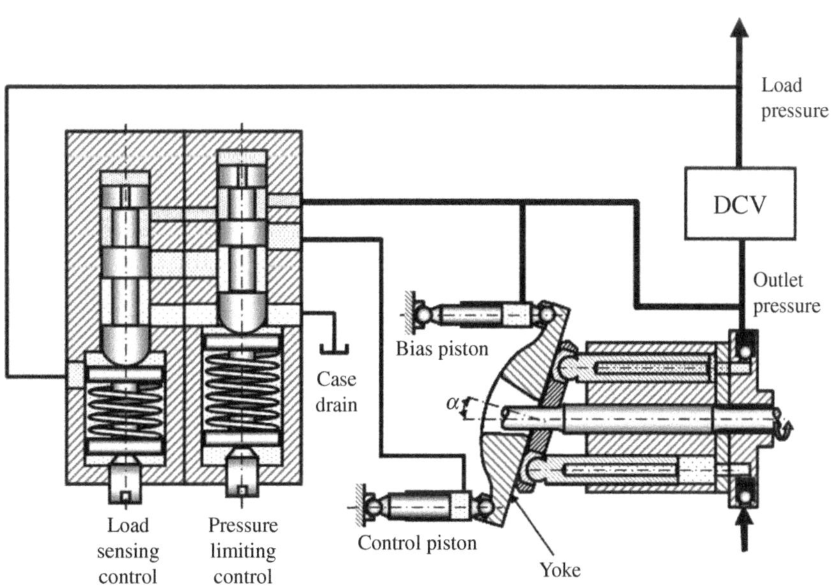


**FIGURE 2.19**
Illustration of the principle of load-sensing with pressure-limiting compensation on a variable-displacement pump.

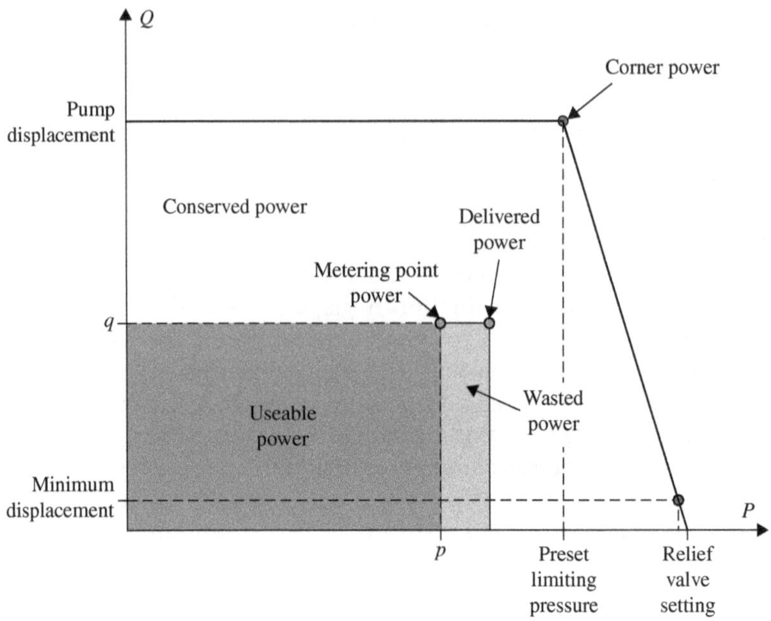


**FIGURE 2.20**
Change of delivered power from a load-sensing with pressure-limiting control on a variable-displacement pump.

yoke control port to further destroke the pump to deliver less flow. The yoke control is then capable of responding to the system pressure directly before it reaches the preset pressure limit. With such a control, the pump can be operated at an outlet pressure that is significantly lower than the line-relief setting by generating only sufficient flow to maintain the needed pressure to drive the load and properly operate control valves. Therefore, the load sensing with pressure-limiting control can save more energy than when the pump is controlled solely by pressure limiting.

Figure 2.20 shows the flow-pressure curve when a pump is controlled using a load sensing with a pressure-limiting compensator. Comparing this curve to the one presented in Figure 2.18, we note that the pressure of discharge flow from the pump can be controlled at a level of system pressure plus a margin pressure instead of the relief pressure. Such a control allows the pump to convert only the needed energy into hydraulic power for driving the load, and makes it operate at higher-energy efficiency. The load sensing with pressure-limiting control is actually a load and flow **dual-parameter control**. The only hydraulic power not used to drive the load in this type of pump is the power needed to maintain a margin pressure required to run the dual-parameter pressure compensator. Similar to a load-sensing-only pressure compensator, this marginal pressure typically ranges from 1.0 to 3.0 MPa.

### 2.2.4 Torque Limiting

While a load-sensing pump offers higher-energy efficiency in fluid power generation by limiting the pressure of the outlet flow to only the needed level, it still requires that the prime mover be sized according to the corner power of the pump. However, many mobile hydraulic systems never require the maximum flow and the maximum pressure at the same time. Instead, these systems are often operated either to drive a light load at a high speed or to move a heavy load at a slow pace, not only for higher-energy efficiency, but more importantly for safe operation. Such a feature makes it possible to reduce the unnecessary power reserve to drive a hydraulic pump by limiting the pump supplying either the maximum flow at a reduced operating pressure or a reduced flow at the maximum operating pressure. A key device to furnish this capacity in a variable-displacement pump is a **torque-limiting compensator**.

As illustrated in Figure 2.21, a torque-limiting compensator often operates in cooperation with pressure-limiting and load-sensing compensators. To support proper torque limiting, a yoke position-sensing piston is required to provide feedback information on the current pump displacement. In this operation, the pressure of pump-discharging flow is fed to a bias piston directly from the pump outlet to the displacement control piston via the compensator assembly. This bias piston pushes the yoke to its maximum angle, corresponding to the maximum pump displacement. The control piston pushes the yoke, reducing its angle to decrease pump displacement when it is energized. Other than pressure-limiting and load-sensing compensation introduced in the previous subsection, the energizing of the control piston is also regulated by torque-limiting compensation. In torque-limiting operations, a displacement-sensing pressure, formed by feeding the outlet pressure of the discharge flow through a sensing clearance on the wall of the yoke position-sensing cylinder, is applied to the head of the torque-limiting spool against a preset torque-limiting spring. When the pump is operating at a large yoke angle, namely, discharging a large flow, the position-sensing piston is largely extended, which will generate a small pressure drop across the piston-sensing clearance and result in a high displacement sensing pressure acting on the torque-limiting spool. In this situation, the torque-limiting spool is very sensitive to any changes in pump outlet pressure. A small increase in the outlet pressure can push the torque-limiting spool to move downward in the configuration depicted in Figure 2.21, which will destroke the pump displacement by feeding the outlet pressure directly to the control piston to decrease the discharge flow. Similarly, when the pump is operating at a small yoke angle, namely, discharging a small flow, the position-sensing

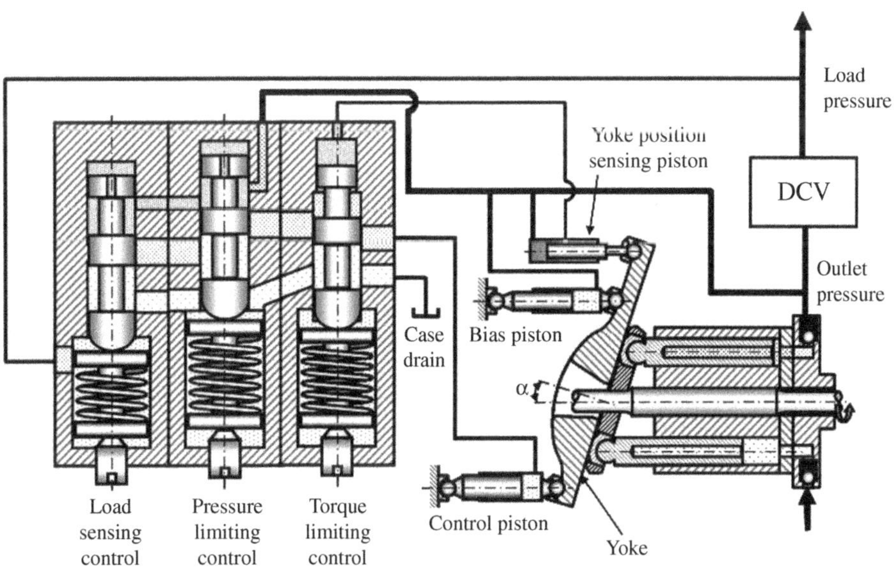


**FIGURE 2.21**
Illustration of the principle of torque-limiting with load-sensing and pressure-limiting compensations on a variable-displacement pump.

piston is mostly retracted, which generates a large pressure drop across the sensing clearance of the position-sensing piston and results in a low displacement-sensing pressure acting on the torque-limiting spool. Under such a case, the torque-limiting spool is pushed by the torque-limiting spring and makes it very insensitive to pressure changes. As a result, the spool connects the path from the control piston to the pressure bleeding port, which will keep the pump operating at its maximum displacement to supply the maximum flow.

Figure 2.22 illustrates the flow-pressure curve when the displacement of a pump is controlled using a torque-limiting compensator. Comparing this curve to the ones presented in Figures 2.18 and 2.20, we observe that the corner power from this pump is no longer a single point, but rather a line of constant values limited by the input torque to the pump in the flow-pressure relationship chart. This means that the pump can either supply the maximum flow at a reduced pressure setting or supply less flow to drive heavier loads when higher pressure is required to push it, all using the same amount of hydraulic power. Such a fact reveals the basic principle of a torque-limiting approach to improve energy efficiency on a variable-displacement pump: either to drive a light load at high speed or to push a heavy load at low speed by efficiently utilizing the limited available energy. This approach is also called a **constant power control**. This torque-limiting pump displacement control approach can also provide overload protection to the prime mover, which implies that it is practical to use a smaller prime mover without losing the capability of driving the load at a proper speed or risking overload stall.

Other than the hydraulic control approaches, the merger of electronic controls and hydraulic systems has made it possible to control pump displacement electronically using electrohydraulic control valves. The major advantage of electronic controls over hydraulic controls is the flexibility it gives in pump control. The electrohydraulic control of a displacement pump is beyond the scope of this textbook.

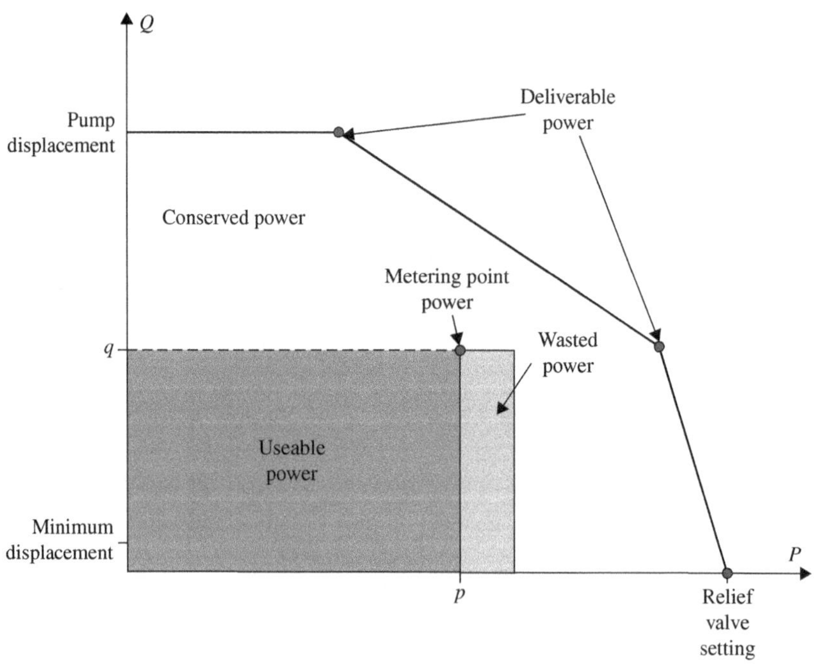


**FIGURE 2.22**
Change of delivered power from a torque-limiting compensated a variable-displacement pump.

## References

1. Akers, A., Gassman, M., Smith, R. *Hydraulic Power System Analysis.* CRC Press, Boca Raton, FL (2006).
2. Burton, A. Developments and trends in hydraulic pump technology. *Power International,* 34: 127 (1988).
3. Cundiff, J.S. *Fluid Power Circuits and Controls: Fundamentals and Applications.* CRC Press, Boca Raton, FL (2002).
4. Dobchuk, J.W., Burton, R.T., Nikiforuk, P. N., Ukrainetz, P.R. Mathematical modeling of a variable displacement axial piston pump. *Proc. ASME Int. Mech. Eng. Cong. & Exp.,* FPST V6: 1-8, Nashville, TN (1999).
5. Eaton Corp., *Load Sensing Principle of Operation.* Eaton Corporation Hydraulic Division, Eden Prairie, MN (1992).
6. Esposito, A. *Fluid Power with Applications (6th Ed.)* Prentice-Hall, Upper Saddle River, NJ, (2003).
7. Goering, C.E., Stone, M.L., Smith, D.W., Turnquist, P.K. *Off-road Vehicle Engineering Principles.* ASAE, St. Joseph, MI (2003).
8. Guan, Z. *Hydraulic Power Transmission Systems* (in Chinese). Mechanical Industry Press, Beijing, China (1997).
9. Harrison, A.M., Edge, K.A. Reduction of axial piston pump pressure ripple. *Proc Instn Mech Engrs: J. Systems and Control Engineering,* 214: 53–63 (2000).
10. Henke, R.W. Understanding pressure compensated pumps. *Hydraulics & Pneumatics,* 38: 80–83 (1985).
11. Hydraulics & Pneumatics. *Fluid Power Basics.* http://www.hydraulicspneumatics.com/200/FPE/IndexPage.aspx. Accessed on November 20 (2006).
12. Kojima, E. Development of a quieter variable-displacement vane pump for automotive hydraulic power steering system. *Int. J. Fluid Power,* 4: 5–14 (2003).
13. Lambeck, R.P. *Hydraulic Pumps and Motors: Selection and Application for Hydraulic Power Control Systems.* Marcel Dekker, New York (1983).
14. Li, Z., Ge, Y., Chen, Y. *Hydraulic Components and Systems* (in Chinese). Mechanical Industry Press, Beijing, China, (2000).
15. McClay, D., Martin, H.R. *The Control of Fluid Power.* John Wiley & Sons, New York, (1973).
16. Manring, N.D. *Hydraulic Control Systems.* John Wiley & Sons, New York (2005).
17. Merrit, H.E. *Hydraulic Control Systems.* John Wiley & Sons, New York (1967).
18. Pease, D.A. *Basic Fluid Power.* Prentice-Hall, Englewood Cliffs, NJ (1967).
19. Pettersson, M., Weddfelt, K., Palmberg, J.-O. Methods of reducing flow ripple from fluid power piston pumps—a theoretical approach. *SAE Transactions: J. Commercial Vehicles,* 100: 158–167 (1991).
20. Vickers, Inc. *Vickers Mobile Hydraulics Manual (2nd Ed.),* Vickers, Inc., Rochester Hills, MI (1998).
21. Yamaguchi, A., Takabe, T. Cavitation in an axial piston pump. *Bulletin of the JSME,* 26: 72–78 (1983).
22. Yeaple, F.D. *Fluid Power Design Handbook.* CRC Press, Boca Raton, FL (1996).
23. Zeiger, G., Akers, A. Torque on the swashplate of an axial piston pump. *Transactions ASME: J. Dynamic Systems, Measurement and Control,* 107: 220–226 (1985).
24. Zhang, Q., Goering, C.E. Fluid power system, In: Bishop, R. (ed.), *The Mechatronics Handbook.* CRC Press, Boca Raton, FL, pp. 10–11~10–14 (2001).

## Exercises

**2.1** Use layperson's language to describe the operating principle of a typical positive displacement pump.

**2.2** Use layperson's language to define the volumetric efficiency of a typical positive displacement pump.

**2.3** Use layperson's language to explain pump cavitation and aeration. List a few practical ways to prevent cavitation and aeration from occurring.

**2.4** Compare fixed-displacement and variable-displacement pumps by listing the main advantages and disadvantages of each.

**2.5** Name two basic designs of gear pumps, and summarize their structural and application characteristics.

**2.6** Name two basic designs of variable-displacement vane pumps, and discuss their structural and application characteristics.

**2.7** What is a pressure-compensated piston pump, and how does it work? What make a load-sensing piston pump more energy efficient?

**2.8** What is the theoretical flow rate discharged from a positive displacement pump illustrated in Figure 2.1 if the diameter of the cylinder is 50 mm, the stroke of the piston is 180 mm, and the piston made 30 reciprocating cycles in one minute?

**2.9** Assume the displacement of a hydraulic pump is 75 $cm^3$. When the pump runs at 1500 rpm and delivers a 100 L/min flow, calculate (a) the theoretical flow the pump can deliver; and (b) the internal flow leakage within the pump under the condition. Also plot the pump volumetric efficiency versus pump speed in the speed range from 1000 to 2000 rpm (assuming the internal leakage remains the same).

**2.10** Calculate the internal flow leakage and the volumetric efficiency of the pump described in Problem 2.9 for cases when it delivers 90 L/min and 105 L/min, and compare the volumetric efficiency curves in speed ranging between 1000 and 2000 rpm.

**2.11** An internal combustion engine puts out 240 $N \cdot m$ torque to drive a hydraulic pump of 50 $cm^3$ displacement to discharge a flow at 25 MPa. Calculate (a) the theoretical torque and (b) the torque losses required to overcome all the resistances. Also plot the pump mechanical efficiency versus discharge pressure in the pressure range from 3 to 30 MPa (assuming the total torque loss remains the same).

**2.12** Calculate the theoretical torque and torque loss of the pump described in Problem 2.10 for cases when it requires 220 and 260 $N \cdot m$ torque to drive the pump. Then calculate and plot the pump mechanical efficiency versus discharge pressure in the pressure range from 3 to 30 MPa.

**2.13** If a 95 $N \cdot m$ torque is required to drive a 0.028 L hydraulic pump operating at 1800 rpm to supply a 45 L/min flow to a system operating at 20 MPa, what is the power efficiency of the pump? What is the relationship of the power efficiency to the volumetric and torque efficiencies of the pump?

**2.14** A vane pump has a cam ring of 85 mm diameter and a rotor of 62 mm diameter, with their centers set eccentrically 9 mm apart. The face width of the vanes is 30 mm. If the actual flow rate discharged from the pump is 105 L/min when the pump is operating at 1800 rpm under its rated pressure, what is the volumetric efficiency of the pump?

**2.15** The full displacement of a variable displacement axial piston pump is 1.6 L, achieved at a 22° swash plate angle. If this pump discharges a 212 L/min flow when operating at 600 rpm with the swash plate set at 6°, what is the displacement of the pump at the condition, and what is the volumetric efficiency under the condition?

**2.16** A balanced vane pump has a cam ring of 50 and 42 mm diameters in a long and short axle, a rotor of 40 mm diameter, and the vane face width of 26 mm. When the pump is operating at 900 rpm to supply pressurized fluid to a system operating at 6.5 MPa, calculate (a) the actual discharge flow rate if the volumetric efficiency is 0.90, and (b) the required mechanical power to drive the pump if the mechanical efficiency of the pump is 0.90 under the operating conditions.

**2.17** What is the theoretical flow rate from a fixed-displacement axial piston pump with a nine-bore cylinder operating at 2000 rpm? Each bore has a 15 mm diameter and all pistons stroke 20 mm.

**2.18** An axial piston pump has seven pistons of 24 mm in diameter evenly arranged in a 72 mm diameter circle. In an operation, the pump needs to discharge flow at 28 MPa. When the swash plate angle is set at 20° and is operating at 1250 rpm, the pump has a volumetric efficiency of 0.97 and a mechanical efficiency of 0.90. What is the pump displacement? What is the actual average discharge rate? What is the required driven torque to the pump?

**2.19** A mobile hydraulic system, normally operating at a system pressure of 10 MPa, uses a 0.02 L pump to obtain a 40 L/min continuous flow supply. If the pump is driven using an 8 kW internal combustion engine, normally operating at 2200 rpm, what are the mechanical, volumetric, and overall efficiencies of the pump?

**2.20** Consider a 0.05 L hydraulic pump that delivers 35 L/min flow to a system operating at 7 MPa. Assuming the pump has 86% overall efficiency and 92% mechanical efficiency, try to specify a prime mover to drive the pump.

# 3

# *Hydraulic Power Distribution*

## 3.1 Hydraulic Control Valves

### 3.1.1 Overview of Hydraulic Valves

Power distribution is the primary function of hydraulic power transmissions, which deliver the right amount of hydraulic power, carried by pressurized flow, to designated actuators in terms of a predesigned strategy within an enclosed system. The core components for constructing an enclosed hydraulic power distribution system are various **hydraulic control valves** and fluid-transporting conductors.

The hydraulic control valves are used to regulate the pressure, flow, and direction of the hydraulic fluid transported within the enclosed system and therefore are often classified into the categories of pressure control, flow control, and directional control valves. However, those categories of valves do not necessarily carry significant differences in their physical configurations. In other words, some valves may be used either as a pressure control, a flow control, or a directional control valve in different applications, with some minor structural modifications or even without any modifications. The functional features for pressure, flow, and directional control valves are discussed in the following section. Here we focus on the configuration features of hydraulic control valves.

According to their configuration features, it is common to classify hydraulic control valves into the two major categories of cartridge valves and spool valves. A **cartridge valve** normally uses a movable **poppet**, shaped either in a ball, a cylinder, or a spool within a constrained space to control the flow passing through the valve. Therefore, it is often also called a **poppet valve**. Figure 3.1 depicts the basic operation principle of a typical cartridge valve. Without loss of generality, the valve illustrated here uses a cylinder-shaped poppet. As shown in the figure, the poppet is pushed by a holding spring on the poppet seat to force the poppet to block the flow path between ports A and B under normal conditions. In operation, the poppet is controlled by two forces: the spring force acting on the top of the poppet and the hydraulic force acting on the bottom of the poppet. The flow path is blocked by the poppet when the spring force is greater than the hydraulic force, and similarly the hydraulic force pushes the poppet open when it is greater than the spring force. Other than the illustrated basic mode of operation, a cartridge valve can also be controlled using a pilot pressure or an electromechanical driver. All modes of operation are discussed in later sections. From the figure, one may also find that the valve cartridge can be thought of as "bodyless" because it requires a supporting body to house the appropriate flow passageways to perform fluid distribution control. **Manifold blocks** are commonly used as the "housing" necessary for cartridge valves and are discussed in detail later in this chapter.

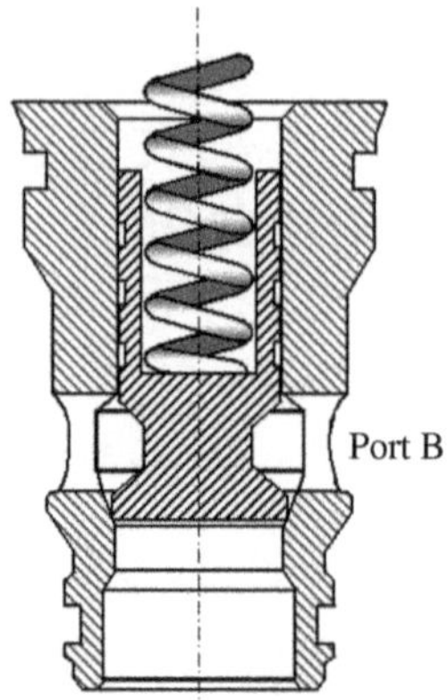


**FIGURE 3.1**
Illustration of the basic operation principle of a typical poppet-type cartridge valve.

There are a wide variety of cartridges, which allow engineers to find an appropriate type of cartridge valve for almost any hydraulic control function, with very few exceptions. The control functions readily available in cartridge configuration today include, but are not limited to, line-relief valves, check valves, sequence control valves, pressure-reducing valves, load control valves, flow control valves, and some specialty valves. Today's cartridge valve technology even allows incorporating two or more functions into single-cartridge housings, such as check and flow control valves, dual-crossover relief valves, and solenoid-operated relief valves.

The cartridge-design approach of hydraulic control valves offers some important advantages. Among them, **flexibility** is probably one of the most attractive features in design practice since the cartridges are normally preassembled as a module and simply need to be mounted into specially designed manifolds on the machine to form different systems. Such a modular feature also grants excellent **serviceability** because a packaged cartridge can be removed and replaced quickly without disturbing any external plumbing. Being **compact** and **lightweight** are other noteworthy advantages of cartridge valves. On the other hand, the convenience of forming valves with different control functions often results in numerous designs of similar valves, which causes challenges in standardization. However, current efforts to create industrial standards for valve design will be helpful in addressing this issue.

Another type of commonly used hydraulic valves are **spool valves**, which use a sliding spool within the valve to control either the pressure, flow, or direction of the fluid passing through. Figure 3.2 depicts the operating principle of a typical spool type valve in controlling flow direction. Without loss of generality, this figure merely illustrates how a sliding spool can be used to control flow in a four-port three-position hydraulic valve. As shown in this figure, this valve often has four ports: the **pump port** (P), **tank port** (T), and **work ports** 1 and 2. The pressurized fluid delivered from the pump port is routed to one of the work ports (port 2 in the illustrated position in Figure 3.2), and bleeding fluid from the other work port (port 1 in this case) is dispatched to the tank port as the spool slides to the left between the passages to open and close flow paths. Often, the flow path between pump port and one of the work ports is called the **pump-to-work** path, and that between the other work port and tank port is called the **work-to-tank** path. Spool valves readily adapt many different spool-shifting schemes, which broaden their use over a wide variety

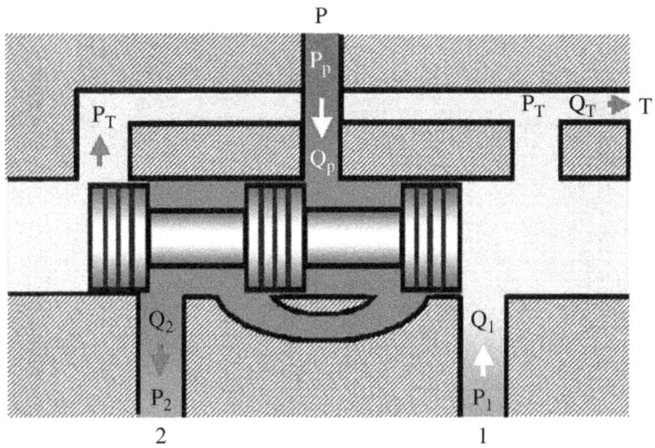


**FIGURE 3.2**
Illustration of operating principle of a typical spool-type hydraulic valve in controlling flow direction.

of applications. The detailed explanation of spool-type valve operation and control principles are discussed in the later relevant sections of this chapter.

Many mobile hydraulic systems require **metering control** or **throttling control** to enable operators to gently accelerate or decelerate a load. To provide such control features, it is common design practice to modify the spools by cutting a few slots, often called **metering notches**, of different shapes and sizes on the **spool shoulders**. Such metering notches provide a gradually changing fluid passage area corresponding to a small change in spool displacement, which in turn results in a fine increase or decrease in fluid flow to gradually increase or decrease the speed of the load movement. Such a beveled or notched edge on the spool is commonly referred to as a **soft-shifting** feature.

### 3.1.2 Fundamentals of Valve Control

In hydraulic power systems, control valves are commonly used to control the pressure, flow rate, and/or direction of the hydraulic fluids in an enclosed system. Other than classifying the valves according to their configuration features, it is also very common to define them according to their functionality features, such as pressure, flow and directional control valves, or based on their control mechanisms, such as on off, proportional, and servo valves.

No matter how it is being classified, a hydraulic control valve achieves the control of flow passing through it by adjusting the flow passage area in the valve. Such an adjustable flow passage area is often referred to as an **orifice area** in engineering practice. Physically, a **hydraulic orifice** is a controllable **hydraulic resistance**. Under the steady-state condition, a hydraulic resistance can be defined as a ratio of pressure drop to the flow rate:

$$R_h = \frac{d(\Delta p)}{dq} \tag{3.1}$$

where $R_h$ is a hydraulic resistance, $\Delta p$ is the pressure drop, and $q$ is the flow rate across a valve orifice.

Control valves make use of many configurations of the orifice to realize various resistance characteristics for different applications. Therefore, it is essential to determine the

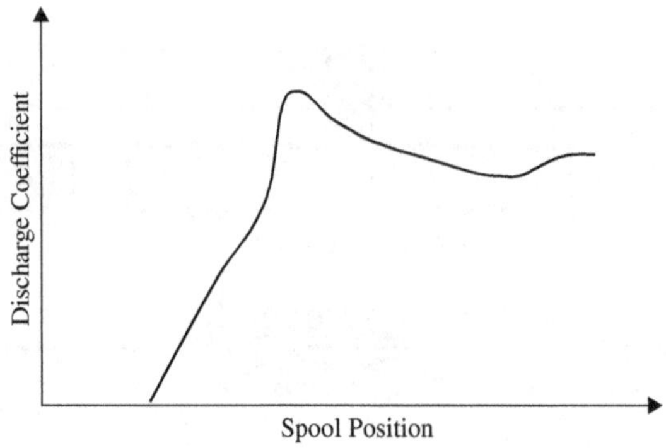


**FIGURE 3.3**
Variation of orifice discharge coefficient versus spool position in a typical spool valve.

relationship between the pressure drop and the flow rate across the orifice. An **orifice equation** is often used to describe this relationship.

$$q = C_d A_o \sqrt{\frac{2}{\rho} \Delta P} \tag{3.2}$$

where $C_d$ is the orifice efficiency, which commonly ranges between 0.6 and 0.8. $A_o$ is the orifice area, and ρ is the density of the fluid.

The pressure drop across the orifice is a system pressure loss. In Eq. (3.2), the **orifice coefficient** plays an important role in determining the amount of flow passing through the orifice and is normally determined experimentally. It has been found that the orifice coefficient varies with the spool position but does not appear to vary much with respect to the pressure drop across the orifice in a spool valve (Figure 3.3). Analysis results obtained from computational fluid dynamics simulation shows that the valve spool and sleeve geometries have little effect on orifice coefficients for large spool displacement. Even though it has been proven experimentally that the orifice coefficient is a variable corresponding to the spool position, it is a very common practice to make this coefficient a constant (often chosen between 0.6 and 0.8) for design calculation because the actual value of the coefficient varies a little when the orifice area surpasses a critical value. Studies show that the orifice opening of a spool valve is greater than this critical value in most normal operations. The orifice area for a spool valve can be calculated using the equation defined as follows:

$$\mathrm{A_O} = \pi d (x - x_0) \tag{3.3}$$

where $d$ is the spool diameter, and $x$ and $x_0$ are the total spool displacement and the spool displacement corresponding to spool deadzone.

When the orifice equation is applied to cartridge valves, use of a relatively higher orifice coefficient is recommended. For instance, the orifice coefficient for a poppet valve is often chosen around 0.8. Another difference in using the orifice equation for analyzing the flow rate through a spool valve and a poppet valve is the determination of the orifice area. Because of the cone structure of a typical poppet, the orifice area can be determined using the following equation:

$$\mathrm{A_O} = \pi d x \sin \alpha \tag{3.4}$$

where $d$ is the poppet diameter, $x$ is the poppet-lifting distance, and $\alpha$ is the angle of the poppet cone.

Control of the valve orifice area is often achieved by controlling the position of the poppet (or the spool in a spool valve). In its normal condition, the poppet is forced on its seat by the spring pushing on the top of the poppet to close the flow path. When pushing to open the path, the opening of the poppet is controlled by the balance of the spring force ($F_S$), the pressure force ($F_P$), and the flow force ($F_F$) acting on the poppet (Figure 3.4). Generally, the spring is always preset with an initial compression to form a nominal force to keep the poppet on the seat. This spring force will be increased as the poppet lifting from its normal position due to such a move further compresses the spring, and it can be determined using this equation:

$$F_S = k(x_0 + x) \tag{3.5}$$

where $F_S$ is the spring force; $k$ is the spring constant; $x_0$ is the initial compression of the spring; and $x$ is the lifting distance of the poppet.

The pressure force acting on a poppet can be determined based on the upstream, downstream, and spring chamber pressures. The determination of the pressure force is closely relevant to the poppet's shape. For instance, the pressure force balance equation for a poppet depicted in Figure 3.4 can be expressed in terms of the upstream, downstream, and spring chamber pressures acting on it as follows:

$$F_P = P_d \frac{\pi d^2}{4} + P_u \frac{\pi\left[D^2 - d^2\right]}{4} - P_s \frac{\pi D^2}{4} \tag{3.6}$$

where $F_P$ is the pressure force' $P_d$, $P_u$, and $P_s$ are the upstream, downstream, and spring chamber pressures, respectively; $d$ is the downstream cavity diameter; and $D$ is the diameter of the upstream cavity and spring chamber as illustrated in Figure 3.4.

Another important force that will act on the poppet is the hydraulic force induced by the flow passing through the poppet, which is a function of the flow rate and fluid velocity passing through the orifice. It can be calculated using the following equation:

$$F_F = \rho q v \cos\alpha \tag{3.7}$$

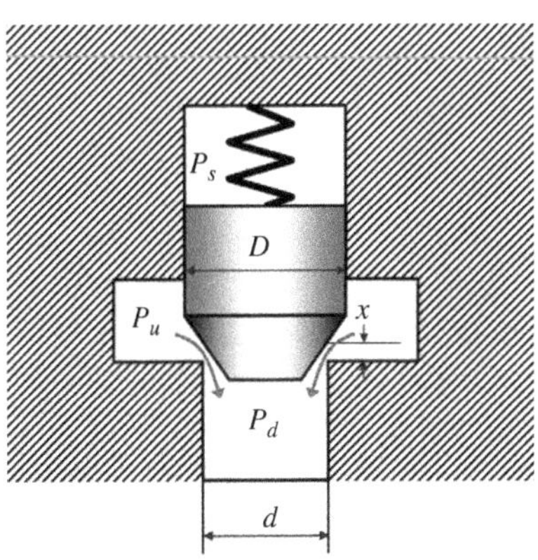


**FIGURE 3.4**
Illustration of the operation principle of a poppet in a typical cartridge valve.

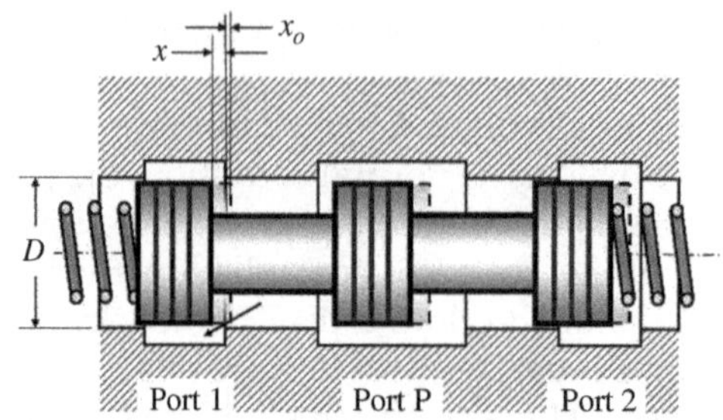


**FIGURE 3.5**
Illustration of a typical spool operation in a spool valve.

where $F_F$ is the flow force $q$ is the flow rate; $v$ is the flow velocity; and $\alpha$ is the flow velocity angle passing through the poppet.

Flow control characteristics of spool valves are similar to those of cartridge valves because spools are also subjected to forces induced by preloaded springs, fluid pressure, and flow forces. It is noteworthy that the pressure force acting on a spool could be either balanced in a direct-actuating valve because of the symmetric spool configuration or unbalanced in a pilot-actuating valve attributed to the pilot-driving force. In the former, an additional mechanical force, such as the actuating force from a solenoid driver or a manual lever, is used to drive the spool to control the orifice opening. In the latter, this task is accomplished by controlling the pilot pressure. As depicted in Figure 3.5, a sliding spool is often centered by two preloaded springs under the balanced spring force to keep the spool in the **central position** (also called **neutral position**) and can be described using the following equation:

$$F_S = k_1\left(x_0 + x\right) - k_2\left(x_0 - x\right) \tag{3.8}$$

where $F_S$ is the spring force; $k_1$ and $k_2$ are spring constants of the left and right springs; $x_0$ is the initial compression of both springs; and $x$ is the spool displacement.

It is common to use identical left and right springs, carrying the same **spring constant** $k$ in a spool-type valve. In this circumstance, Eq. (3.8) can be redefined as follows:

$$F_S = 2kx \tag{3.9}$$

It can always be assumed that the total pressure force acting on a directly actuated spool is zero because of the symmetric distribution of the pressure on a spool. The flow forces acting on the spool can be calculated using Eq. (3.7) with the **flow velocity angle** $\alpha$ being normally taken as 69°.

The theoretical equations introduced here may need some modifications to more accurately represent the force balance and flow control characteristics for various types of hydraulic valves in terms of their specific design features. In practice, it is common to use a variety of specifications to describe design configurations and performance characteristics of hydraulic control valves.

One very important specification for spool-type valves is the **valve lap**, defined as the distance a spool has to move before forming any perceptible degree of valve opening. All spool valves can be classified either as underlapped, zero-lapped, or overlapped valves. As illustrated in Figure 3.6, an **underlapped spool valve** has always-open flow passages formed between the spool and valve ports at its neutral position because the

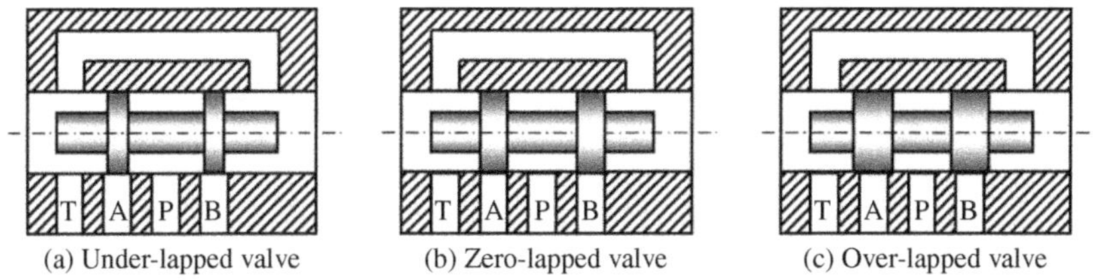


**FIGURE 3.6**
Illustration of the principle of valve laps in a conceptual four-way spool valve. (a) Under-lapped, (b) zero-lapped and (c) overlapped designs.

narrow spool sections cannot completely block the ports (Figure 3.6(a)). This group of valves belongs to the category of **normal-open valves**. When the width of spool sections is designed to be the same size as the valve port, those sections can just block the ports to close the valve (Figure 3.6(b)); therefore, this group of **zero-lapped valves** can be sorted into the normal-close valve category. However, the zero-lap closing is insufficient to stop the flow from leaking through the clearance between the spool and the port. This inevitably results in a significant amount of internal leakage in this type of valve. To reduce such internal leakage, another type of valve, the **overlapped valves**, is used. These valves, which have a certain length of spool section, overlap to cover the port at its neutral position as depicted in Figure 3.6(c) and are commonly used in applications requiring normally close features.

Often, we can call the edge of the spool section forming a flow passage at a port the **valve land**. As depicted in Figure 3.6(b), a critically normal closed four-way spool valve has four individual lands. All lands change simultaneously, with two going into an open state and the other two moving into a closed state, as the spool shifts. Figure 3.7 shows the changes of two orifice areas of the open ports of typical four-way valves with different types of spool laps as the spool moves within the valve.

The arrangement of lands follows the rule that while two of the lands are used to connect the hydraulic supply passages, the other two are used to connect the return route. The former two are commonly called the **powered lands** and the latter two the **return lands**. All those lands can be physically placed at different locations on the spool, which makes each land carry its own opening pattern. Normally, the return lands should be opened slightly ahead of their corresponding powered lands to avoid forming a pressure spike induced by temporarily shutting off the passage for the returning flow.

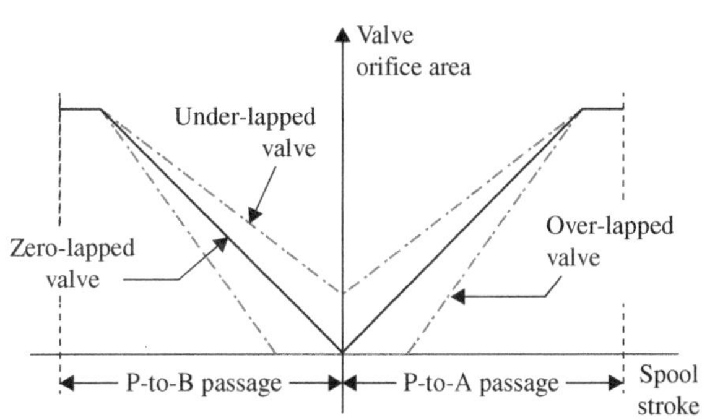


**FIGURE 3.7**
Typical valve orifice area versus valve stroke relationships for different types of spool laps.

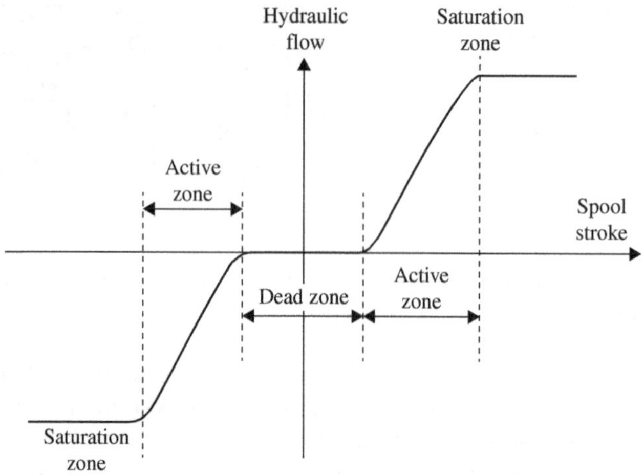


**FIGURE 3.8**
Typical valve-transform curve for a generic spool valve operating under constant pressure.

When a spool shifts to any direction in an operation, two lands will open their flow passages and the other two will be overlapped to close their flow passages, which means that there are only two active lands, a powered land and a return land, in a particular operation.

A **valve transform** curve, presented in a form of the flow passing through the valve at different **spool strokes**, is commonly used to disclose the open pattern of a valve. As presented in Figure 3.8, a typical valve transform curve of an overlapped spool valve normally consists of five characteristic zones within three categories: the **dead zone**, two **modulating zones** (also called the **active zones**), and two **saturation zones**. This valve transform curve reveals several key flow control characteristics of the **response delay**, the **modulation sensitivity**, and the **flow capacity** on both directions of the spool stroke.

The valve dead zone is embraced by two **cracking points** of the valve at where the valve just begins to open its flow passage. Within this zone, shifting the spool will result in no flow passing through the valve due to the overlap. This implies that a zero-lapped valve or an underlapped valve carries an ignorable dead zone or even no dead zone. The dead zone is a very important valve parameter because a substantial dead zone could make a significant impact on response characteristics in both flow and pressure control. Therefore, a dead zone compensator is often used in those systems.

The modulating zones are the actual operating ranges of a valve. A fundamental parameter used to describe the control characteristics of a valve is the **flow gain** of the valve, defined as the change in output flow with respect to the change in spool stroke expressed using the following equation:

$$G_q = \frac{\Delta q}{\Delta x} \tag{3.10}$$

where $G_q$ is the flow gain, $\Delta q$ is the flow rate increment passing through the valve, and $\Delta x$ is the spool stroke increment.

Ideally, a hydraulic valve should have a constant flow gain over the entire modulating range of the valve, and we often call this valve a **linear valve**. In fact, many of the

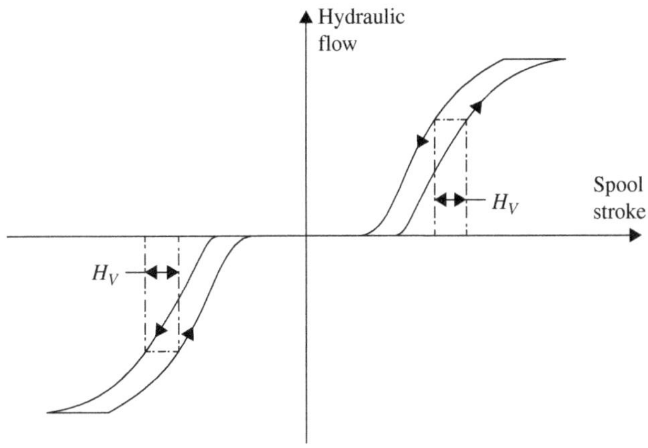


**FIGURE 3.9**
Conceptual illustration of valve hysteresis between spool-stroking and destroking motions.

hydraulic valves do not carry a linear gain in their modulating range and therefore are **nonlinear valves**. The flow gain is not a timely constant parameter determined only by the valve opening because it is also strongly related to the pressure drop across the valve. Engineers often use an **average flow gain** as the design parameter for many applications.

Another important parameter revealing the control characteristics of a valve is the **hysteresis** of the valve. The hysteresis of a valve is defined as the point of widest separation between the flow gain curves with increasing input relative to that with decreasing input, as measured along a horizontal line (Figure 3.9). Physically, it means that the flow-passing capacity through a certain valve opening is lower when the spool is stroking to open the valve than when it is destroking to close. This phenomenon could have a significant impact on flow control performance if the hysteresis were not compensated properly.

The saturation is often induced by the structural limitation of a spool valve under which the orifice area will stay the same after the spool is stroked past this limit point. Therefore, the flow saturation value in a valve transform can be used to estimate the **rated flow** of the valve under a specific operating pressure. This parameter gives the essential information for selecting valve size.

**Example 3.1: Cartridge-Type Valve Opening and Flow Rate**

Assume the poppet diameter of a cartridge-type flow control valve as shown in Figure 3.4 is 20 mm and the angle of poppet cone is 45°. When the poppet is lifted 3 mm by 1.5 MPa of upstream pressure, what will be the flow rate passing through the valve if the downstream of the valve is connected to the pressurized reservoir with a 0.3 MPa back pressure (assume valve orifice coefficient of 0.8 for cartridge-type valve and fluid density of 850 $kg \cdot m^{-3}$ for hydraulic fluids).

a. The orifice area of the valve can be calculated using Eq. (3.4):

$$
\begin{aligned}
A_o &= \pi dx \sin\alpha \\
&= 3.14 \times 2.0 \times 0.3 \times \sin(45°) \\
&= 1.33\left(cm^2\right)
\end{aligned}
$$

b. The flow rate passing through this valve can be calculated using Eq. (3.2):

$$Q = A_o C_d \sqrt{\frac{2}{\rho}\Delta P}$$

$$= 1.33 \times 10^{-4} \times 0.8 \times \sqrt{\frac{2}{850} \times 1.2 \times 10^6}$$

$$= 5.65 \times 10^{-3}\left(m^3 \cdot s^{-1}\right) = 339\left(L \cdot \min^{-1}\right)$$

DISCUSSION 3.1: An orifice area of a valve is determined by the geometric configuration of the valve flow control element within a valve body, as does the orifice coefficient. The flow rate controlled by the valve can be reasonably estimated using an orifice equation.

### 3.1.3 Pressure Control Valves

A hydraulic power transmission system employs the pressurized flow to transport the energy to perform desired work. Therefore, **pressure control** of the working fluids is one of the most basic control functions in hydraulic power transmission. Different types of pressure control valves are designed to perform a variety of functions, from keeping systems safely operating below a maximum allowable pressure limit to allowing only certain pressure fluids into a particular branch in a circuit. The commonly used pressure control functions include pressure limiting, reducing, sequencing, balancing, and releasing. Most pressure control valves are normally closed valves, except for pressure-reducing valves, which are normally open. In principle, all pressure control valves are operated using the balance between the pressure force(s) and the preload spring force(s) on the poppet or the spool to control the orifice opening to achieve the pressure control goal by means of controlling the pressure drop across the valve.

The **relief valve**, designed to maintain a maximum allowable pressure for a system, is commonly used in hydraulic systems to safeguard system operation. We often define the pressure at which a relief valve first opens as the **cracking pressure**—the pressure at which the valve releases full flow the **full-flow pressure**—and we call the difference between full-flow and cracking pressures the **pressure override** (Figure 3.10).

The pressure override in relief valves is caused mainly by the increased pressure force requirement for enlarging the valve opening. An ideal relief valve should have a small pressure override to attain consistent operation characteristics. Pressure relief valves of either direct-acting or pilot-operated are designed to achieve different pressure overriding characteristics.

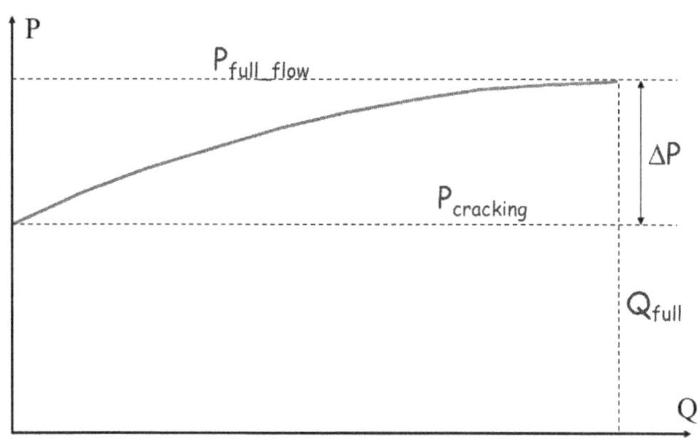


**FIGURE 3.10**
Illustration of the pressure override characteristic in a typical pressure relief valve.

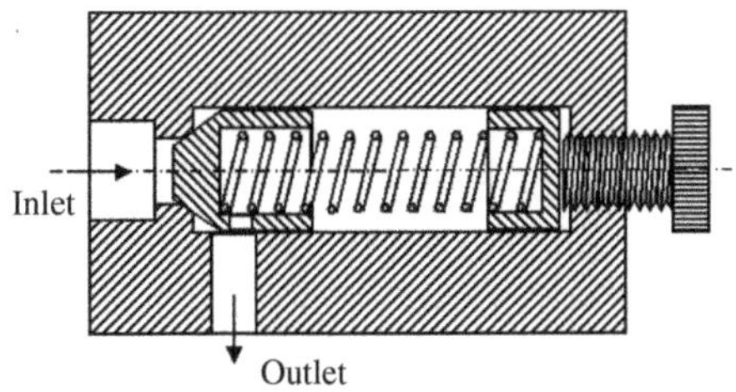


**FIGURE 3.11**
Illustration of the operating principle of a typical direct-acting pressure-relief valve.

**Direct-acting relief valves** often use a poppet, a ball, or a spool to perform the pressure control. Figure 3.11 depicts the operating principle of an adjustable, normally closed poppet-type pressure relief valve. Just like a typical cartridge valve, it is also common to use the poppet as the control element in a typical relief valve. In the depicted basic relief valve, the pressure force in the inlet side is acting on the poppet against the spring force preloaded by compressing the spring in terms of the maximum allowable operating pressure of the system. If the pressure force is smaller than the spring force, the spring pushes the poppet on to its seat to close the flow path, and the valve holds the pressure. When the pressure force exceeds the spring force acting on the poppet, it pushes the poppet open to let the excess fluid bleed from the system, avoiding excessive pressure build-up. The higher the pressure, the more the valve will open, which will allow more flow to bleed from the relief valve until the valve is widely open at full-flow pressure.

Direct-acting valves are generally used for small flow applications. This type of valve normally has negligible leakage below the cracking pressure and responds rapidly after surpassing the cracking pressure, which makes these types of valves ideal for relieving shock pressures. The rapid responding characteristics will also make the valve open or close quickly when the pressure exceeds or drops below the cracking pressure and results in a frequent alteration between opening and closing. This feature also makes this type of valve suitable for use as safeguard valves to prevent damage caused by high-pressure surges or to relieve pressure caused by thermal expansion. In addition, the differential between cracking and full-flow pressure on direct-acting poppet valves is normally high, which, together with the high frequency of open–close alterations, makes this type valve not recommended for precise pressure control.

The major shortcoming of direct-acting relief valves is the high differential between cracking and full-flow pressures, namely, **pressure inconsistency**. To solve this problem, a pilot-operated design is often used. As shown in Figure 3.12, a pilot-operated relief valve operates in two stages: a **pilot relief** and **main relief** stage. Each stage performs by using a separate

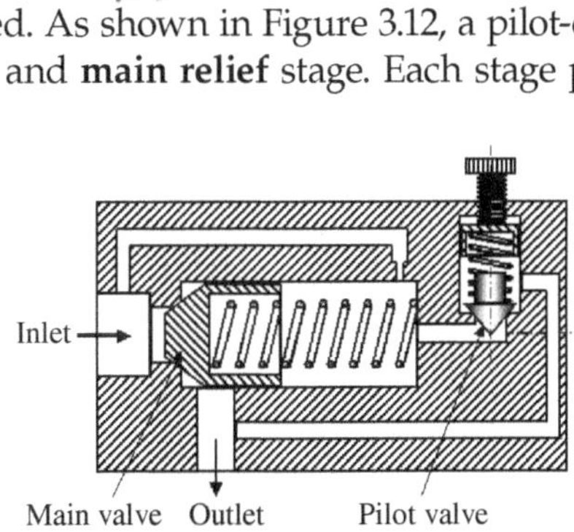


**FIGURE 3.12**
Illustration of the operating principle of a typical pilot-operated relief valve.

pressure-acting valve. At the pilot relief stage, a small, spring-biased relief valve (often built into the main relief valve as illustrated in Figure 3.12) acts as a trigger to control the main relief valve when the system pressure reaches the cracking pressure. The system pressure can be connected either directly from the valve inlet port or from a remote point carrying the system pressure. Under normal conditions, the system pressure is below the cracking pressure, the main spring chamber of the relief valve keeps the same pressure as the system, and the pilot spring chamber holds no hydraulic pressure. Under these conditions, the pilot spring holds the pilot valve against the system pressure to keep the valve closed. The main spring chamber pressure and the spring, with the pressure providing the main force, keep the main valve closed. The spring is used mainly to push the valve back to the closed position when external forces are removed. When the system pressure exceeds the cracking pressure preset for the pilot valve, this pressure will force the pilot valve to open, which in turn quickly reduces the pressure in the main spring chamber due to a large pressure drop formed by the orifice restriction between the pilot pressure source and the main spring chamber. The release of the main spring chamber pressure removes the main resistance force that acts on the back of the main valve, allowing the system pressure to act on the front of the valve to push the valve open.

A pilot-operated relief valve usually has much less pressure override than a direct-acting counterpart because the former uses a much softer spring than the latter. This feature allows a pilot-operated relief valve to be set at a much higher cracking pressure than a direct-acting one for the same full-flow pressure. It is not uncommon for a pilot-operated relief valve to be set at a cracking pressure at 90% of the full-flow pressure, while a direct-acting valve is normally set at a much lower level (Figure 3.13). This higher cracking pressure can effectively improve system efficiency because less pressurized fluid is discharged during the releasing process. While a pilot-operated relief valve can maintain a system operating at a more constant pressure, its response is generally slower than that of a direct-acting counterpart, mainly because of the two-stage procedure in pressure release. In addition, the cost is much higher to manufacture pilot-operated relief valves than direct-acting ones because of their complexity in structure.

**Pressure-reducing valves** are practical components for maintaining a lower pressure in secondary or branch circuits in a hydraulic system. Because of their unique function in pressure control, this type of valve is normally open and inserted in between the main circuit and the branch circuit. The main application of pressure-reducing valves is to use pressurized fluid supplied from one hydraulic pump to drive multiple actuators, at different operating pressures. The basic principle of a pressure-reducing valve is to form a

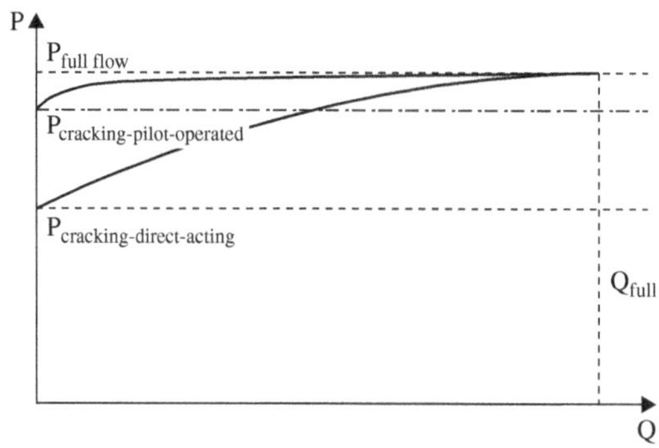


**FIGURE 3.13**
Performance comparison between typical direct-acting and pilot-operated relief valves.

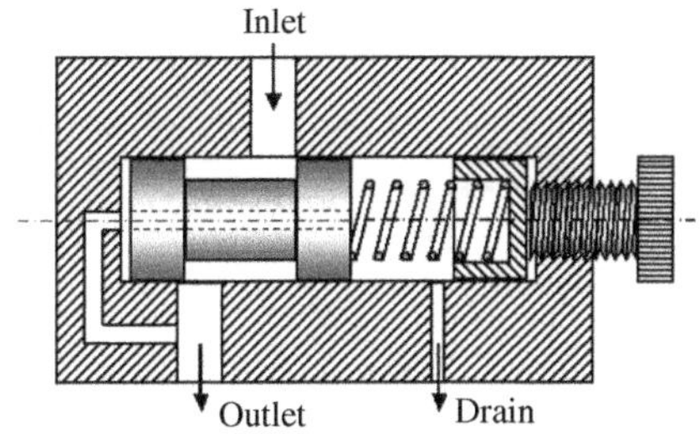


**FIGURE 3.14**
Illustration of the operating principle of a typical direct-acting constant-pressure-reducing valve.

controlled pressure drop across the valve to provide a lower pressure to a downstream branch. Similar to relief valves, there are also the two types of direct-acting and pilot-operated pressure-reducing valves.

To maintain a lower pressure in a second circuit means to limit the maximum pressure available in that circuit regardless of the pressure changes in the main circuit by adjusting the valve opening and also prevent any backflow generated by the workload getting into the main circuit by closing the valve. To offer such a function, pressure-reducing valves commonly use a spool to adjust the orifice opening, as illustrated in Figure 3.14. During the operation, an adjustable spring holds the valve at a normally open position. The spool will sense the pressure at the outlet port. As the downstream pressure increases, a larger hydraulic force will act on the left side of the spool, as illustrated in the figure, which pushes the spool to close the orifice partially against the spring force. This partially opened port will allow enough fluid to flow passing the valve with a certain pressure drop, to maintain the downstream circuit at a constant operating pressure. Compared to a relief valve which senses the pressure at upstream, a pressure-reducing valve senses the pressure signal from downstream and therefore operates in reverse fashion from a relief valve.

When the downstream pressure exceeds a preset value, this pressure will push the spool to close the path completely. In such a case, the unavoidable internal leakage in this spool valve could cause a pressure build-up in the branch circuit. To prevent such a potentially damaging situation from occurring, it is common to design a bleeding orifice at the centerline of the spool, as illustrated in Figure 3.14, to provide a drain passage for bleeding the leakage to the reservoir.

Figure 3.15 depicts the principle of a pressure-reducing valve to keep a constant pressure at the branch circuit regardless of the system pressure, as long as pressure in the main circuit is higher than that in the branch. Other than this **constant-pressure-reducing**

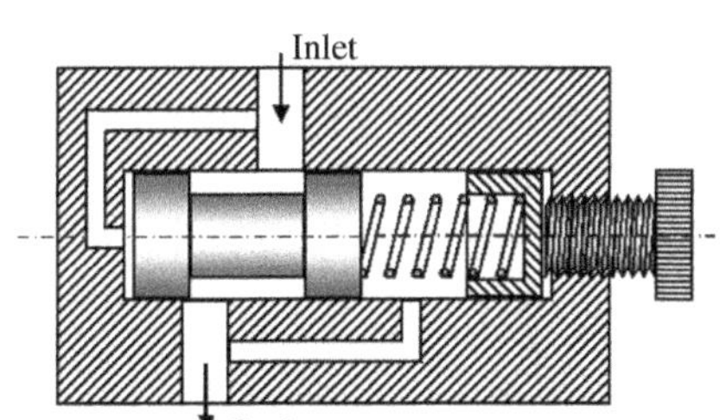


**FIGURE 3.15**
Illustration of the operating principle of a typical direct-acting fixed-pressure-reducing valve.

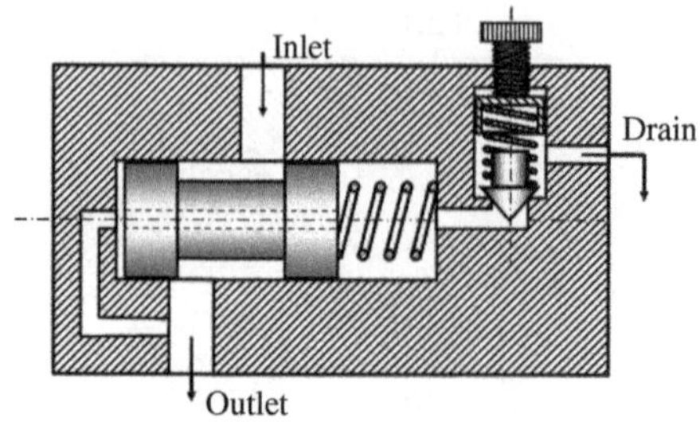


**FIGURE 3.16**
Illustration of the operating principle of a typical pilot-operated pressure-reducing valve.

control, it is also possible to achieve **fixed-pressure-reducing** control to keep a fixed pressure reduction for the downstream regardless of the upstream pressure. As depicted in Figure 3.15, this fixed-pressure-reducing valve operates by balancing the force exerted by upstream pressure against the sum of the forces exerted by downstream pressure and the spring. Because the pressurized areas on both sides of the spool are equal, the fixed reduction is that exerted by the spring.

Similar to relief valves, the pressure control performance of direct-acting pressure-reducing valves is also affected by the high differential between the cracking pressure and the target pressure attributing to the stiffness of the spring. The piloted-operated approach offers similar advantages to achieve better control performance over the direct-acting valves for the spool being balanced hydraulically by downstream pressure at both ends (Figure 3.16). A small pilot relief valve, usually built into the main valve body, releases the fluid in the main spring chamber to the tank when its pressure (the same as the downstream pressure) reaches the pilot valve spring setting. This fluid drain will cause a pressure drop across the spool due to orifice effect and will form a pressure differential to shift the spool toward its closed position against the light main spring force. To avoid a significant amount of energy loss from the pilot operation, the pilot valve releases only enough fluid to position the main valve spool to control the flow through the main valve to accomplish the pressure-reducing function. When the downstream pressure is high enough during a portion of the operation cycle, the main valve will be fully closed to stop flow supply to the circuit, and any internal leakage from the high-pressure end to the main spring chamber will be returned to the reservoir through the pilot-operated relief valve.

Pilot-operated pressure-reducing valves generally have a wider range of spring adjustment than their direct-acting counterparts and can generally provide more repetitive accuracy in pressure control. However, these types of valves are also much more contamination sensitive because the long orifice paths are easily blocked, which will consequently cause the main valve to fail to close properly.

**Sequence valves** are a type of pressure control device used to control more than one actuator in separate branches operating in a predefined order of sequence. To perform this function, a typical sequence valve is a three-way valve, and normally open to the primary branch. As depicted in Figure 3.17, a typical three-port direct-acting sequence valve basically resembles a direct-acting pressure-reducing valve except for having a secondary port. This valve regulates the sequence following the control logic that an adjustable spring pushes the spool to a position which fully opens the port to the primary branch and completely closes the port to the secondary branch. With a normally open position, a sequence valve supplies fluid freely to the primary branch to perform its first function until the pressure reaches the designed setting for switching functions.

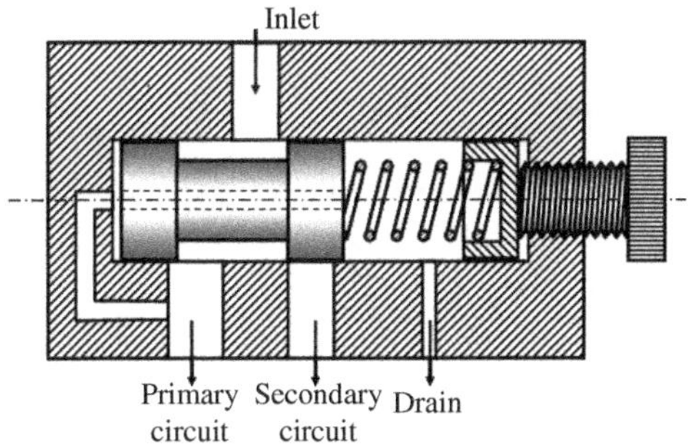


**FIGURE 3.17**
Illustration of the operating principle of a typical three-port direct-acting sequence valve.

After that, the valve spool will be pushed by the rising pressure to right against the spring force to open the secondary branch. The sequence valve then opens the path to the secondary branch and permits the pressure fluid flowing into this branch to perform the second function.

Sequence valves sometimes use a check valve to provide an additional bypass to release excess flow from the secondary to the primary branch. As depicted in Figure 3.18, the typical sequencing-control function is provided only when the flow is from the primary to the secondary branch. However, in some special applications, it is desirable to provide a path allowing the flow from the secondary branch back to the primary branch during a retraction motion. A check valve (introduced in the next section) added between the secondary and primary branches will serve this function. Sequence valves can also be operated remotely.

**Counterbalance valves** are used primarily to maintain a set pressure in part of a circuit to counterbalance a weight or an external force to keep the load from free-falling. Many off-road vehicles use this type of valve as a speed-limiting control device in various applications, often inserted in the line connecting the cylinder rod-end port and directional control valve with its primary port connected to the cylinder and the secondary port connected to the valve. As depicted in Figure 3.19, a counterbalance valve stops flow from its primary port to its secondary port unless the pressure at the primary port overcomes the preset spring force, which is normally set slightly higher than the pressure required to keep the load from free-falling. A check valve is often used to provide an additional

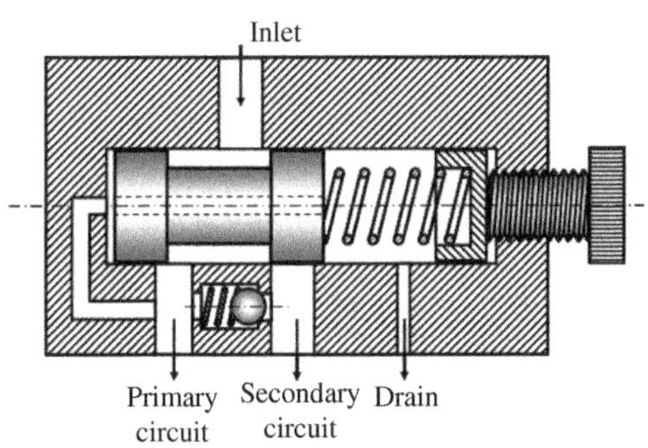


**FIGURE 3.18**
Illustration of the operating principle of a typical direct-acting sequence valve that allows a reverse flow from secondary to primary branch.

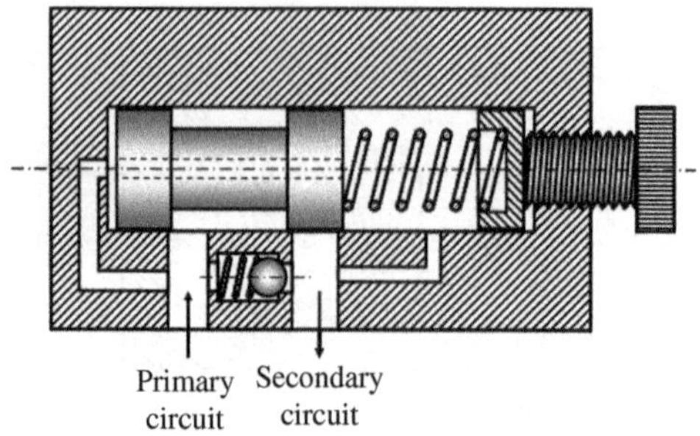


**FIGURE 3.19**
Illustration of the operating principle of a typical counterbalance valve integrated with a returning bypass check valve.

path to reverse the fluid so that it flows freely from the secondary to the primary branch. Similar to other types of pressure control valves, counterbalance valves can also be operated remotely.

Comparison of the structural features between the sequence and the counterbalance valves, depicted in Figures 3.17 and 3.19, shows that the counterbalance valve can be modified from a sequence valve by blocking its inlet port. Another important difference is that counterbalance valves are usually drained internally to the secondary port, while the sequence valve is normally drained to the reservoir. However, during the counterbalancing operation, the secondary port is normally connected to the reservoir, and so is the drain path for the spring chamber. Similarly, during the reversing operation, the check valve will connect the secondary port and the drain path to the reservoir via the check valve.

**Unloading valves** are another important type of pressure control valve and are often used to unload pumps by means of releasing pump-discharging flow directly to the tank at a low pressure after the system has reached sufficient pressure to hold the load. Structurally, unloading valves are very similar to counterbalance valves, except that the feedback pressure to the unloading valve is normally sensed remotely where the load is located as compared to being sensed at the primary port in a counterbalance valve (Figure 3.20). As depicted in the figure, the preset spring keeps the valve closed under its normal condition. When a high-pressure external pilot signal is transmitted to the opposite end of the valve spool through the pilot port, it will push the spool against the adjustable spring to open the valve, allowing dumping of the discharge flow.

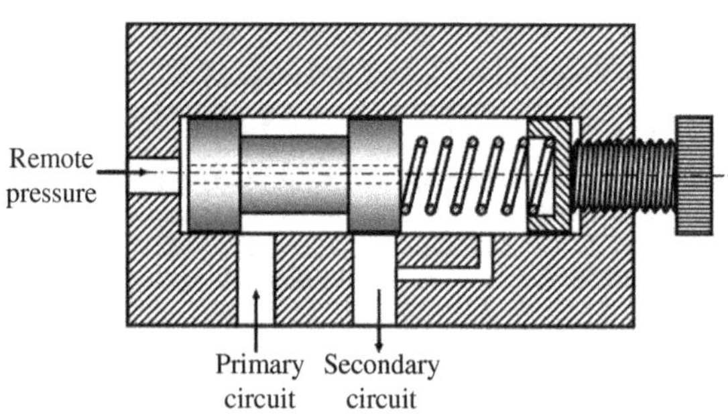


**FIGURE 3.20**
Illustration of the operating principle of a typical unloading valve.

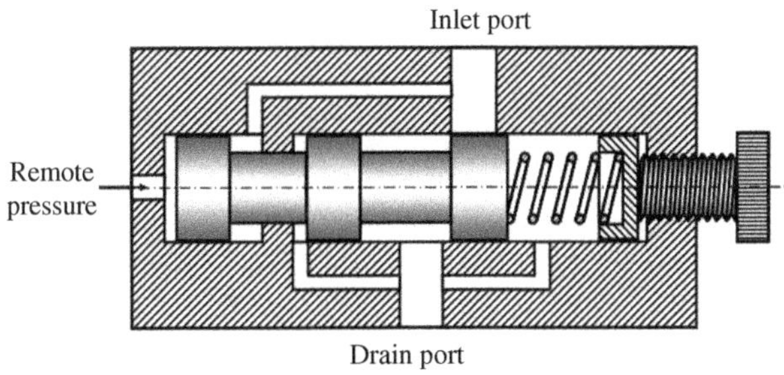


**FIGURE 3.21**
Illustration of the operating principle of a typical unloading valve for accumulator circuits use.

Unloading valves are also commonly used in accumulator circuits to unload the pump after the accumulator has been fully charged. For this application, some necessary modifications are required. As depicted in Figure 3.21, the valve is normally pushed closed by the spring while charging the accumulator. After the accumulator is charged, the system pressure is transmitted through the remote pressure port and pushes the unloading valve spool against the spring to open the draining path to unload the pump. Every time pressure in the accumulator drops below a preset level by the spring, the spring will push the spool to close the draining path, and the charge and unload cycle repeats.

Unloading valves are also made with a pilot to control the main valve, as are any of the other types of pressure control valves, to reduce the pressure override margin.

### 3.1.4 Directional Control Valves

To deliver the energy to the place where useful work is performed is a fundamental function of hydraulic power transmission. Hydraulic systems commonly utilize different types of directional control valves to supply the pressurized fluids to targeted users in a proper route. In other words, direction control does not primarily control the amount of energy being delivered but directs the energy transfer to the proper place at the proper time in a hydraulic system. Therefore, directional control valves can be thought of as switches in hydraulic systems that make the desired connections by means of directing the high-energy input flow to the actuator inlet and provide a return path for the lower-energy fluids back to the reservoir.

To provide such a switching function alone, a basic directional control valve is often operated under a "bang-bang" control mode, in which the valve either fully opens or fully closes a flow path, usually within an instant, to rapidly provide and discontinue the flow supply. Such a control mode makes a typical directional control valve a discrete valve that can only be shifted from one discrete position to another, such as forward, backward, or neutral. While this control mode satisfies the basic functional requirements for hydraulic control, it will certainly induce large pressure surges, causing fluid hammer effects and resulting in very jerky operations under many conditions. To solve this problem, more sophisticated designs of directional control valves, such as proportional directional control valves, which control both the flow direction and rate at the same time, have been widely applied in many mobile hydraulic systems. While still belonging to the direction control category, proportional directional control valves will be introduced in a later section to comprehensively explain their operating principles on direction-flow dual controls. In this section, we focus on the operating principles of a few basic directional control

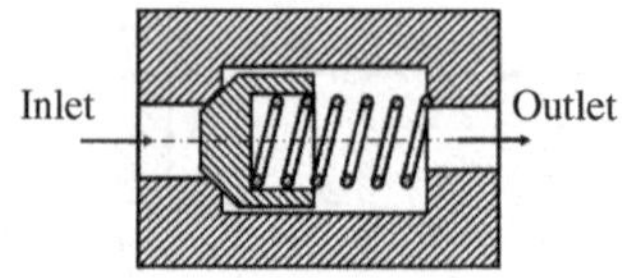


FIGURE 3.22
Illustration of the operating principle of a typical check valve.

valves commonly used in many hydraulic systems, such as check valves, shuttle valves, and multiple port/position directional control valves.

**Check valves** are the simplest type of directional control valve. Basically, a check valve is designed to permit the fluid to flow freely from one port (the inlet port) to another (the outlet port) and prevent reversible flow. As illustrated in Figure 3.22, a typical check valve uses a spring to keep the valve at a normally closed position, which keeps the fluid from flowing unless the inlet pressure acting on the poppet overcomes the spring force. To avoid too much energy loss in overcoming the spring force, a light spring is often selected for holding the poppet. When the fluid attempts to flow backwards from the outlet port to the inlet, the fluid pressure will push the poppet, along with the spring force, to securely close the valve.

In some applications, we may want to allow the fluids to flow backwards if a specifically defined operation condition is satisfied. For example, if we want to allow an actuator installed in the downstream of the check valve to operate reversibly, we will have to open this check valve to permit the fluid in this circuit to bleed. To offer such a function, a **pilot-operated check valve** (Figure 3.23) is always used to provide the **undercondition valve-open control**. As illustrated in this figure, a pilot-operated check valve operates in the same way as a basic counterpart during the normal operating conditions. However, when the pilot port of the valve receives a high pressure, this pilot pressure will drive the pilot actuator to push the poppet open to bleed the returning flow through the **forced-open** check valve from the outlet port to the inlet. Pilot check valves are often used to lock hydraulic cylinders in position.

In some hydraulic systems, one can find that a special type of check valve is being used which connects three lines: two upstream lines and one downstream. These types of check valves are called **shuttle check valves** (often simply called **shuttle valves**). Functionally, a shuttle valve will selectively permit the flow from the upstream circuit of higher pressure to pass through the valve and meanwhile block the other upstream circuit. For example, under the condition depicted in Figure 3.24, this shuttle valve connects the path between inlet A port and the outlet port due to a higher pressure at the inlet A port; this higher pressure pushes the ball to block inlet B port. When the pressure of inlet B circuit increases to exceed that in inlet A circuit, the higher pressure will then push the ball to block the

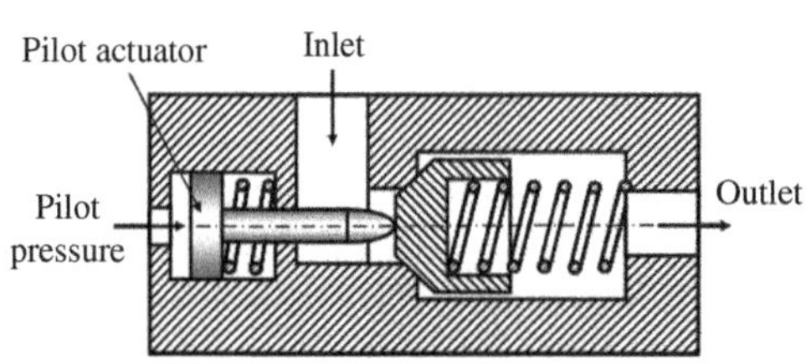


FIGURE 3.23
Illustration of the operating principle of a typical pilot-operated check valve.

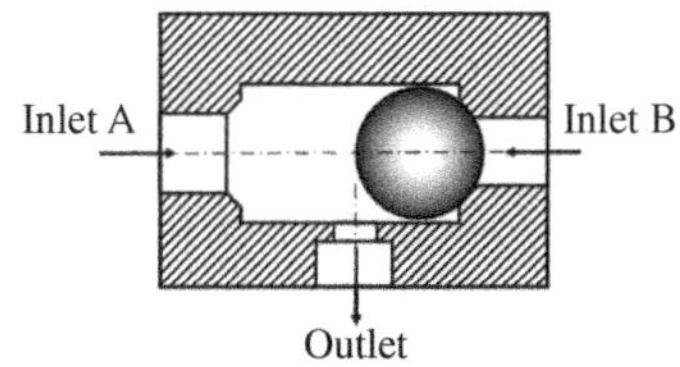


**FIGURE 3.24**
Illustration of the operating principle of a typical shuttle check valve.

A port and connect the inlet B port to the outlet. Shuttle valves are commonly used in circuits where the higher of two pressures is to be sensed or to be connected, such as in load-sensing circuits and hydrostatic transmission circuits.

Extending the concept of direction control from check valves, the two primary characteristics for directional control valves are the total number of ports (also called **ways**) and the total number of flow states (also called **positions**) a valve can attain. **Valve ports** provide paths for the fluid to be transported to or from other components, and **directing states** refer to the number of distinct flow-passing states the valve can provide. When defined based on these two characteristics, directional control valves are commonly sorted into classes such as two-way two-position, three-way two-position, four-way two-position, four-way three-position, and six-way three-position. Directional control valves of more than six ways can also be found in some applications. Figure 3.25 depicts the typical structural features of a few commonly seen spool-type directional control valves and their corresponding ISO symbols. For a **three-way two-position valve** as depicted in Figure 3.25(b), the **P-port** receives pressurized fluid from the pump, the **A-port** is

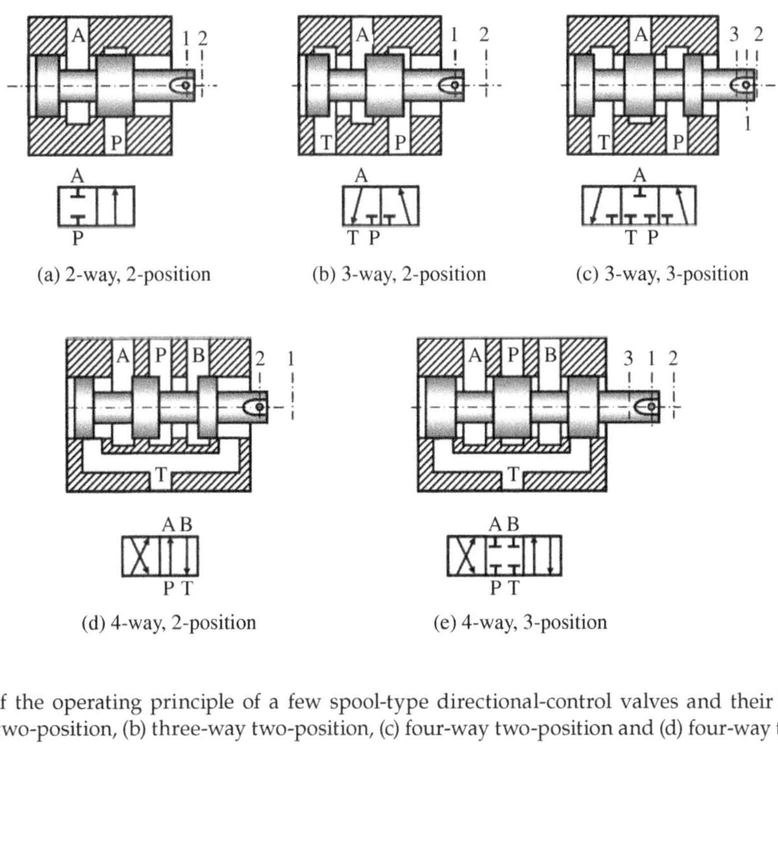


**FIGURE 3.25**
Illustration of the operating principle of a few spool-type directional-control valves and their ISO symbols. (a) Two-way two-position, (b) three-way two-position, (c) four-way two-position and (d) four-way three-position designs.

connected to the actuator to transport flow to or from the actuator, and the **T-port** routes the returning fluid back to the reservoir. The **four-way three-position valve** represented in Figure 3.25(e) can be shifted to any of three discrete positions and is one of the most commonly used directional control valves in mobile hydraulic systems. As shown, all ports of this valve are blocked when the spool is located at the neutral position, so no fluid will flow under this state. Therefore, it is often called a **normally closed valve**. Shifting the spool to the right of the valve routes the pressurized fluid from pump port to A-port (**P-to-A**) and leads the returning fluid from B-port back to the reservoir (**B-to-T**). Similarly, shifting the spool to the left of the valve connects the **P-to-B** and **A-to-T** fluid pathways to support desirable operations at the hydraulic actuator. When the spool returns to the center position, the valve again blocks all flow.

Spool-type valves are widely used in mobile hydraulic systems because these types of valves can easily be shifted to two, three, or more positions to route fluid between different combinations of inlet and outlet ports. While using this type of valve, the flow-directing state at their neutral position differentiates the control characteristics of a hydraulic system. In terms of their flow-directing states, there are four most commonly applied neutral position designs: closed center, open center, tandem center, and float center (Figure 3.26). All of these designs will connect flow pathways of P-to-A and B-to-T when shifting the spool to the left and will connect P-to-B and A-to-T when shifting to the right.

When a **closed-center valve** is used, all the ports are under a normally closed condition (Figure 3.26(a)), and the hydraulic system will hold the pressure set either by the line-relief valve or by the load-sensing control on the hydraulic pump. This type of valve is commonly used in hydraulic systems with a variable-displacement pump, often in load-sensing systems. An **open-center valve** offers a normally open condition to all ports

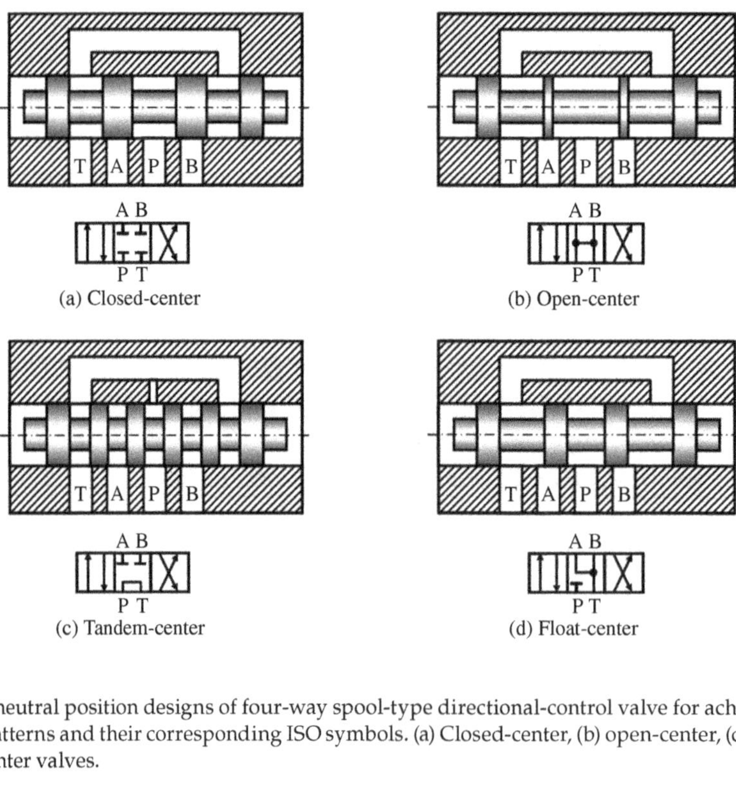


**FIGURE 3.26**
Four common neutral position designs of four-way spool-type directional-control valve for achieving different flow-routing patterns and their corresponding ISO symbols. (a) Closed-center, (b) open-center, (c) tandem center and (d) float center valves.

(Figure 3.26(b)), which means all the ports in this valve are connected at its neutral position. Since both the pump port and working ports are connected to the tank port, the one carrying the least resistance, the pump flow is then directed to the tank and the actuator is unable to support any load. Open-center valves are often used in hydraulic systems with a fixed-displacement pump. Similarly, often used in hydraulic systems powered using a fixed-displacement pump, the **tandem-center valve** (Figure 3.26(c)) connects the pump and tank ports to permit discharge flow return to the tank freely while it blocks both working ports to keep the pressure on both sides of an actuator to hold the load while setting the valve at its neutral position. Functionally opposite to the tandem-center valve, a **float-center valve** (Figure 3.26(d)) connects both working ports to the tank port to permit the working flow to freely return to the tank, but it blocks the pump port from the tank port to prevent pump flow from being drained to the tank at the neutral position. Float-center valves are also often used in hydraulic systems powered by a variable-displacement pump.

As explained in Section 3.1.2, when stroking a spool within the valve body, it will change the relative positions of its lands and valve body edges, and such position changes will in turn form different flow paths in performing the designed proportional flow control. Taking the tandem-center valve as an example, let us see how different flow paths are formed and changed corresponding to the stroking of the spool. As depicted in Figure 3.27, when the spool is at its neutral position (normally at its center position), it blocks flow from the pump to both working ports of A and B (P-to-A and P-to-B), as well as from both working ports to the tank (A-to-T and B-to-T). But it has the flow path of pump to tank (P-to-T) fully open to bypass the pump, discharging flow directly back to the tank to lower the pump-discharge pressure. In direction controlling, the spool is stroked to a desired position for a specific operation, say $x$ to the right direction as illustrated in Figure 3.27 as an example, which makes the P-to-T path partially closed and the P-to-A and B-to-T paths partially open. It should be pointed out that the B-to-T path must always be opened slightly ahead of the opening of P-to-A path in order to reliably release the fluid in the back-chamber to avoid excess back pressure from building up. By switching the spool to the opposite direction, this valve can change the controller hydraulic actuator to move in reverse.

All of the aforementioned valves can be used to form multistage, mostly two to four stages, hydraulic valves, either packaged as sectional valves or enblock valves or their

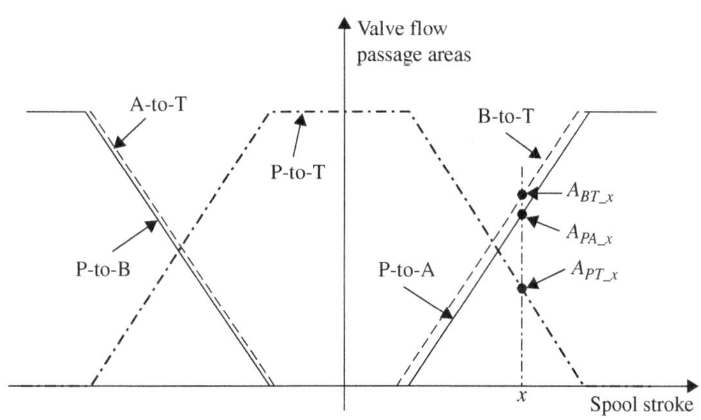


**FIGURE 3.27**
Conceptual illustration of valve transforms of a typical tandem-center proportional directional-control valve.

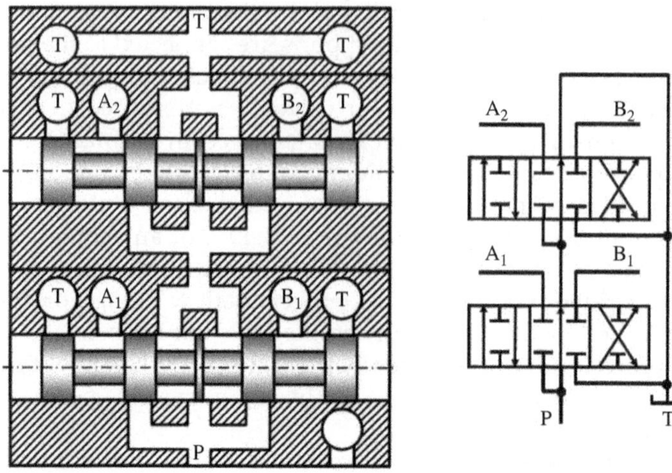


**FIGURE 3.28**
Illustration of the operating principle of an exemplar two-stage directional-control valve.

combinations in terms of the application requirements. Figure 3.28 illustrates the operating principle of a two-stage spool-type directional control valve package, which is integrated using two six-way directional control valves as a sectional valve and can be found in many mobile hydraulic systems. Such a two-stage valve can be used to control two branch circuits with a priority assigned to the upstream circuit. As shown in the figure, both sectional valves used to construct this two-stage valve package are closed-center valves, with a normally open bypass route to supply the inlet fluid to downstream at their neutral position. When both sectional valves are at their neutral position, the pump flow will take the bypassing route and go directly back to the tank without doing any work. When the spool in the upstream valve moves away from its neutral position, say to the right without loss of generality, the valve will connect the paths between the P-$A_1$ ports (the pump port to the $A_1$ work port) and the $B_1$-T ports (the $B_1$ work port to the tank port) and close the bypassing route. Therefore, *all* the pump flow is used to drive the upstream circuit (circuit 1) actuator to do the desired work. Because all flow is supplied to the upstream circuit in this case, the downstream circuit will not get any flow and thus cannot do any work. In other words, downstream actuators can only perform work when the upstream valve is set at its neutral position. Such a design gives the upstream circuit a **higher priority** in getting flow supply to perform the work. **Priority control** is commonly used in multistage systems. If the sectional valves used to construct this multistage valve are proportional valves, a **partial priority** function could also be realized, when the upstream circuit demands only a portion of the pump flow and the remaining flow could be used to drive the load carried by the downstream circuit.

### 3.1.5 Flow Control Valves

Flow control valves are used to regulate the amount of flow supplied to a branch circuit to control the actuator speed. Often, such flow regulation is accomplished through adjusting orifice areas, either by direct-acting or pilot-operating means. While there are numerous ways to control the flow, most of them can be categorized as one of three basic types: (1) noncompensation flow control, (2) pressure-compensated flow control, and (3) flow-dividing control.

As the energy-carrying medium, the pressure flow transport rate determines the energy transfer rate, and flow control valves play a dominant role in controlling energy distribution in hydraulic power transmission. For the rate of energy transferred to an actuator to equal the speed of the work being done, which can be determined by multiplying the amount of hydraulic force used to move a load by the distance this load is being moved per unit of time, a flow control valve will control the energy transfer rate at the actuator by regulating the flow rate passing through the valve. Controlling the flow in a hydraulic system is to regulate the fluid volume per unit of time, namely, the **volumetric flow rate**, through a valve. Two other ways of flow rate measurement, the **mass flow rate** and the **weight flow rate**, are used in engineering calculations. The volumetric flow rate is very convenient for calculating the linear speeds of piston rods or the rotational speeds of motor shafts. The mass flow rate is often used to calculate inertia forces during periods of acceleration and deceleration. The weight flow rate is normally used in the calculation of fluid power using English units of measure. This textbook uses volumetric and mass flow rates in all calculations using SI units.

A noteworthy feature of **noncompensation flow control valves** is that the flow rate passing through the valve changes with the pressure drop between the upstream and downstream of the valve. The simplest effective method of noncompensation flow control is probably the use of orifices, which is also the most basic pressure control device. As illustrated in Figure 3.29, a simple orifice can be simply placed in a flow transport line. Such an orifice will create a pressure drop to increase the resistance in the line, which will force more flow to turn into the other parallel branches to achieve the goal of flow control in the particular line. If there is only one branch in a system, the use of an orifice in the line does not provide a flow control capability. Instead, it will only generate a higher resistance to deliver the same rate of flow through the line until the pressure reaches the line-relief setting at which a portion of the flow will be relieved as a bypass through the relief valve.

Like their fixed counterparts, adjustable orifices are also among the popular methods for noncompensation flow control in many applications. Among all different designs of adjustable orifices, **needle valves** are probably the most commonly used. As depicted in Figure 3.30, needle valves are designed to finely control the orifice area by turning the needle to adjust the opening of the valve. Needle valves are often used as **meter-in** or **meter-out** flow control in many hydraulic systems. Because it is always associated with a large energy loss in controlling flow using either fixed or adjustable orifices, these types of flow control valves are generally used only on low-power systems or on temporal rate control applications.

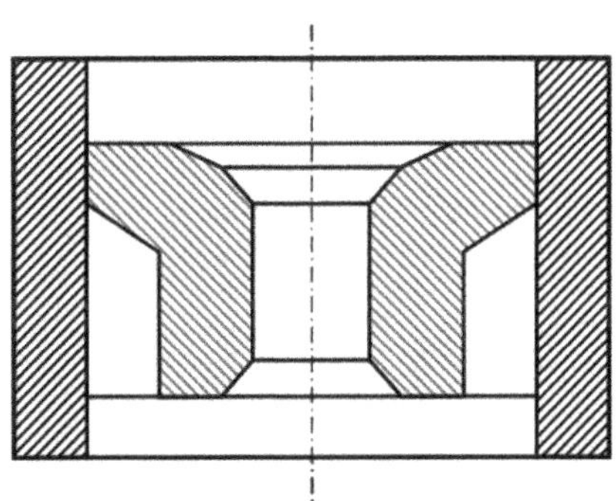

**FIGURE 3.29**
Schematic illustration of a simple fixed orifice used in hydraulic lines.

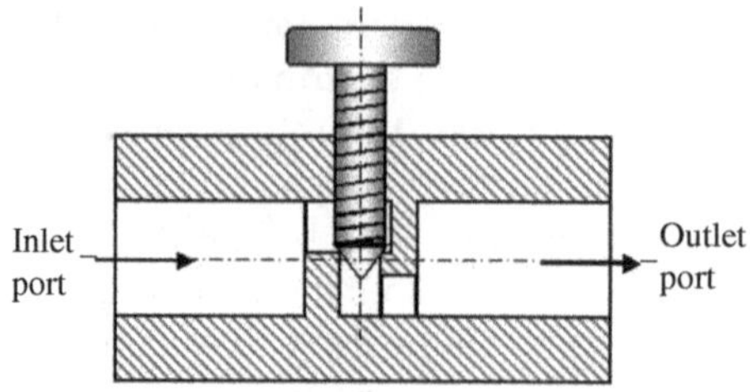


**FIGURE 3.30**
Schematic illustration of a typical needle valve.

The flow rate passing through a noncompensated flow control valve can be determined using the orifice equation defined by Eq. (1.9). Importantly, the equation provides only the general form of equation for flow control and offers a satisfactory estimation of the flow rate passing through a sharp-edged round orifice on a disk installed in a pipeline. Since there are many different designs using the orifice shape or size, such as a ring-shaped gap between a spool and valve body or a long hole, various empirical equations have been formulated based on test data to provide a more accurate flow rate estimation to achieve more precise flow controls.

In many applications, a branch circuit often requests a consistent flow supply regardless of the variations in either the system pressure or the load pressure. Noncompensated flow control valves will be unable to satisfy this performance requirement. In such a case, a **pressure-compensated flow control valve** is desirable to offer automatic flow control capability in responding to system and load pressure changes. As illustrated in Figure 3.31, a typical pressure-compensated flow control valve uses an adjustable pilot relief valve placed in serial with the main valve. When more flow than needed is supplied to the controlled branch, it will increase the pressure on the outlet flow due to the incompressibility of hydraulic fluids, which in turn will push the serially placed relief valve open to release the back pressure on the compensator to maintain a constant flow rate under varying system and load pressures. This type of function is accomplished by bleeding the excess flow from a drain port. A typical pressure-compensated flow control valve can achieve between 3 and 5% flow control accuracy.

A typical pressure-compensated flow control valve operating under a principle of bleeding the excess flow to maintain a constant flow supply will inevitably result in a significant amount of energy loss from the bled flow. To improve energy efficiency, an alternative design for pressure compensation is to limit the flow getting into the circuit using an

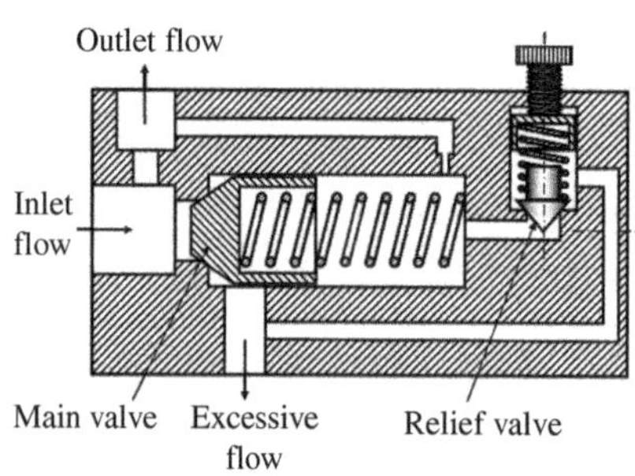


**FIGURE 3.31**
Illustration of the operating principle of a typical pressure-compensated flow-control valve.

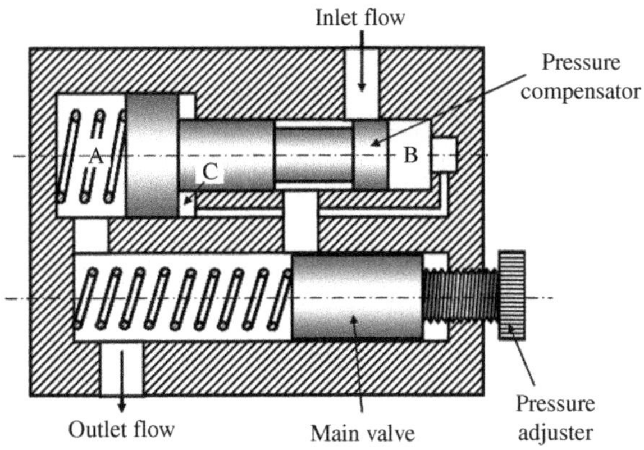


**FIGURE 3.32**
Illustration of the operating principle of a typical alternative pressure-compensated flow-control valve.

adjustable orifice on its main valve (Figure 3.32). As illustrated in this figure, this alternative design uses two spools, one serves as the pressure compensator and the other as the main valve, to accomplish the no-bleeding flow control. During operation, the pressure compensator is used to keep the pressures in the spring chamber (Chamber A) and in the opposite chamber (Chamber B) at a constant difference. This constant difference between two opposite chambers is achieved by a force balance, which is expressed by the following equation:

$$F_S + p_A A_c = p_B A_c \tag{3.11}$$

where $F_S$ is the spring force acting on the pressure compensator spool, $p_A$ and $p_B$ are fluid pressures in chambers A and B, respectively, and $A_c$ is the cross-sectional area of the spool in chamber A, because the pressure at chambers B and C are the same. The total pressure-bearing area on the nonspring side is the same as that on the spring-acting side.

According to Eq. (3.11), the pressure difference between $p_A$ and $p_B$ is determined by the stiffness of the compensator spring. In a case where the load pressure (namely, the outlet pressure) is increased for any reason, the pressure in chamber A will also increase, which will generate a large pressure force acting on the spool to push open the pressure compensator orifice more. The larger opening at the pressure compensator will raise the pressure in the chamber between the pressure compensator and the flow control valve, and consequently keeps the pressure drop across the flow control valve constant. Because the orifice area across this valve is also a constant during an operation (for it can only be manually adjusted by the pressure adjuster as shown in Figure 3.32), the flow rate passing through this valve is therefore a constant according to the orifice equation.

Because the viscosity of hydraulic fluid varies with temperature, the flow supplied to a controlled circuit using a flow control valve will often drift with temperature changes. To compensate for the outlet flow inconsistency induced by temperature variations, it is also possible to add temperature compensators to a flow control valve to adjust the orifice openings in response to temperature changes. This is often done in combination with adjustments to the control orifice for pressure changes as well.

The other basic category of flow control valve is the **flow-dividing control valve**, often called the **flow divider**. As implied by its name, flow dividers are designed to split flow supplies from one pump equally or proportionally to two circuits that may be operating

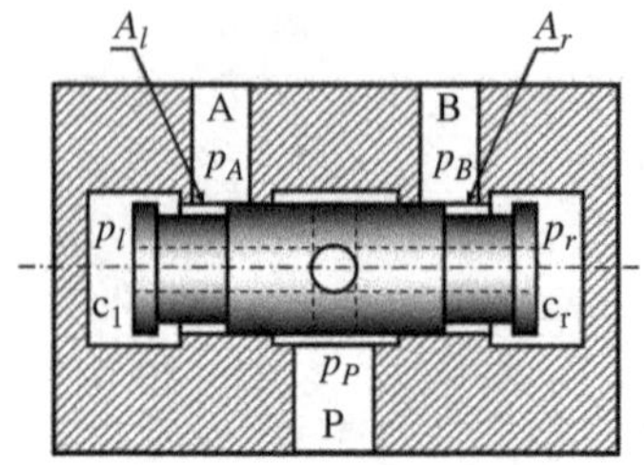


**FIGURE 3.33**
Illustration of the operating principle of a typical spool-type linear flow divider.

at different pressures. The portion of flows supplied to each circuit is generally predetermined by the design of the valve. As depicted in Figure 3.33, the main component in a flow divide is the sliding spool with a fluid path in the middle. The flow entering the valve from the P port is split into two streams in the path and is supplied to fluid chambers at both ends of the spool. The fluid is then transported to actuators in both circuits via the left and the right flow-passing orifices, $A_l$ and $A_r$, formed by the sliding spool and valve body as shown in Figure 3.33. Because the sliding spool is balanced by the pressure acting on both ends, it will maintain an equilibrium pressure until pushed to shut down the flow path at one end. Therefore, within its normal operational range, the equation is always satisfied:

$$p_l = p_r \tag{3.12}$$

where $p_l$ and $p_r$ are pressures in the left- and right-side pressure chambers of the flow divider.

Substitute the relationship presented in Eq. (3.12) to orifice equations as expressed in Eq. (3.2) to determine the flow transported to both ends of the flow divider. It has:

$$\frac{Q_A}{Q_B} = \frac{A_l}{A_r} \tag{3.13}$$

where $Q_A$ and $Q_B$ are the split flows supplying to the left and right side of the divider, and $A_l$ and $A_r$ are the orifice areas at the left side and right side of the sliding spool.

Equation (3.13) indicates that the flow split ratio is proportional to the orifice area ratio of the left- and right-side pressure chambers. If the orifices are identical, the flow is split in a 50:50 ratio. To maintain constant flow supplies to both circuits in a case when one of the actuators (let us assume in the left-side circuit) carries a heavier load than the other (in the right-side circuit in this case), a higher working pressure at port A will result. The higher $p_A$ will then increase $p_l$, and a higher $p_l$ will in turn push the spool sliding to the right. The rightward sliding of the spool will reduce the orifice area of the left pressure chamber to working port A, which in turn will reduce $p_l$ until it reaches a new equilibrium state between $p_l$ and $p_r$.

It is practically impossible to keep the split flow exactly constant because of the imperfect sensitivity and the responsive time delay of the spool motion to load variations. One should also keep in mind that a flow divider uses the principle of controlling pressure drop to control the consistency in flow split, and any pressure drop that converts hydraulic energy to heat is lost. Therefore, use of a flow divider will always lower power transmission efficiency.

### 3.1.6 Electrohydraulic Control Valves

While the conventional hydraulic control valves, normally actuated by either mechanical or pilot drivers, can satisfactorily accomplish many required control tasks in various applications, electrohydraulic control valves, constructed by integrating electromechanical drivers on hydraulic control valves, can offer more accurate and responsive controls to perform these functions. For example, in conventional hydraulic control valves, a mechanical or pilot driver is used to convert the operator's maneuvering force applied on a control lever either directly through a mechanical linkage or through the hydraulic pressure to shift the flow-directing element, such as the sliding spool, from one position to another. While operating a mobile hydraulic system, a manual maneuvering action is always subjected to the impact of motion or vibration, which often results in difficulty holding the control lever accurately at the desired position and consequently affects the control accuracy of fluid power delivery. Electrohydraulic control valves, by using the muscle of the hydraulic power and the accuracy of electrical controls, can provide enhanced functionalities to control hydraulic systems. The most commonly used electromechanical drivers for hydraulic control valves are **proportional solenoid drivers** and **servo drivers**.

As the actuating devices in electrohydraulic valves, electromechanical drivers are designed to convert the received electrical signal to mechanical force to drive the control element, often a poppet or a spool, in a valve to perform the required hydraulic control functions. Although electrohydraulic controls can be used either to direct pressurized fluid to drive a load or to stroke a pump to generate the flow to do the same, the direct mechanical actions in electrohydraulic valves are almost always performed within valves. In general, an electrohydraulic proportional valve consists of the two core elements of a **proportional solenoid** or **servo driver** and a controllable metering area formed by a movable hydraulic control element (either poppet or spool) and the valve body. Many electrohydraulic valves also use an optional element, an electronic position-feedback device, to gather feedback information to achieve more accurate control.

In terms of the ways electrohydraulic control valves perform their control actions, these valves can be classified as either **on–off control valves** or **proportional control valves**. On–off control valves generally use solenoid drivers to actuate the spool shifting between two or three positions to change flow-directing states controlled by those valves. Because of its ability to generate a force to directly push a spool, a solenoid valve is also called a **force motor**. Proportional control valves can accurately adjust the valve-opening using either solenoid drivers or servo drivers regulated by sophisticated electronic controllers. Electrically controlled proportional valves help simplify hydraulic circuitry by reducing the number of components a system may require, while, at the same time, substantially increasing system control accuracy and efficiency.

The most commonly used electromechanical drivers in mobile hydraulic systems are proportional solenoid drivers, including traditional air-gap type and wet-armature-type solenoid drivers. Figure 3.34 depicts a traditional air-gap-type solenoid driver in which a coil is used to generate a magnetic field while energized with a current and a metal plunger is placed in the center of the coil to convert the magnetic potential into mechanical force to actuate the valve spool through a push pin. Because there are no permanent magnets in a solenoid, the plunger normally rests partially out of the solenoid frame pushed by the spring acting on the opposite side of the valve spool when the solenoid is not energized. The separation between the plunger and the base of the frame is called the **air gap**. When energized, the magnetic field will flow through both the plunger and the air gap. Because the air has higher resistance to the magnetic flux than the metal plunger and

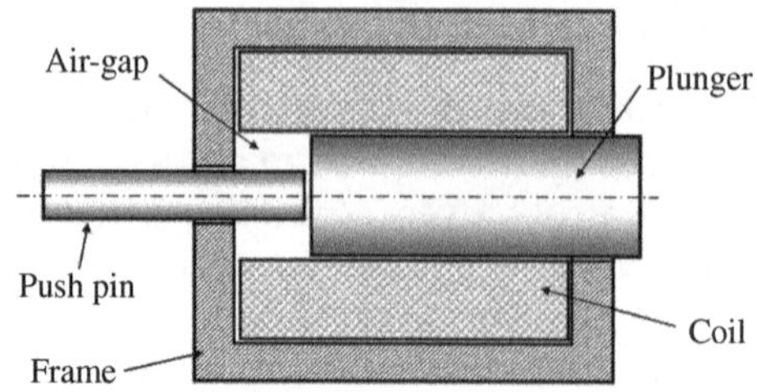


**FIGURE 3.34**
Schematic illustration of a traditional air-gap type proportional solenoid driver.

weakens the magnetic field, the metal plunger pulls in to fill the air gap and results in a stronger magnetic field. This phenomenon indicates that the solenoid has the minimum driving force when the plunger is out and reaches its maximum value as the plunger is fully filled the air gap at this position. Such a feature could induce a major problem when using solenoid drivers to actuate a spool valve because it requires the highest force to drive the spool to overcome all the resistance, including flow forces, drags, and static friction, at the very beginning of the spool stroke. As stated earlier, a solenoid driver has the lowest force potential at this point. Another important feature of this type of solenoid driver is that an energized coil normally generates a relatively constant force regardless of plunger position when the current is constant. Therefore, it is possible to make the solenoid generate more force by supplying a higher current at the beginning of the stroke to drive the spool to overcome the heavier load.

An alternative design of solenoid driver for hydraulic system use is the **wet-armature solenoid**. Depicted in Figure 3.35, a wet-armature solenoid typically consists of three functional elements: the coil, the tube, and the plunger (also called the **armature**). In this type of solenoid, a thin-wall tube is specially designed to tolerate fluid coming into the chamber of the plunger without the worry of shorting out the electrical circuit by the hydraulic fluids. The plunger is always shorter than the tube and fits in the tube very loosely to give it sufficient space to stroke back and forth, driving the valve spool throughout the entire length of travel.

Solenoid drivers can be energized using either AC power or DC power. AC solenoids are commonly used on stationary hydraulic systems for industrial applications and are energized directly using electrical current. Mobile hydraulic systems normally use DC solenoids, which are actuated using a **pulse-width modulation** (PWM) driver. PWM is

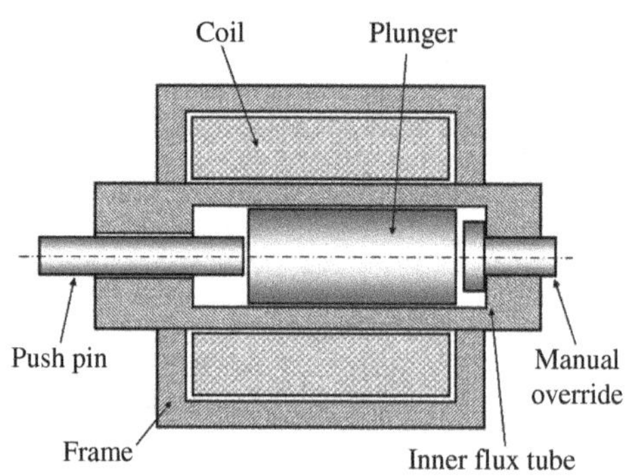


**FIGURE 3.35**
Schematic illustration of an innovative wet-armature type proportional solenoid driver.

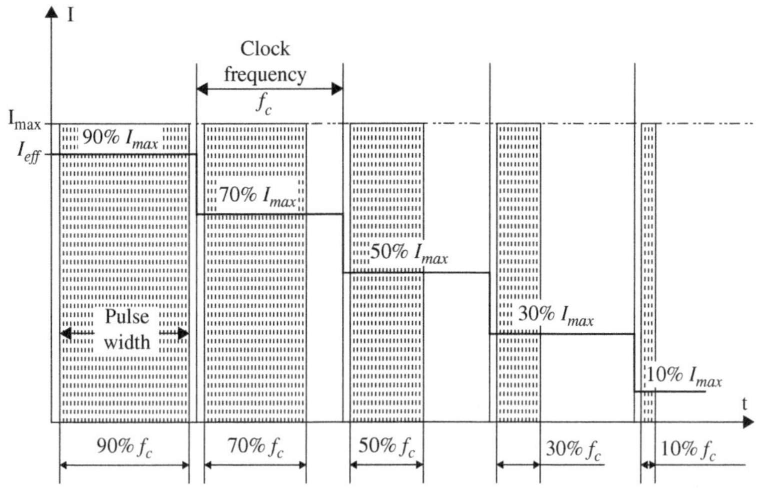


**FIGURE 3.36**
Illustration of the principle of pulse-width modulation (PWM) signals and their equivalent currents.

a modulation technique generating variable-width pulses to represent the amplitude of an analog input signal (see Figure 3.36). PWM is widely used in **switch-mode** power supplies that convert AC power to DC to energize DC solenoid devices. The input power is provided as stepped current pulses of a constant clock frequency, $f_c$, with fixed amplitude. Normally, the pulse is maintained between 10 and 90% of the interval, with the duration of the pulse within an interval termed the **pulse width** and the formation of such a pulse called the **modulation**. The average current (also called the **effective current**) carried by those pulses is the available current to energize the solenoid. The solenoid drivers used in North America typically operate at PWM frequencies from 33 to 400 Hz and in the rest of the World often at a higher range of frequency. If the PWM frequency is sufficiently low, it could automatically provide a mechanical dither to minimize stick-induced hysteresis. However, if the frequency is too low, it could result in noticeable pulsations in the hydraulic system and deteriorate the system performance.

Practically, there are limits on the stroke force and distance a solenoid drive can generate. Typically, a proportional solenoid driver consumes 5 to 40 W power to exert enough force to perform a 0.02 to 1.00 N·m work. This feature implies that solenoid drivers cannot directly shift valves requiring a high driving force. Technically, it is possible to design larger solenoid drivers. However, a larger valve will not only consume more electrical power to drive the spool, but may also result in substantial heat build-up, which can noticeably deteriorate the performance of a hydraulic valve. The solution to provide a large actuating force on a big valve spool with longer stroke distance is the use of small and low-power direct-acting solenoids in combination with pilot pressure, by which the solenoid controls a pilot flow to create a high force to shift the main spool of the valve. These types of solenoid valves are often called **pilot-operated** solenoid valves.

Direct-acting solenoid valves use solenoid drivers to directly actuate the main spool. Such a design can be added to virtually any small-size hydraulic valve. There are three fundamental solenoid arrangements on a direct-acting valve in terms of the configuration of a solenoid driver being installed on a valve: a two-position single-solenoid valve, a two-position double-solenoid valve, and three-position double-solenoid valve. Figure 3.37 shows the operating principle of a typical two-position valve directly actuated using a

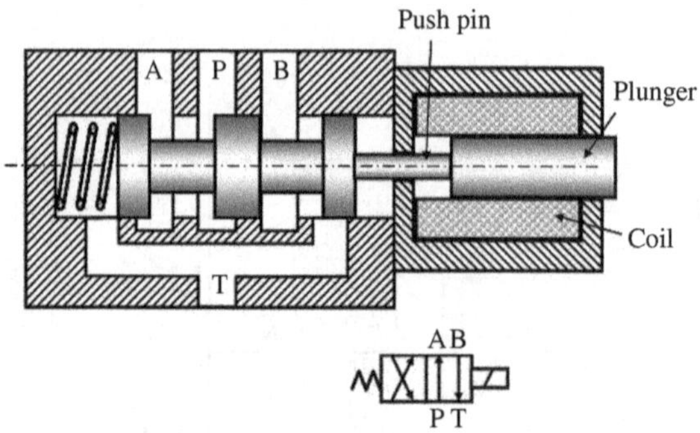


**FIGURE 3.37**
Illustration of the operating principle of a typical two-position single-solenoid controlled electrohydraulic valve.

single solenoid. When the solenoid is not energized, the spring pushes the spool all the way to the right, which opens the flow paths from ports P-to-A and B–to-T. After the solenoid is energized, the plunger pushes the spool against the spring to the left, which will switch flow passes to ports P-to-B and A-to-T. Again, as soon as the solenoid is deenergized, the spool will move back to its original position due to the reposition force provided by the compressed spring. Depending on the applications, the function of these types of valves can easily be changed to get normally open P-to-B and A-to-T paths by exchanging the position of the solenoid and the spring. In this kind of arrangement, the solenoid driver always pushes the spool when it is energized and therefore is a one-direction control arrangement.

Many spool-type electrohydraulic control valves require bi-directional controllability on the spool. To provide such a capability, double-solenoid arrangements are widely used on electrohydraulic control valves. As illustrated in Figure 3.38, a three-position double-solenoid valve typically uses two solenoids to actuate the spool and uses two springs to keep the spool at its neutral position when neither solenoid is energized.

Similarly, the solenoid driver can be installed on virtually any type of hydraulic control valve. However, as we pointed out earlier in this section, there is a practical limitation on the amount of force a solenoid driver can generate to drive a spool, and there are many valves that require a much higher force to actuate. To solve this problem, a **solenoid-controlled pilot-operating design,** as illustrated in Figure 3.39, has been widely adopted by industry. As depicted in this figure, a typical pilot-operated valve uses a small solenoid-actuated pilot valve to direct the pilot flow to actuate the large spool of the main valve.

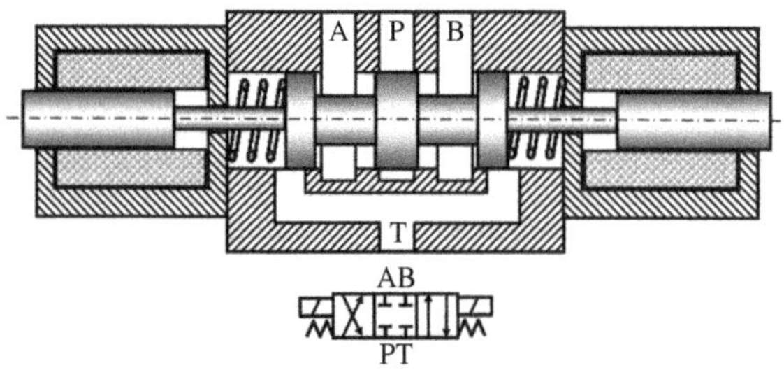


**FIGURE 3.38**
Illustration of the operating principle of a typical three-position double-solenoid controlled electrohydraulic valve.

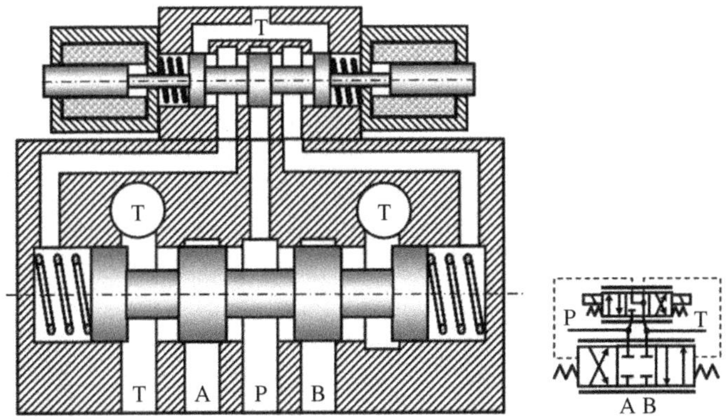


**FIGURE 3.39**
Illustration of the operating principle of a typical double-solenoid controlled pilot-operated electrohydraulic valve.

While different spool designs often have their specialized functions corresponding to a certain spool position in the valve, they all allow the same rule of exchanging flow paths by connecting different ports through shifting spool positions. We can explain such a rule without loss of generality based on the design depicted in Figure 3.39. At its neutral position, both solenoids attached on both ends of the pilot valve are not energized, and both the pilot spool and the main spool are centered by the springs acting on both ends of the spools. When one solenoid, for example, one on the left side, is energized, the solenoid driver pushes the pilot spool to the right to open the pilot paths of P-to-pilot_left and pilot_right-to-T to direct the pilot pressure to the left_end_chamber of the main valve to push the main spool to shift to the right to open the main flow paths of P-to-B and A-to-T. When the solenoid is deenergized, both valves will again be centered by a pair of balance springs.

When proportional solenoids are used to drive a spool against a set of balanced springs, the resultant spool displacement, consequently the **valve-opening areas** controlled by the spool, is proportional to the current driving the solenoids. In other words, the solenoid drivers can make the spool stop at an arbitrary intermediate position rather than only at the ends of the spool stroke. Therefore, such electrohydraulic control valves are also often called the **proportional control valves**. While all types of electrohydraulic valves can be modified into proportional control valves, the most commonly used ones are **proportional directional control valves**. A typical electrohydraulic proportional directional control valve remains the base function of direction control, as discussed in Section 3.1.4, with the additional function of controlling the flow proportional to the stroke of the spool. Using the general form of a valve transform introduced in Figure 3.8, we can create a different form of valve transform to represent the characteristics of a valve-opening area control on a typical four-way proportional directional control valve as shown in Figure 3.40. From these valve transforms, one may notice that at the same spool stroke the valve-opening area from the working port to the tank port always opens before the pump to working port does. Such a feature is often specially designed to bleed the fluid in the back chamber of the hydraulic actuator to avoid forming a blocked space in the flow return line, which could induce pressure surges at the beginning of an operation and cause instability in system operations.

Solenoid drivers are commonly used in many mobile hydraulic systems due mainly to their simple structure, high reliability, relative insensitivity to operating environment,

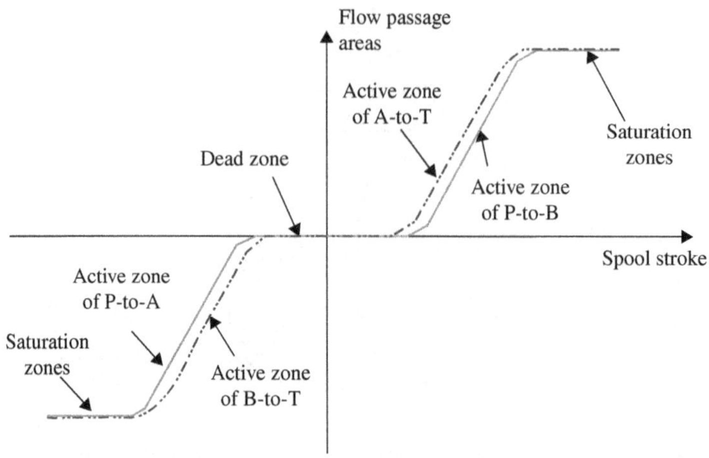


**FIGURE 3.40**
The valve-opening area form of valve transforms for a typical four-way proportional direction control valve.

and low cost. However, solenoid drivers are inherently slow in response: the typical frequency response for solenoid valves is normally less than 10 Hz. Such a slow reaction feature will also cause difficulty in achieving high accuracy in valve control, which is critical in many industry applications. The servo valves, in contrast, can quickly and accurately react to input commands (exceeding 100 Hz), and therefore are often used in hydraulic systems requiring accurate and prompt controls. The term **servo** traditionally leads people to think of mechanical feedback controls. Because a servo driver typically generates a rotary force in a very small arc in driving a spool and can be operated in two directions by simply reversing the current flow, it is also called a **torque motor**. As depicted in Figure 3.41, a servo driver is built using two permanent magnets, two coils, and an armature (with a flapper-feedback spring element connected at the center). Before being energized, the armature is held at the midpoint in between the coils by

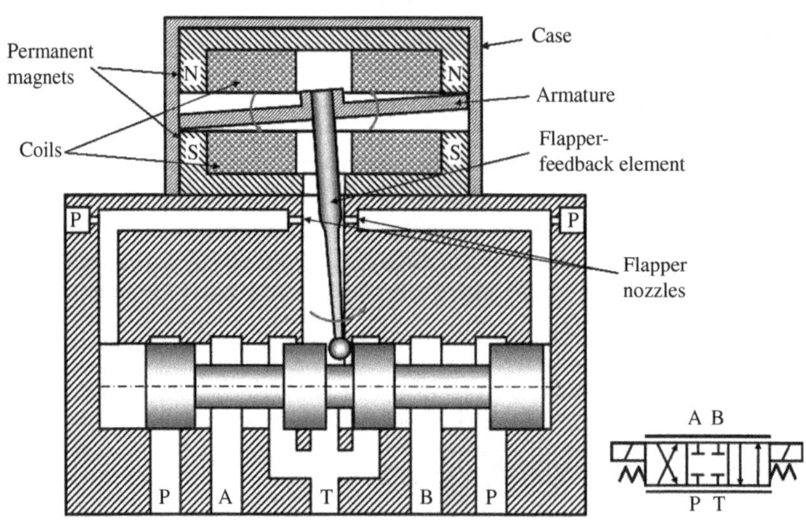


**FIGURE 3.41**
Illustration of the operating principle of a typical servo-type electrohydraulic control valve.

the permanent magnets, which in turn keeps the spool in the neutral position to have the valve normally closed. When it is energized, the two coils form two electromagnets to rotate the armature a small arc either clockwise or counterclockwise in terms of the direction of the input current, which in turn creates a torque to actuate the valve spool to stroke either left or right via the flapper-feedback element to open different paths. This bi-directional actuating capability offers the servo driver push-pull functionality and allows a valve to use only one electrical driver to perform all valve control actuations. Under the condition depicted in Figure 3.41, the servo drive turns the armature counterclockwise to shift the spool rightward via the flapper-feedback element. One consequence of this rightward rotation of the flapper-feedback element is opening the left-side flapper nozzle, which allows a release in the pressure in the left-side back-chamber of the spool. Since the right-side back-chamber is blocked in this case, the pressure acting on the right side of the spool (the same as the pump port pressure) will push it leftward back to the neutral position as soon as the driver is deenergized.

Other than response time, electrical power consumption is another important difference between proportional solenoids and servo drivers: a solenoid drive requires a much higher input of electrical power to actuate a valve than a servo driver. Some other apparent differences are that proportional solenoids, in general, require a substantially greater stroke and operate with greater hysteresis than their servo counterparts.

### 3.1.7 Programmable Electrohydraulic Valves

Electrohydraulic proportional control valves are widely applied for motion control in mobile hydraulic systems. A typical valve of this type uses a sliding spool to regulate the direction and amount of fluid passing through the valve. For different applications, this sliding spool is often specially designed to provide the desired flow control characteristics. Therefore, spool-type electrohydraulic control valves are generally not interchangeable even if they are exactly the same size and have the same flow capacity. Consequently, it is inconvenient and costly to manufacture, distribute, and service. To provide a solution to those problems, a programmable electrohydraulic control valve, integrated using a set of individually controllable generic all-purpose proportional control valves, can use a software to change valve functions and/or characteristics to make it interchangeable for different applications.

As depicted in Figure 3.42, a typical programmable electrohydraulic valve is normally constructed using five proportional electrohydraulic control valves and an electronic controller. In the depicted configuration, valves 1 and 2 are connected by the pump to cylinder ports A or B to provide equilibrium paths of P-to-A and P-to-B, and valves 3 and 4 are connected to cylinder ports A or B to the tank to provide equilibrium paths of A-to-T and B-to-T in a conventional directional control valve. In addition, valve 5 connects the pump and tank directly and provides a dual function of line relief and an equilibrium port of P-to-T in a conventional directional control valve. All five individually controllable normally closed valves working together under the depicted condition provide a **closed-center** mode on the programmable valve equilibrant to a conventional closed-center four-way directional control valve. By setting the initial condition differently on those individually controllable valves, we can easily realize all other basic functions of conventional four-way proportional valves, including the **open-center**, **tandem-center,** or **float-center valve**, using this programmable valve. Each of the composing valves is controlled separately in terms of a set of predefined control logics for each operation mode. Table 3.1 summarizes the control logic for realizing these valve function modes on the integrated programmable valve.

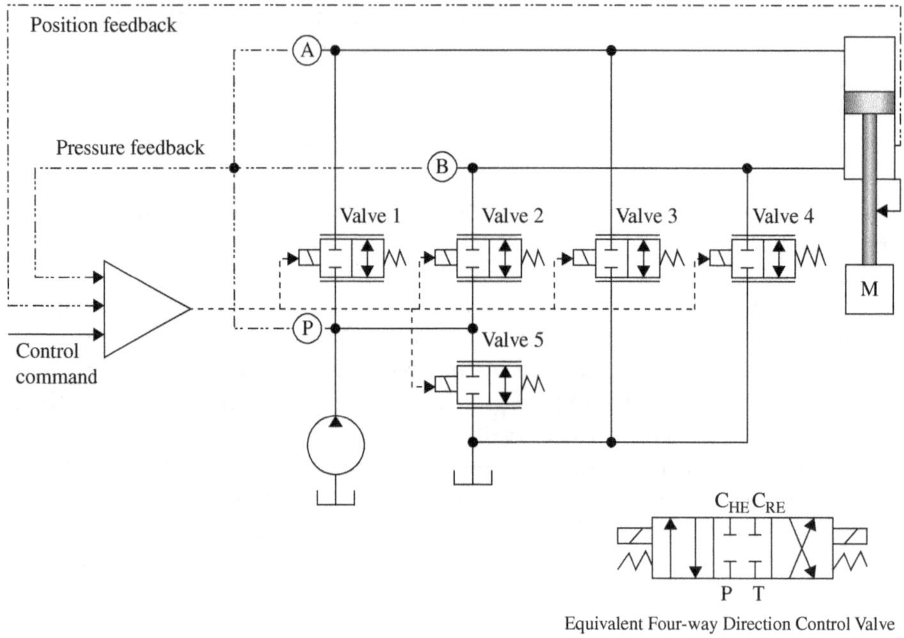


**FIGURE 3.42**
Illustration of the operating principle of a programmable electro-hydraulic control valve under the "closed-center" mode.

With a proper logic on–off control on all five valves, the programmable valve is capable of realizing all basic functions listed in Table 3.1. For example, in a conventional tandem-center or closed-center directional control valve, the working ports are normally closed to hold the pressure in actuator chambers, with the pump and tank ports either normally open or normally closed. To fulfill this function, a programmable valve keeps valves 1 to 4 closed to hold the cylinder chamber pressure and fully opens valve 5 to bleed the flow to the tank either at low pressure (tandem-center function) or when the system pressure exceeds a preset relief level (closed-center function). In conventional open-center directional control valves, all ports are normally connected. To fulfill this function, the generic valve keeps all valves open. Similarly, to provide a float-center function, the programmable valve needs to open valves 3 and 4 to release pressure in both working chambers of the actuator. In both cases, valve 5 will be opened only when the system pressure exceeds a preset relief level. Notably, modulation control is required to realize proportional control functions.

**TABLE 3.1**

Control Logics for Realizing Multiple Valve Functions using the Integrated Programmable Electrohydraulic Control Valve.

| Valve Function | Valve 1 | Valve 2 | Valve 3 | Valve 4 | Valve 5 |
|---|---|---|---|---|---|
| **Open-center** | Normally open | Normally open | Normally open | Normally open | Normally open |
| **Closed-center** | Normally closed | Normally closed | Normally closed | Normally closed | Line-relief |
| **Tandem-center** | Normally closed | Normally closed | Normally closed | Normally closed | Normally open |
| **Float-center** | Normally closed | Normally closed | Normally open | Normally open | Line-relief |

### 3.1.8 Select Appropriate Control Valves

The design of a hydraulic system often starts at the calculation of the load and the motion of hydraulic actuators it needs to drive. For example, when linear actuators are planned to be used in an application, we often start the design process by selecting the diameter of the cylinders and the length of cylinder stroke in terms of the load to be driven. Then, the required flow rate is determined based on the time of motion required for the application. The sizing of cylinders should also have an adequate dynamic response to meet acceleration and deceleration needs, which usually require calculating the desirable system pressure. As a critical component in a hydraulic system, the selection of a hydraulic control valve is often done after the flow rate and load pressure are determined.

Because the basic function of hydraulic control valves is distinctively determined by the category, as discussed in the previous sections, and the size of those valves needs to match the system flow rate requirement, the selection of control valves is actually limited to operating characteristics and valve-actuating means. As discussed in the previous section, the most commonly used electrohydraulic valves are actuated either by proportional solenoid drivers or by servo drivers. Mobile hydraulic systems often choose proportional solenoid valves because they are less susceptible to contamination and, maybe most importantly, because they are less expensive.

The pressure drop of typical proportional valves is normally rated at 1000 kPa, whereas the corresponding servo valves can be seven times higher at 7000 kPa. On the other side, the high-pressure drop could significantly raise the flow-passing capacity over a servo valve. Research results indicate that the flow rate passing through a valve can be increased 1.4 times when the pressure drop over the valve is double. Another very important factor in choosing servo valves is that those valves generally respond much faster than proportional solenoid counterparts mainly because of the force available to shift the spool. A solenoid valve normally uses the linear force to directly push the spool, overcoming the spring-centering force, and sometimes even needs to push an inline **linear variable differential transformer** (LVDT). In comparison, a servo valve often relies on a rotary force to actuate merely a small pilot spool and can have a faster and more linear response on the main spool. However, the precise pilot control loop often causes servo valves to be more susceptible to contamination, which often drives the price up. In many applications, especially in mobile hydraulic applications, such features have steered people away from servo valves and toward choosing proportional solenoid valves.

In some cases, proportional solenoid valves may be unable to provide sufficient power to overcome the resistance induced by Bernoulli forces caused by high flows, which could result in momentarily losing controllability to the valve. A commonly applied method to solve this flow force problem is the use of a multiple-stage valve, using a large hydraulic force supplied by a pilot flow to control the position of the main spool. The main drawback of this approach is the higher cost and slower response. However, when a valve is large, the pilot flow can provide sufficient force to effectively move the main spool faster than a solenoid driver alone can do. In such cases, pilot valves could enhance the main valve response performance to actuate the main spool quickly.

Proportional valves, servo or solenoid driven, move the spool to positions proportional to the input the control signals to the driver. Ideally, a proportional control valve should have a linear spool gain for flow control. However, the flow passing through those valves is not necessarily in proportion to the input control signal, but relies heavily on the design of the spool. Therefore, making a correct spool choice is critical for optimizing system performance. The most commonly used spool type in a proportional valve is probably the **overlapped** spool. The unique feature of this type spool is having a **deadband**, also called **zero gain**, in which no flow will pass through the valve regardless of the level of the input control signal. This type of spool is designed to reliably seal the ports to reduce internal leakage and hold the system status when the valve is at its neutral position. On the other hand, the deadband also induces a slow response to small correction signals, which makes such valves a poor choice for both position and pressure control applications. Other than the deadband, it is also very common for spools to carry a **dual gain**, often formed by the niches being cut on the land of the spool corresponding to small correction signals and by the spool land as the signal exceeds a threshold value. For manual systems, such a dual-gain spool can offer fine control on position or speed by means of metering control at small spool strokes and allows fast response control by providing a lot of flow at large spool strokes. While such control nonlinearity does not really cause a problem in manual systems, it does bring some difficulty in achieving high tracking accuracy in either position or speed control for many electrohydraulic systems. Fortunately, the advancement in electrohydraulic control technologies, such as feedforward-proportional-integral-derivative (PID) and fuzzy controls, can satisfactorily overcome those difficulties.

**Example 3.2: Pressure Drop and Flow-Passing Rate in a Typical Valve**

Assume a spool-type proportional directional control valve as shown in Figure 3.38 has a 25 mm diameter spool with a 1 mm deadzone on each direction that the spool moves. When the total spool is 4 mm by 1.0 MPa of pressure drop across the valve, what will be the flow rate passing through the valve? How much will the flow rate change if the pressure drop is increased to 2.0 MPa under the same condition (assume a valve orifice coefficient of 0.65 for a cartridge-type valve and fluid density of 850 $kg \cdot m^{-3}$ for hydraulic fluids)?

a. The orifice area of the valve can be calculated using Eq. (3.3):

$$\begin{aligned} A_o &= \pi d(x - x_o) \\ &= 3.14 \times 0.025 \times (0.004 - 0.001) \\ &= 2.36 \times 10^{-4} (m^2) \end{aligned}$$

b. The flow rate passing through this valve at 1.0 MPa pressure drop is:

$$\begin{aligned} Q_1 &= A_o C_d \sqrt{\frac{2}{\rho} \Delta P} \\ &= 2.36 \times 10^{-4} \times 0.65 \times \sqrt{\frac{2}{850} \times 1.0 \times 10^6} \\ &= 7.44 \times 10^{-3} (m^3 \cdot s^{-1}) = 446 (L \cdot \min^{-1}) \end{aligned}$$

c. The flow rate passing through this valve at 2.0 MPa pressure drop is:

$$\begin{aligned} Q_2 &= A_o C_d \sqrt{\frac{2}{\rho} \Delta P} \\ &= 2.36 \times 10^{-4} \times 0.65 \times \sqrt{\frac{2}{850} \times 2.0 \times 10^6} \\ &= 1.05 \times 10^{-2} (m^3 \cdot s^{-1}) = 630 (L \cdot \min^{-1}) \end{aligned}$$

DISCUSSION 3.2: The results proved that with the same valve-opening area, the flow rate passing through a spool valve increases 1.4 times as the pressure drop across the valve doubles.

## 3.2 Hydraulic Manifolds

### 3.2.1 Overview of Hydraulic Manifolds

As introduced in the previous sections, many hydraulic control valves are manufactured and distributed in the form of cartridges. While such cartridge formations provide the industry with more flexibility in using generic components to design their hydraulic systems, often required are specially designed hydraulic manifolds to host multiple cartridge valves in a centralized location. Some high-cost valves, such as servo and proportional valves, are also often mounted on a manifold to have the fluid conductors connected to the more structurally durable manifold to protect the expensive valves. Therefore, the other major benefits of hydraulic manifolds include lower cost, more compact design, less leakage, and simpler maintenance.

By definition, hydraulic manifolds are simply blocks of metal with drilled flow paths to connect various ports. Other than flow paths, a typical hydraulic manifold often has some pressure channels drilled to intersect with the main flow paths so that a pressure gage or sensor can be installed. According to their design, hydraulic manifolds come in either modular-block and single-piece designs. A **modular-block** design generally supports one valve and contains only internal paths for one-valve functions. It normally needs to connect a series of similar modular blocks to build a complete system. In comparison, **single-piece** manifolds are designed to support all associated valves by providing all the paths and channels needed to form a stand-alone system.

From numerous engineering practices, it has been proven that the use of manifolds could not only cut the installation space by one-third, which is an attractive feature for mobile hydraulic systems due to the very limited space, but also reduces assembly and installation costs from 30 to 50%. In addition, both types of manifolds have their distinctive advantages; the best suited for a particular application will depend on a variety of factors, including specific function, cost, space, and system endurance.

### 3.2.2 Modular-Block Manifolds

Modular manifolds are normally manufactured on blocks of cast iron, aluminum, or steel to support one valve, which makes it relatively easy to design cartridge-interchangeable and mounting-identical manifolds for building different types of modulated valves, such as pressure, directional, flow, and proportional control valves. In common engineering practices, most modular manifolds are designed in a **ready-to-install** style and can be bench-assembled horizontally and stacked. Figure 3.43 illustrates the cross section of a simple modular manifold commonly used to build some specific control valves, such as flow control valves.

### 3.2.3 Single-Piece Hydraulic Manifolds

In many applications, more than one cartridge valve is installed in one manifold to form a custom-packaged valve group to provide complete functionality. Those valve packages are often installed on a single-piece valve manifold, which offers some very attractive advantages to mobile applications, such as the compactness and centralized valve installation. Figure 3.44 shows an example of a single-piece valve manifold for installing five cartridge valves to form a four-way individually controllable programmable valve, which carries all the necessary flow passages for the support of an integrated programmable valve performing all designed functions.

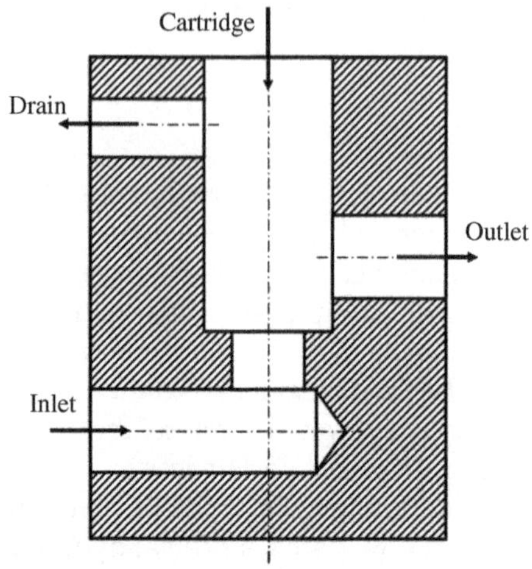


**FIGURE 3.43**
Configuration illustration of an exemplar cross section of simple modular manifolds.

Typically, single-piece manifolds can be designed in two basic styles: either the laminar or the drilled metal block. In a **laminar-type manifold**, several layers of metal plates having proper flow paths machined on or through them are stacked to form a complete network of flow paths customer-designed for specific applications. Because the internal paths can be machined in many shapes and sizes, laminar-type manifolds have virtually no limit to composing the number or size of valves to be mounted on it. With a proper plate-stacking technique, laminar-type manifolds can also be used in systems with operating pressures. **Drilled-type manifolds**, by comparison, can also be custom-designed for specific applications. The only difference is that the network of flow paths is all drilled, which causes some limitations because the drilled paths must be straight.

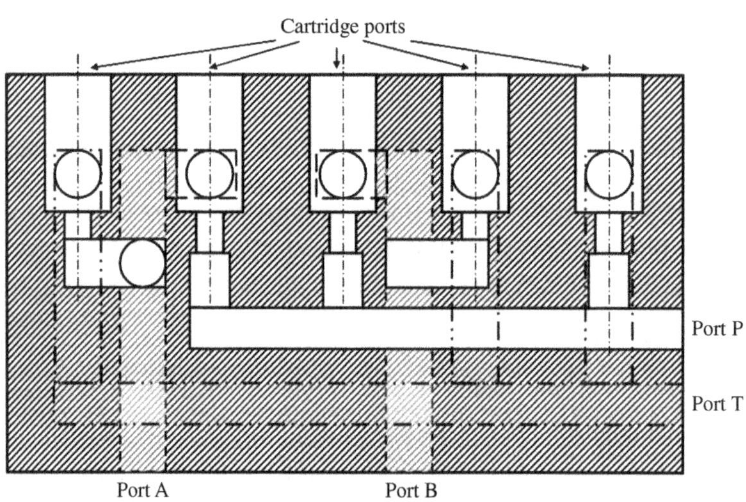


**FIGURE 3.44**
Configuration illustration of the cross section of a specially designed single-piece manifold for installing five cartridge valves.

## 3.3 Hydraulic Lines

### 3.3.1 Major Components of Hydraulic lines

While hydraulic control valves play a key role in quantitatively determining the amount of energy being transmitted for performing the desired work, a hydraulic system still needs to have **hydraulic lines** to complete the energy transmission process. Hydraulic lines use conductors or connectors to create a contained flow path for transporting pressure flows to the places where hydraulic power is used. Designed for different applications, hydraulic conductors can often be categorized into one of three basic kinds: pipes, tubing, and hoses. Both **hydraulic pipes** and **tubing** are rigid conductors made of metal materials. The major difference between the two is that pipes in general use threaded fittings, and tubing often uses flared or flareless fittings. In comparison, **hydraulic hoses** are a type of flexible conductor, often made of oil-resistant synthetic rubber or thermoplastic material and can be bent easily during installation or even in use. Mobile hydraulic systems commonly use only hydraulic hoses and tubing for pressure flow delivery.

### 3.3.2 Hydraulic Hoses

Flexibility is the basic and unique feature of hydraulic hoses and enables them to be positioned in the most efficient or convenient places by bending hoses around corners and running through tight spaces or varying gaps. Also, mainly because of this flexibility feature, a hydraulic hose can withstand movements or vibrations very well and can absorb pressure surges, which are some of the common problems encountered in placing hydraulic lines in mobile hydraulic systems.

Modern hydraulic hoses typically consist of at least three parts: an inner tube that carries the fluids, a reinforcement layer, and a protective outer layer. The **inner tube** must have some flexibility. Due to the direct contact with the hot hydraulic oils, it is often made of synthetic rubber or other oil-resistant thermoplastic material. The **reinforcement layer** consists of one or more sheaths of braided wire, spiral-wound wire, or textile yarn, and can therefore be categorized into two types of hose, **wire-reinforced** and **fabric-reinforced**, to make hoses stronger. The **outer layer** can be either weather-, oil-, or abrasion-resistant, depending on the type of environment the hose is designed for, to achieve an endurable useful life. Figure 3.45 shows six examples of commonly used hydraulic hoses on mobile systems with different numbers of layers. All but one of those six types is fabricated in a traditional three-layer design, with the only difference being the number of reinforcement layers. The last type, however, uses a **polytetrafluorethylene (PTFE)** inner tube reinforced with a single layer of metal wires, often made of stainless steel, braided outside to provide an electrically conductive feature for preventing electrostatic charge build-up.

Hydraulic hose is specified by its **inside diameter** (ID), and this dimension does not change with the type of hose design illustrated in Figure 3.45. This means that different thicknesses in hose design will only change the **outside diameter** (OD). The hose pressure rating is specified in working and burst values. For standardizing mobile system applications, the Society of Automotive Engineers (SAE) has issued a series of industrial standards, such as J517 Hydraulic Hose and J1273 Recommended Practices for Hydraulic Hose Assemblies, to provide guidelines for design, fabrication, selection, installation, and maintenance of hydraulic hoses and hose assemblies for mobile system applications. The SAE J517 standard specifies all hydraulic hoses using a 100R series number, from SAE 100R1 to

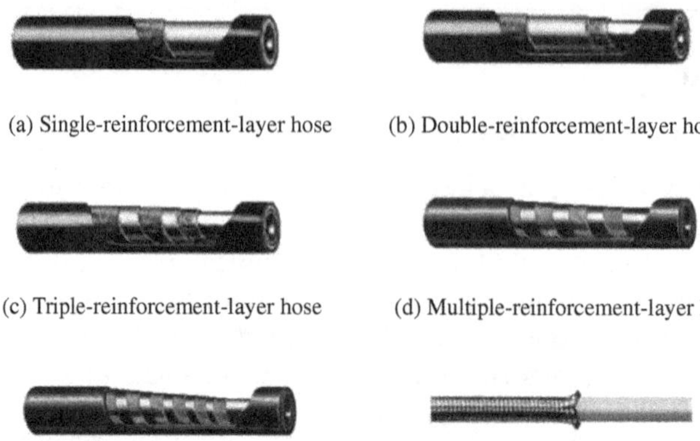

(a) Single-reinforcement-layer hose (b) Double-reinforcement-layer hose

(c) Triple-reinforcement-layer hose (d) Multiple-reinforcement-layer hose

(e) Multiple-heavy-reinforcement-layer hose (f) Metal-reinforcement-layer hose

**FIGURE 3.45**
Illustrations of a few representative typical structures of hydraulic hose of different fabrications.

100R17, with additional letters (A, AT, B, or BT) to indicate some of the special features of those hoses. Among them, 100R1 is a steel wire-reinforced, rubber-covered hose suitable for delivering petroleum-based fluids or water of medium pressure (less than 20 MPa) under a temperature of 100°C. By using one more layer of steel reinforcement, 100R2 can be used to deliver high-pressure (up to 34.5 MPa) fluids. When reinforced using four layers of spiral steel wire 100R9 and 100R10, hoses can be used to deliver higher-pressure fluids than 100R2. When the reinforcement layer is increased to six layers, 100R11 hoses can be safely used to deliver high-pressure fluids of up to 86.2 MPa. Reinforced with two-layer fiber braid, 100R3 and 100R5 can deliver lower pressures than 100R1 and 100R2, but offer a superior flexibility to allow a smaller bend radius. With only one layer of fiber braid, 100R6 is normally used in low-pressure and very small bend radius applications. To meet the special requirements for hydraulic systems of heavy duty and high impulse, it is strongly recommended to use hydraulic hoses (100R12, 100R13, and 100R15) specially designed for such applications. 100R7 and 100R8 are, respectively, the normal-pressure rating and high-pressure rating thermoplastic hoses. 100R14, with its inner layer made of PTFE material, can safely be used in high-temperature (over 200°C) applications. 100R16 and 100R17 are compact hydraulic houses, which can replace 100R1 and 100R2 in applications where larger hoses can hardly be installed. 100R4 is a special type of wire-inserted hydraulic hose and is often used in low-pressure applications such as the suction line in a hydraulic system.

The SAE standard also recommends including size information of the hose by dashing the size number. For example, a serial number of 100R2AT-8 represents a two-wire, type AT hydraulic hose of 1/2 inch ID with a maximum operating pressure of 24.1 MPa. Similarly, a serial number of 100R4-32 represents a 2 inch ID suction hose with a maximum operating pressure of 0.7 MPa, and a serial number of 100R14B-16 represents a PTFE tube, electrical conductive hydraulic hose of 7/8 inch ID with a maximum operating pressure of 5.5 MPa. For ultra-high-pressure systems, it is often recommended to select 100R10 or 100R11 series hoses.

In general, the materials for fabricating flexible hoses have limited heat resistance capability. Placing hoses near a heat source will not only degrade the life of the hoses but may pose some safety hazards as well. Research indicates that an increase of 10°C above the maximum ambient temperature rating of a hose may cut its expected life in half. Moreover, the use of thermoplastic materials in fabricating hoses can effectively increase

**FIGURE 3.46**
Configuration illustration of three commonly used hydraulic hose fittings.

the temperature tolerance and allow such hoses to be used at higher-temperature environments, with no degradation in performance. The multilayer structure furnishes hydraulic hoses with an insulation effect to retain heat, which can also help bring the hydraulic fluid to operating temperature quickly, and therefore offers an advantage for mobile equipment operating in cold environments. However, no matter which type of hose is used, it is strongly recommended to avoid placing the hoses at locations where it will be in direct contact or be in close proximity to an ignition or heat source, such as engine exhaust system components, for safety concerns.

To use hydraulic hoses to form a fluid power delivery line network, we must use some types of reliable **hose connectors**, often called **hose fittings**, to couple hoses with other hydraulic elements. Three types of hydraulic fittings (Figure 3.46) are available for different applications, either reusable or nonreusable. The reusable ones are often screw-together or bolt-together, and nonreusable ones are normally crimp or swage (Figure 3.47).

In many mobile hydraulic systems, it may be required to connect different types of hydraulic actuators to perform different work and to connect and disconnect hydraulic lines frequently. A type of special fitting, the **quick-disconnect coupling**, is often used in those applications. There are over a dozen common designs of quick-disconnect couplings; among them the **double shutoff lock-ball-type coupling** is probably the most popular quick-disconnect coupling used on mobile hydraulic systems. As shown in Figure 3.48, this type of coupling uses a group of balls locked by a sliding collar to secure the connection between the female and male connectors. In disconnecting or connecting the coupling, one needs to manually pull the sliding collar to release the locking balls. A spring-loaded shutoff valve is installed in both mating halves to block the flow passage. The flow will not commence unless these valves are pushed open by each other after the coupling is connected as depicted in the figure. Once the couplings are connected, releasing the sliding collar will force the balls against a locking groove on the outside diameter of the male connector and return to the secured connecting status.

There are other lock designs for quick-disconnect couplings, such as **roller-lock couplings** and **pin-lock couplings**. These quick-disconnect couplings work under the same principle as the ball-lock couplings, with the only difference being the locker designs. Another type of commonly used quick-disconnect coupling is the **flat-face coupling**, which employs a poppet-style shutoff valve on each mating half to prevent air ingression during coupling to

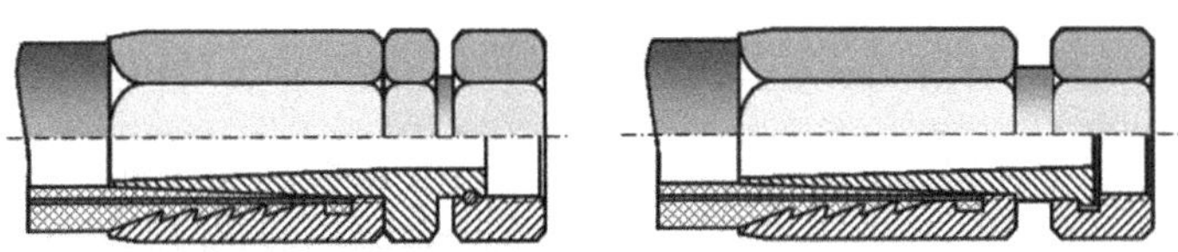


**FIGURE 3.47**
Configuration illustration of typical (a) reusable and (b) non-reusable fittings (straight) for hydraulic hoses.

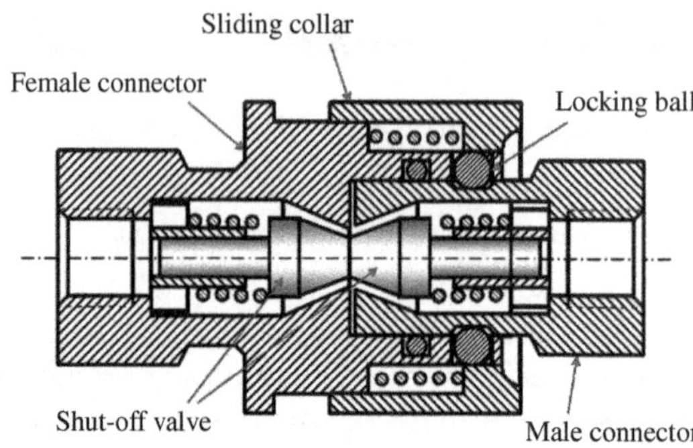


**FIGURE 3.48**
Configuration illustration of a typical pair of quick-disconnect coupling for hydraulic hoses.

achieve a **no-spill coupling**. These couplings are also designed for minimum-flow restriction to achieve the goal of minimizing a pressure drop during operation. Other commonly used quick-disconnect couplings include, but are not limited to, **bayonet, ring-lock,** and **cam-lock couplings**. One can often find their application features in the specifications provided by the manufacturers.

### 3.3.3 Metal Tubes and Pipes

Compared to hydraulic hoses, metal tubes and pipes not only exhibit a longer service life with a lower cost, but more importantly, they offer better heat dissipation and often a higher working pressure rating (≥41.4 MPa) than most hydraulic hoses. The main advantage of tubes over pipes is that the tube can be easily bent, cut, and connected, with more sizes and materials available for different applications. Because of these advantages, a metal tube is often used in mobile hydraulic systems when solid metal hydraulic lines are required. Pipe lines will mostly be selected over tube lines for large hydraulic systems with long, permanent straight runs.

One major difference of metal tubes from hydraulic hoses is that the tube has a fixed outside diameter (OD). Since their inside diameters vary with the thickness of the wall, the nominal diameter of tubes is specified by its OD. The choice of wall thickness is based on a combination of tube strength and flow capability, often in terms of the following two equations:

$$p = \frac{2\sigma\delta}{D_o} \tag{3.14}$$

$$q = v_t A \tag{3.15}$$

where $p$ is the allowed maximum working pressure; $\sigma$ is the allowable metal stress; $\delta$ is the tube thickness; $D_o$ is the nominal diameter; $A$ is the cross-sectional area of a tube; $q$ is the flow; and $v_t$ is the maximum allowable flow velocity within a tube.

Steel tubes can be used in different ways, including straight runs, bent runs, and coiled runs, to offer maximum flexibility for connection. **Straight runs** connect two hydraulic elements directly. The only concern in design of a straight run is to make sure the tube will not be subjected to extraordinary mechanical axial forces during the entire operation cycle to avoid buckling the tube. Normally, if a tube is sturdily installed and is mechanically fastened at both ends, a **buckling effect** may cause a bursting failure on a tube. Bent and coil runs of steel tube often require careful design to make the fabrication practical. The problem that often arises is how to figure out the exact bends and determine the exact length of the tube before bending. The procedure usually starts with a system layout, followed by adding the fitting points (in a three dimensional space), analyzing the available space, and choosing the optimum paths. A practical criterion for an optimal path design is to use as many straight-line segments and as few bends as possible.

Although a metal tube in general has better heat-dissipating capability than flexible hydraulic hoses, the increasing use of load-sensing pumps in mobile hydraulic systems has significantly reduced heat generation during operation, which consequently makes heat dissipation less of a concern. In addition, the development of stronger thermoplastics for making hydraulic hoses has resulted in new generations of lighter-weight, high-pressure hydraulic hoses, narrowing the gap between the heat dissipation and higher-pressure rating advantages of steel tubes over hoses. Both reasons have led to less need to use metal tube on modern mobile hydraulic systems.

The rigid nature of tubing does introduce a couple of crucial disadvantages, however, when compared to flexible hoses: they must be shaped using sophisticated bending equipment and often require special fittings and considerable labor to install and they have a fast wear-out rate if involved in relative motion between the tube and other metal surfaces. Considering that the operational environment of mobile equipment, the vibration—the main source of relative motion between tube and other solid material surfaces—is often very severe, some hybrid tube-hose assemblies are often used, with tubes being used at only a few select locations not suitable for using hydraulic hoses. For example, to install a boom attachment on an agricultural tractor, manufacturers often use flexible hydraulic hoses to connect the implement cylinders to corresponding ports of the control valve, and they use metal tubes only on places that can be reliably fixed, such as the connector between the rod end and the cap end of the cylinders.

Similar to hose lines, tube lines can use either **all-metal fittings** or **O-ring type fittings** to securely confine hydraulic fluids within the lines. The all-metal relies on metal-to-metal contact, and the O-ring compresses elastomeric materials to form a high-pressure seal to contain pressurized fluids. Different from connecting fittings to hydraulic hoses, tube fittings can either be welded, which is often done by sliding fitting sockets onto the tube and then welding them into position for economical mass production, or tapered, which normally uses the stress generated by forcing the tapered threads of the male half of the fitting into the female half or component port, to form reliable tube fittings. Tube threads have inherent disadvantages for sealing high-pressure systems, such as having a tendency to leak when it is either insufficiently tightened or overtightened, as well as having a tendency to loosen when exposed to vibration and wide temperature variations. Because pipe threads are tapered, repeated assembly and disassembly will only aggravate the leakage problem even more by distorting threads.

To avoid tube damage during the assembly/disassembly process, it is common to use either flared or flareless fittings (Figure 3.49) in mobile hydraulic systems. The **flared-fitting** technique is suitable for connecting thin-wall tubes. As depicted in Figure 3.49(a), a sleeve is placed on the tube before it is flared. When the **fitting nut** is tightening, the sleeve absorbs the twisting friction induced by the nut turning to deliver only the axial force against the flared tube, forming a positive seal between the flared tube face and the fitting body. A common recommendation is to use 37° or 45° flare fittings for thin-wall to medium-thickness tubes in systems with operating pressures up to 20.7 MPa. This flare-fitting technique can provide satisfactory tube connections for hydraulic systems operating at temperatures

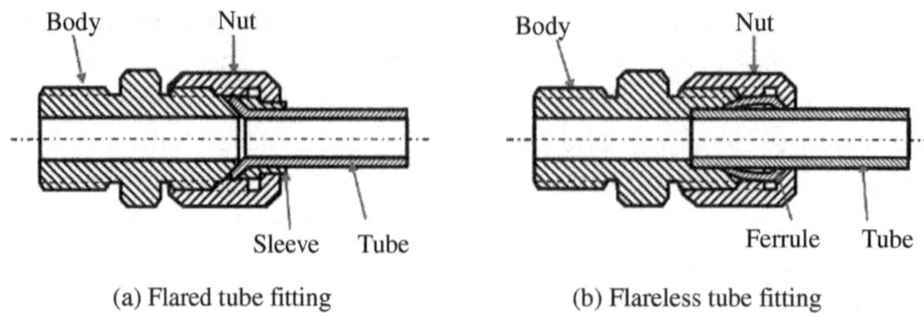


**FIGURE 3.49**
Configuration illustration of commonly used (a) flared and (b) flareless tube fittings.

from –50 to 200°C. Because it is more compact than most other fittings and can easily be adapted to metric tubing, the three-piece type of flared tube fitting is probably the most commonly used tube-fitting method used in mobile hydraulic systems. In addition, it is also readily available and is one of the most economical methods.

Because thick-walled tubing is difficult to flare, **flareless fittings** are recommended for these applications. As depicted in Figure 3.49(b), when tightening the nut of a typical flareless fitting onto the body, it will draw a ferrule into the body, which compresses the ferrule around the tube, consequently forcing the ferrule to contract and penetrate the outer circumference of the tube to create a positive seal. Due to such a feature, flareless fittings must be used with medium- or thick-walled tubes. Flareless fittings can handle average fluid working pressures to 20.7 MPa and are more tolerant of vibration than other types of all-metal fittings. This fitting technique is gaining wider acceptance because it requires minimal tube preparation.

In many mobile hydraulic systems, tubes are often combined with hoses in forming hybrid hydraulic lines. Because hose length can be increased as much as 2% when heated or decreased as much as 4% when pressurized, a **leak-proof seal** functioning reliably under vibrations and shock impulses must be established whenever a hydraulic hose and steel tube are connected. Therefore, it is crucial to make hydraulic hoses slightly longer than the actual distance between two connections in hybrid lines in order to allow expansion and contraction of the hose during operation.

### 3.3.4 Designing Hydraulic Lines

Since mobile hydraulic systems use more hoses than tubes/pipes, this textbook focuses on introducing the design of hydraulic lines using hydraulic hoses. Similar principles can be applied to selection and size of metal tubes and pipes.

Hydraulic hoses have a finite service life, and a number of factors will influence hose service life, including but not limited to (1) flexing hoses under a recommended minimum bend radius; (2) twisting or pulling the hose during operating cycles; (3) exposing the hose to working pressure above its rated value; (4) operating at temperatures above the rating of the hose; and (5) connecting hoses and fittings in a way not recommended as compatible by the manufacturer. To best design a hydraulic line for mobile applications, a seven-step design procedure, often called a **STAMPED** procedure, is recommended to select proper hoses and associated fittings in terms of the size (*S*), temperature (*T*), application (*A*), materials (*M*), pressure (*P*), ends (*E*), and delivery (*D*).

The selection of hose **size** is normally determined by fluid velocity. A hose must have a large enough size to deliver the fluids without a significant amount of energy loss induced by excessive fluid velocity. If the hose is too small, it will cause more serious consequences other than the energy loss, including but not limited to excessively high fluid temperature and, system noises and vibration. On the other hand, if the hose is too large, a larger radius is needed to bend the hose, which will not only make it more difficult to install it, but more importantly require more space to place it, which is often a major challenge for mobile hydraulic systems.

Since fluid velocity is the main factor to be considered in hose sizing, the size of hose should be specified using its inside diameter (ID). In sizing the hose ID, the design procedure normally starts with determination of the maximum fluid rate to be delivered, followed by identification of the functionality of the line in the system and selection of the maximum allowable fluid velocity in those lines. After all those design parameters are determined, the minimum hose ID can be calculated using the following equation.

$$ID = \sqrt{\frac{4Q}{\pi v_c}} \tag{3.16}$$

where *ID* is the inside diameter of the hose, *Q* is the maximum flow rate to be delivered, and $v_c$ is the critical velocity of the flow allowed to be transported in specific lines.

It is recommended that hydraulic lines be sized according to the following **maximum flow velocity** for different line segments:

1. Pump inlet lines: $0.6\ \text{m}\cdot\text{s}^{-1}$;
2. Pressure lines: $4.6\ \text{m}\cdot\text{s}^{-1}$ for low pressure (lower than 3.4 MPa)
   $6.1\ \text{m}\cdot\text{s}^{-1}$ for medium pressure (between 3.4 and 20.7 MPa)
   $7.6\ \text{m}\cdot\text{s}^{-1}$ for high pressure (higher than 20.7 MPa);
3. Return lines: $4.6\ \text{m}\cdot\text{s}^{-1}$.

The **pump inlet line** (also called the **suction line**) is the line used to connect the pump to the reservoir. The reason for a very slow flow recommendation for this line is that the flow is often delivered under a slight vacuum pressure in this line, and a high velocity will dramatically increase the vacuum-induced cavitation, thereby resulting in a high risk of damaging the pump. A special type of hose (SAE 100R4) is designed exclusively for inlet line use. Whenever possible, it is recommended to use a pump inlet line equal to or larger than the size of the pump inlet port being plumbed. In addition, the pump inlet line should be placed as straight as possible, with a minimal number of fittings, and that it be security sealed. By comparison, the **return lines** are used to direct the return flow back to the tank. One major difference between pump inlet lines and return lines is that the latter often needs to hold a backpressure to push returning fluid, overcoming the line resistance. A high-efficiency circuit always keeps the pressure drop in return lines low, and a low-pressure rating hose is often selected for return line use.

The size calculated using the above equation is the minimum inside diameter, and if the value is a nonstandard dimension for hose inside diameter, the next larger standard size hose should be used. The use of the next smaller standard size hose, even if the calculated size is only a little bit larger, will result in a velocity that is noticeably higher than the recommended level because the flow velocity is inversely proportional by the square of the inside diameter of the hose selected.

Sizing a hose often includes determining the length of the hose. As pointed out in the previous section, the hose length may be decreased by 4% when pressurized in operation due to the pressure-induced expansion in the diameter. If a hose segment is designed without making the hose length longer than the actual distance between the two connections to compensate for the hose shortening, the operating pressure will cause the hose to be stretched and consequently lead to a reduced service life.

The next design consideration in selecting a hose is its operating **temperature** range. All hoses are rated with a maximum working temperature ranging from 95 to 150°C based on the fabricating materials. Exposure to operating temperatures higher than the rated value for a long period of time will result in hoses losing their flexibility, or even in charcoaling of the hose. When a system is expected to operate at an excessively high temperature range for a considerable period of time, it is strongly recommended to select hoses made of special materials, such as the PTFEinner layer hoses (SAE 100R14 hose) for safe use in high-temperature (over 200°C) applications. Most of the time when talking about the operating temperature, it means the fluid temperature inside the hoses. The external temperature, also called the environmental temperature, will be of concern only when the hoses are exposed to some heat sources emitting excessively high temperatures. Long-term exposure to high external and internal temperatures concurrently will considerably reduce the service life of those hoses. Insulation sleeves are sometimes used to shield the hoses from being cooked by excessive heat sources, protecting their service life.

The **application** condition, mainly the routing plan of hydraulic lines, is another important issue to be considered. The rule-of-thumb for designing a good line routing plan is to place the high-pressure lines parallel to machine contours whenever possible. This practice could help reduce line lengths, minimize the number of bends, protect the hose from being physically damaged, and provide easier serviceability. One important issue needing special attention in design for application is to make sure that the selected hoses meet bend radius requirements because bending a hose in smaller radius than recommended will likely injure the hose reinforcement and therefore reduce the service life.

In selecting hydraulic hoses, it is mandatory to check the compatibility of hose-fabricating **materials** being used with the fluids. Operating temperature and fluid contamination are two of the most common factors that could affect the chemical compatibility of the hose and the fluid. While most hydraulic hoses are compatible with **petroleum-based fluids**, the recent adoption of **biodegradeable fluids**, especially in European countries may present a problem for some hoses.

Hose-rated working **pressure** is another critical parameter that needs to be chosen correctly to allow the system to operate safely, even under pressure spikes that are often significantly greater than the normal working pressure; if such pressure spikes exceed the rated working pressure of a hose, it will significantly shorten the service life of the hose.

Properly coupling all the line segments is an essential task in hydraulic system design. One fundamental requirement is having **leak-free connections**. The use of proper fittings on both **ends** of the hose segments will provide a necessary structural assurance for designing satisfactory leak-free connections. As introduced in the previous sections, two general types of couplings, the permanent and the quick disconnect, are available for connecting hydraulic hoses. The permanent type is available for most rubber and thermoplastic hoses and offers a wide range of dependable low-cost connections. The quick-disconnect type is generally more complicated and expensive, but it is much more convenient for applications requiring frequent connecting and disconnecting on many mobile hydraulic systems during field operations.

The last but not the least factor to be considered in a hydraulic line system design is the **deliverability** of the selected components in a timely manner. The preferred design is the one that does not use components with limited deliverability for the best serviceability of the system.

**Example 3.3: Sizing Hydraulic Hoses**

For a hydraulic system to deliver a maximum of 600 $L \cdot min^{-1}$ fluid to drive a heavy load up to 15 MPa, try to properly size pump input, pressure, and return hoses for the system (assume the standard size hoses are 5, 10, 15, 20, 25, and 30 cm in ID).

a. By applying Eq. (3.16), pump input hose can be sized in terms of a maximum allowed flow velocity of 0.6 $m \cdot s^{-1}$:

$$\begin{aligned} ID_{pump\_inlet} &= \sqrt{\frac{4Q}{\pi v_c}} \\ &= \sqrt{\frac{4 \times \frac{0.6}{60}}{3.14 \times 0.6}} \\ &= 0.146(m) \approx 15(cm) \end{aligned}$$

b. The hose size for a pressure line up to 20.7 MPa can be determined in terms of a maximum allowed flow velocity of 6.1 m/s using the same equation:

$$\begin{aligned} ID_{pressure\_line} &= \sqrt{\frac{4Q}{\pi v_c}} \\ &= \sqrt{\frac{4 \times \frac{0.6}{60}}{3.14 \times 6.1}} \\ &= 0.046(m) \approx 5(cm) \end{aligned}$$

c. The hose size for a return line is determined based on allowed flow velocity of 4.6 $m \cdot s^{-1}$:

$$\begin{aligned} ID_{return\_line} &= \sqrt{\frac{4Q}{\pi v_c}} \\ &= \sqrt{\frac{4 \times \frac{0.6}{60}}{3.14 \times 4.6}} \\ &= 0.053(m) \approx 10(cm) \end{aligned}$$

DISCUSSION 3.3: The maximum allowed velocities are different for hoses of different use. In engineering design practice, a hose-sizing nomograph is used for determining the inside diameter of a hose in terms of the maximum recommended velocity and the flow rate to be delivered.

### 3.3.5 Hose Routing and Installations

While the **STAMPED** design approach provides an optimized procedure for selecting proper hoses and associated fittings for an application, an adequate design on hose routing can result in higher reliability and better serviceability. Because of its flexibility, hydraulic hoses can be easily routed over, under, around, or through a series of obstacles lying on the optimized path of the line and are widely used in applications, allowing relative motion between elements being connected using the hose. To achieve an adequate design on hose routing and installation, it is generally recommended to follow a few commonsense rules for properly placing the hoses:

1. Allow the bend, on a large radius, to be greater than seven times the hose outside diameter (OD).
2. Keep the bend at least six times the size of the hose OD away from the fitting.
3. Avoid placing hoses where they are directly impacted by external forces.
4. Avoid placing hoses where they are directly exposed to excessive heat sources.
5. Avoid placing hoses where they will be extremely bent or twisted by the relative motion of the connected elements.

A good design should also always have some slack to relieve any hose stretches and be fixed properly to prevent kinking during the operation. Figure 3.50 depicts a few examples of good design guidelines accompanied by a few representative poor designs commonly seen for those cases. One of those guidelines is always to make a hose slightly (5% or more) longer than needed. Otherwise, the pressure-induced hose elongation and contraction will strain the reinforcement wires in a hose cut exactly to fit and easily lead to an earlier failure, especially at the hose-to-coupling interface (Figure 3.50(a)).

Another important guideline for good design is always to route a hose in a large bending radius and avoid bending the hose smaller than its minimum allowed bending radius (Figure 3.50(b)). In addition, it is strongly recommended that hoses be bent only in one plane if it is possible to avoid twisting its wire reinforcement, which consequently would reduce the expected service life of the hose. When it is unavoidable to bend a hose in two

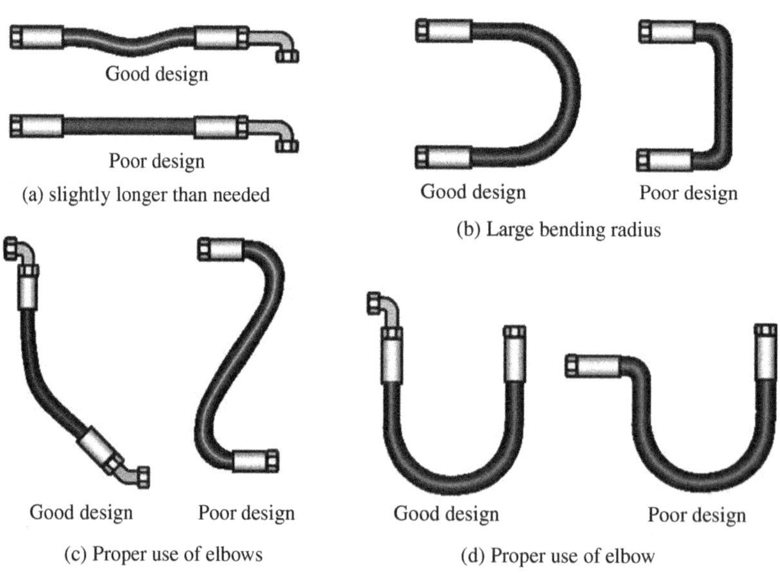


**FIGURE 3.50**
Examples of a few good and poor designs of hose routing and installations. (a) Straight, (b)-(d) bent installations.

different planes, use of a hose clamp to fix the hose in between bends with sufficient length on both sides of the clamp can effectively relieve the twist-induced strain on the reinforcement wires in the hose.

Whenever sharp turns are necessary for placing the line, it is preferable to use elbow fittings rather than to bend hoses, as illustrated in Figure 3.50(c) and (d). When one end of the hose has to be installed on a moving component, such a movement will often cause the hose to become twisted or bent. To avoid the motion-induced hose twist or bend, one can make a good design by using a **swivel joint** (also called **live swivel**) to replace a standard **swivel fitting** to accommodate relative motion between the hose and the component to which it is connected.

Most hydraulic hoses are wire reinforced, which makes them electrical conductors. For equipment that may be used near electrical lines or where hoses may be in close proximity to flammable solutions that could be ignited by static electricity discharged from the hoses, nonconductive hoses should be used. In addition, neatly planning the hose routing and carefully choosing hose end fittings, especially for cases with multiple lines running together or close, does not only help prevent tangling, twisting, and rubbing together (which can cause abrasive wear), but also make maintenance and troubleshooting easier.

When cutting hose to fit the required length, it is important to make sure that no hose debris gets into the finished hose assembly, as such debris could induce serious wear and damage to sensitive components in a hydraulic system. Debris may also be produced while crimping the hoses, as the crimping process will squeeze the coupling onto hose surfaces which often scratches some residual materials off the hose or gets into the hose. Improper storage of hoses without securely sealing the hose ends may also result in a wider variety of contaminants, such as dirt, water, foreign materials, and other types of contaminants, migrating into the hoses. Therefore, all hoses should be cleaned thoroughly before being put into service. Cleaning hoses is normally an inexpensive process; such an economical process could potentially avoid big expenses by preventing hydraulic systems from being damaged by all types of contaminants carried in a dirty hose.

## 3.4 Energy Losses and Heat Generation in Power Distribution

During the hydraulic power distribution process, a certain amount of energy carried by the fluid will be unavoidably lost to overcome all kinds of resistance to pressure fluid transporting. Those resistances may be represented in the forms of pressure drop and/or mechanical friction, and almost all the energy lost in overcoming those resistances is converted into the form of heat. While a portion of the heat is dissipated into the environment, a considerable amount of heat will always be transmitted to the fluid and result in a noticeable temperature rise. When the fluid temperature exceeds a certain level (70°C for most hydraulic fluids) over a long period of time, it may often induce a series of negative consequences in hydraulic systems, including but not limited to accelerating the fluid deterioration process (such as oxidation and/or insoluble gum formation), shortening the service life of sealing components, increasing leakage due to the lower viscosity, and even causing excessive component wear attributed to the loss of lubricity. To avoid overheating, most hydraulic systems have a limit on their maximum allowable operating temperatures. For mobile hydraulic systems, the maximum allowable operating temperature is commonly set at 80°C.

At any moment, the total amount of heat generated in a hydraulic system always equals the sum of the heat dissipating to the environment and the heat warming up the fluid. As discussed in Section 1.3, the energy is carried by the pressure fluids in a hydraulic system, and the amount of energy delivered in the pressure fluids can be determined according to the amount of fluids being delivered under a certain pressure. Since the amount of heat generated within a fluid power system is actually the amount of energy lost from the pressure fluid within this system, the energy loss is quantitatively determined by the pressure drop of fluid passing through the system by using the following equation:

$$q_t = Q\Delta P \tag{3.17}$$

where $q_t$ is the heat-generating rate within the hydraulic system, $Q$ is the flow rate, and $\Delta P$ is the pressure drop of the flow passing through the system of interest.

As mentioned earlier, a portion of the generated heat will be dissipated to the environment as natural cooling to the hydraulic system. The problem is that another considerable portion of the heat will be transmitted to the fluid and result in a noticeable fluid temperature rise. Assume that the initial operating temperature of the fluid is $T$ and that the surrounding environment temperature is $T_e$ for a hydraulic system. During an operation, a portion of the generated heat will raise the fluid temperature a $\Delta T$ after a short time interval, $\Delta t$, with another portion being dissipated to the environment. The energy balance for the heat-transfer process can be defined as follows:

$$q_t \Delta t = \sum c_i m_i \Delta T + \sum k_i A_i \left( T + \frac{\Delta T}{2} - T_e \right) \Delta t \tag{3.18}$$

where $c_i$ is the specific heat and $m_i$ is the mass of the relevant materials (such as the fluid, and lines and valves, the reservoir, the pump and actuator, etc.) in the hydraulic system, $k_i$ is the heat-dissipating coefficient, and $A_i$ is the heat-dissipating area.

For a very short time interval, $\Delta t \to 0$, Eq. (3.18) can be rewritten in a form of differential equation, and solving this differential equation results in a new equation capable of indicating the fluid temperature-rising pattern:

$$T = T_e + (T_0 - T_e) e^{-\frac{\Sigma k_i A_i}{\Sigma c_i m_i} t} + \frac{q_t}{\Sigma k_i A_i} \left( 1 - e^{-\frac{\Sigma k_i A_i}{\Sigma c_i m_i} t} \right) \tag{3.19}$$

where $T_0$ is the initial temperature of the fluid at time instant $t = 0$.

From the temperature-rising pattern represented by Eq. (3.19), it can be found that the fluid temperature will increase as the time duration increases. Theoretically, the temperature will reach its maximum value when the duration is infinitively long, determinable using the following equation:

$$T_{\max} \approx T_e + \frac{q_t}{\Sigma k_i A_i} \tag{3.20}$$

This equation indicates that the fluid temperature could reach a maximum value of $T_{\max}$ after a sufficiently long time of operation. After that, the heat generated in the hydraulic system will be equivalent to the heat dissipated from the system to the environment. This equation also reveals an important fact that the possible maximum temperature of the fluid is affected by both the amount of heat generated during the operation and the heat-dissipating capacity (namely, the value of $\sum k_i A_i$) of the system. Generally speaking, the higher the heat-dissipating capacity, the lower the possible maximum operating temperature of the fluid will be. Therefore, to control the maximum operating temperature under a certain allowable threshold, one very effective method is to improve the heat-dissipating capacity of the system; some of the effective methods are the use of heat exchangers or cooling fans for cooling the fluids.

**Example 3.4: Equilibrium Fluid Temperature**

When a hydraulic system delivers 450 L/min fluid from the pump to the actuator, it results in a total pressure drop of 750 kPa. How much heat will be generated in the power-transporting process? What will be the equilibrium fluid temperature if the overall heat-dissipating coefficient is 125 $J \cdot m^{-2} \cdot °C^{-1}$ and the total heat-dissipating area is 0.75 $m^2$ (assume the environment temperature is 15°C)?

a. Based on Eq. (3.17), the total heat being generated is:

$$\begin{aligned} q_t &= Q\Delta P \\ &= \frac{450 \times 10^{-3}}{60} \times 750 \times 10^3 \\ &= 5625\left(N \cdot m \cdot s^{-1}\right) = 5.625(kJ) \end{aligned}$$

b. From Eq. (3.20), we have:

$$\begin{aligned} T_{eq} &\approx T_e + \frac{q_t}{\sum k_i A_i} \\ &= 15 + \frac{5625}{125 \times 0.75} \\ &= 75(°C) \end{aligned}$$

DISCUSSION 3.4: The equilibrium fluid temperature can be determined by the amount of heat generated in the system, the environment temperature, and the system heat-dissipating capacity.

## References


1. Aardema, J.A., Koehler, D.W. System and method for controlling an independent metering valve, *United States Patent No. 5960695* (1999).
2. Akers, A., Gassman, M., Smith, R. *Hydraulic Power System Analysis*. CRC Press, Boca Raton, FL (2006).
3. Anderson, W.R. *Controlling Electrohydraulic Systems*. Marcel Dekker, New York (1988).
4. Book, R., Goering, C.E. Programmable electrohydraulic valve. *SAE Transactions: J. Commercial Vehicles*, 108: 346–352 (1999).
5. Caputo, D. Manifolds simplify fluid power systems. *Plant Engineering*, 56: 40–44 (2002).
6. Caterpillar, Inc. *Basic Hydraulic Valves*, Caterpillar, Inc., Peoria, IL (1983).
7. Cui, P., Burton, R.T., Ukrainetz, P.R. Development of a high speed on/off valve. *SAE Transactions: J. Commercial Vehicles*, 100: 312–316 (1991).

8. Esposito, A. *Fluid Power with Applications (6th Ed.)*. Prentice-Hall, Upper Saddle River, NJ (2003).
9. Fales, R. Stability and performance analysis of a metering poppet valve. *Int. J. Fluid Power*, 7: 11–17 (2006).
10. Goering, C.E., Stone, M.L., Smith, D.W., Turnquist, P.K. *Off-road Vehicle Engineering Principles*. ASAE, St. Joseph, MI (2003).
11. Guan, Z. *Hydraulic Power Transmission Systems* (in Chinese). Mechanical Industry Press, Beijing, China (1997).
12. Hedges, C.S. *Industrial Fluid Power (3rd Ed)*. Womack Educational Publications, Dallas, TX (1988)
13. Henke, R. Proportional hydraulic valves offer power, flexibility. *Control Engineering*, 28: 68–71 (1981).
14. Henke, R.W. Electrohydraulic proportional control valves. *Hydraulics & Pneumatics*, 38: 20–32 (1985).
15. Hu, H., Zhang, Q. Realization of programmable control using a set of individually controlled electrohydraulic valves. *International Journal of Fluid Power*, 3: 29–34 (2002).
16. Hu, H., Zhang, Q. Development of a programmable E/H valve with a hybrid control algorithm. *SAE Transactions: J. Commercial Vehicles*, 111: 413–419 (2002).
17. Hu, H., Zhang, Q. Multi-function realization using an integrated programmable E/H control valve. *Applied Engineering in Agriculture*, 19: 283–290 (2003).
18. Hutchison, E.A., Krone, J.J. Displacement controlled hydraulic proportional valve. *US Patent No. 5350152* (1994).
19. Hydraulics & Pneumatics. *Fluid Power Basics*. http://www.hydraulicspneumatics.com/200/FPE/IndexPage.aspx. Accessed on November 20 (2006).
20. Keller, G.R. *Hydraulic System Analysis*. Penton Media Inc., Cleveland, OH (1985).
21. Kong, X., Shan, D., Yao, J., Gao, Y. Study on experiment and modeling for the multifunctional integrated valve control system. *Proc. Int. Conf. on Intelligent Mechatronics and Automation*, pp. 455–459, Chengdu, China (2004).
22. Krone, J.J., Lunzman, S.V., Devier, L.J. Hydraulic flow priority system. *US Patent No. 5560387* (1996).
23. Krone, J.J., Zhang, Q. Method and apparatus for determining a valve transform. *US Patent No. 5784945* (1998).
24. Li, Z., Ge, Y., Chen, Y. *Hydraulic Components and Systems* (in Chinese). Mechanical Industry Press, Beijing, China (2000).
25. McClay, D., Martin, H.R. *The Control of Fluid Power*. John Wiley & Sons, New York (1973).
26. Mack, D.C., Hutchison, E.A., Szentes, J.F., Zimmermann, D.E. Hydraulic control valve having a centering spring device. *US Patent No. 5316044* (1994).
27. Manring, N.D. *Hydraulic Control Systems*. John Wiley & Sons, New York (2005).
28. Merrit, H.E. *Hydraulic Control Systems*. John Wiley & Sons, New York (1967).
29. NFPA, *Fluid Power Training: Basic Hydraulics*. NFPA, Milwaukee, WI (2000).
30. Norvelle, F.D. *Electrohydraulic Control Systems*. Prentice-Hall, New York (2000).
31. Pease, D.A. *Basic Fluid Power*. Prentice-Hall, Englewood Cliffs, NJ (1967).
32. Qing, G., Burton, R., Schoenau, G. Dynamic response of flow divider valves. *J. Fluid Control*, 19: 20–42 (1988).
33. Reed, E.W., Larman, I.S. *Fluid Power with Microprocessor Control: An Introduction*, Prentice-Hall, New York (1985).
34. Ruan, J., Pei, X., Li, S. Two-dimensional digital directional control valve. *Chinese Journal of Mechanical Engineering* (in Chinese) 36(3): 86–89 (2000).
35. Stoecker, W.F. *Design of Thermal Systems*. McGraw-Hill, New York (1989).
36. Stringer, J. *Hydraulic Systems Analysis: An Introduction*. John Wiley & Sons, New York (1976).
37. Valenti, M. Improving hydraulic performance with intelligent valves. *Mechanical Engineering*, 118: 56–60 (1996).
38. Viall, E.N., Zhang, Q. Spool valve discharge coefficient determination. *Proc. 48th National Conference of Fluid Power*, pp. 491–496, Chicago (2000).
39. Vickers, Inc. *Principles of Proportional Valves*. Vickers, Inc., Rochester Hills, MI (1987).

40. Vickers, Inc. *Vickers Mobile Hydraulics Manual (2nd Ed.)*. Vickers, Inc., Rochester Hills, MI (1998).
41. Wang, L., Chen, Y., Lu, Y. Numerical study on the axial flow force of a spool valve. *Proc. Proc. ASME Int. Mech. Eng. Cong. & Exp.*, FPST V5: 177–183, Anaheim, CA (1998).
42. Welty, J.R., Wicks, C.E., Wilson, R.E. *Fundamentals of Momentum, Heat, and Mass Transfer (3rd Ed.)*. John Wiley & Sons, New York (1984).
43. Wu, H.W., Lee, C.B. Influence of a relief valve on the performance of a pump/inverter controlled hydraulic motor system. *Mechatronics*, 6: 1–19 (1996).
44. Yeaple, F.D. *Fluid Power Design Handbook*. CRC Press, Boca Raton, FL (1996).
45. Yuan, Q, Li, P.Y. Self-calibration of push-pull solenoid actuators in electrohydraulic valves. *Proc. ASME Int. Mech. Eng. Cong. & Exp.*, FPST V11: 269–275, Anaheim, CA (2004).
46. Zahe, B., Prinsen, T., Schultz, M. A new type of pressure relief valve: The "soft relief" valve. *Proc. 48th National Conference of Fluid Power*, Pp. 481–490, Chicago (2000).
47. Zahe, B., Prinsen, T. Soft ventable relief valve. *US Patent No. 20060201554* (2006).
48. Zhang, Q. Hydraulic linear actuator velocity control using a feedforward-plus-PID control. *Int. J. Flexible Automation and Integrated Manufacturing*. 7: 275–290 (1999).
49. Zhang, Q. A generic fuzzy electrohydraulic steering controller for off-road vehicles. *Proc Instn Mech Engrs: J. Automobile Engineering*, 217: 791–799 (2003).
50. Zhang, Q., Goering, C.E. Fluid power system. In: Bishop, R. (ed.), *The Mechatronics Handbook*, CRC Press, Boca Raton, FL, pp: 10-11 ~ 10–14 (2001).
51. Zhang, Q., Meinhold, D.R., Krone, J.J. Valve transform fuzzy tuning algorithm for open-center electrohydraulic systems. *J. Agric. Engng. Res.*, 73: 331–339 (1999).

## Exercises

**3.1** Use a layperson's language to explain the two major categories of hydraulic control valves in terms of their configuration features.

**3.2** Use a layperson's language to explain the three major categories of hydraulic control valves in terms of their control features.

**3.3** Use a layperson's language to explain the three methods for actuating directional control valves.

**3.4** Use a layperson's language to explain the concept of cracking pressure.

**3.5** What is the primary function of a line-relief valve, and how does it work?

**3.6** What is the primary function of a pilot-operated check valve, and how does it work?

**3.7** What is a shuttle valve, what functionality does it provide to a hydraulic system?

**3.8** What is a sequence valve and what functionality does it provide to a hydraulic system?

**3.9** What are the major differences between an open-center and closed-center type of directional control valve?

**3.10** If two direct-acting relief valves, one of which opens at 5 MPa and the other at 4 MPa, are connected in parallel in between the main pressure line and the tank, what will be the system pressure? How about if those two valves are connected in series?

**3.11** Estimate the flow rate passing through a cartridge-type flow control valve that has a flow passage area of 25 $mm^2$ and a 1.5 MPa pressure drop crossing the valve. (Assume a valve orifice coefficient of 0.8 and fluid density of 850 $kg \cdot m^{-3}$ for petroleum-based hydraulic fluids.)

**3.12** The collected test data obtained from a flow control experiment on a spool-type valve showed that the flow rate through this valve was 35 $L \cdot min^{-1}$ when the flow passage area at the valve was 22 $mm^2$ and the pressure drop crossing the valve was 1.8 MPa. If the testing fluid used in this experiment is petroleum-based hydraulic fluid (assume the fluid density of 850 $kg \cdot m^{-3}$), estimate the effective orifice coefficient for this case.

**3.13** Compute the power loss across a check valve if the pressure drop across this valve at a full flow capacity of 90 $L \cdot min^{-1}$ is 800 KPa.

**3.14** A line-relief valve has a poppet of 3.6 $cm^2$ in area on which the fluid pressure acts. If a spring with spring constant of 300 $kN \cdot m^{-1}$ is compressed for 0.9 cm to form the initial holding force for keeping the poppet closed in normal condition, calculate (a) the cracking pressure of the valve; and (b) the full flow pressure it requires to lift the poppet for 0.3 cm for fully releasing the pump flow.

**3.15** A line-relief valve has a poppet of 3.6 $cm^2$ in area on which the fluid pressure acts and uses a spring with spring constant of 300 $kN \cdot m^{-1}$ to hold the poppet closed. If lifting the poppet for 0.3 cm to release the full flow, the system pressure should be no more than 15% higher than the cracking pressure; what should be the initial compression of the spring?

**3.16** A pressure-reducing valve is used to control the operating pressure in a branch. Assume that this branch requires having a constant level of supplying flow pressure at 9.2 MPa regardless of the flow rate being supplied. If the system keeps a 10 MPa pressure, how large a valve opening on the pressure-reducing valve is required to get 40 $L \cdot min^{-1}$ supplying flow from the system (assume the fluid density of 850 $kg \cdot m^{-3}$ and the valve orifice coefficient of 0.65)?

**3.17** Assuming we need to get a supplying flow pressure at 5 MPa using the same pressure-reducing valve as described in Problem 3.16, even with the same valve opening, how much flow can the branch get if the system pressure remains at 10 MPa? If we want to get the same amount of flow as in Problem 3.16, what should be the valve opening?

**3.18** A hydraulic system delivers a maximum of 500 $L \cdot min^{-1}$ fluid to drive a heavy load up to 25 MPa. Try to properly size pump input, pressure, and return hoses for the system. (Assume the standard size hoses are 2, 3, 4, 5, 10, 15, 20, 25, and 30 cm in ID.)

**3.19** The total pressure drop in a hydraulic system is 850 kPa when it delivers 600 $L \cdot min^{-1}$ fluid to the actuator. How much heat will be generated during the power-transporting process? How much higher will the equilibrium fluid temperature be over the environment if the total surface area of the hydraulic lines is 0.25 $m^2$, with an overall heat-dissipating coefficient of 80 $J \cdot m^{-2} \cdot °C^{-1}$, that of the fluid reservoir is 0.55 $m^2$ with a heat-dissipating coefficient of 148 $J \cdot m^{-2} \cdot °C^{-1}$, and that of all other components, such as the pump, cylinder, and valves, is 0.15 $m^2$ with a heat-dissipating coefficient of 105 $J \cdot m^{-2} \cdot °C^{-1}$?

**3.20** A hydraulic system that delivers 300 $L \cdot min^{-1}$ fluid will have a total line resistance of 900 kPa. Can this system work properly in a place with an environment temperature of 35°C if the total heat-dissipating area is 0.65 $m^2$ and the overall heat-dissipating coefficient is 125 $J \cdot m^{-2} \cdot °C^{-1}$? How about if the environment is changed to 25°C? (Assume the maximum allowable operating temperature is 80°C.)

# 4

# *Hydraulic Power Deployment*

## 4.1 Hydraulic Power Deployment Components

### 4.1.1 Hydraulic Actuators

Hydraulic actuators, linear or rotary, are the primary types of hydraulic power deployment components. The primary function of hydraulic actuators is to convert the potential energy, carried by fluid pressure and flow rate, to mechanical power in the form of force and velocity to drive the load.

### 4.1.2 Principle of Hydraulic Actuating

Hydraulic actuators can commonly be classified into two categories: **linear actuators** (often called **hydraulic cylinders**) and **rotary actuators** (also called **hydraulic motors**). Because of the significant dissimilarity in their structures, the operating principles of linear and rotary actuators have some fundamental differences. This section first discusses the basic operating principles of hydraulic cylinders and then introduces the basic principles of hydraulic motors.

Hydraulic cylinders are a line of linear actuators that deploy hydraulic power to do work by converting the hydraulic potential energy into mechanical power in linear motion. As illustrated in Figure 4.1, the output motion from a **single-rod double-action cylinder** is limited to a linear reciprocal motion, and so is the exerting force. In such a cylinder, the force and velocity for extension and retraction are different due to the difference in the effective area of the piston in the **cap end** (also called the **head end**) and the **rod end** of the cylinder. In **extension operation**, the pressure flow is imported to the cap end of the cylinder, which has an effective piston area determined by the cylinder bore size. Corresponding to a certain inlet flow rate $Q_1$ and pressure $p_1$ to a cylinder, the converted velocity and force on the cylinder rod can be calculated using the following two equations:

$$v_1 = \frac{4}{\pi D^2} Q_1 \tag{4.1}$$

$$F_1 = \frac{\pi D^2}{4} p_1 - \frac{\pi \left( D^2 - d^2 \right)}{4} p_2 \tag{4.2}$$

where $v_1$ is the extending velocity and $F_1$ is the exerting force of the rod; $D$ and $d$ are cylinder bore and rod diameters; $p_1$ and $p_2$ are inlet and outlet pressures; and $Q_1$ and $Q_2$ are, respectively, inlet and outlet flow rates of the fluids.

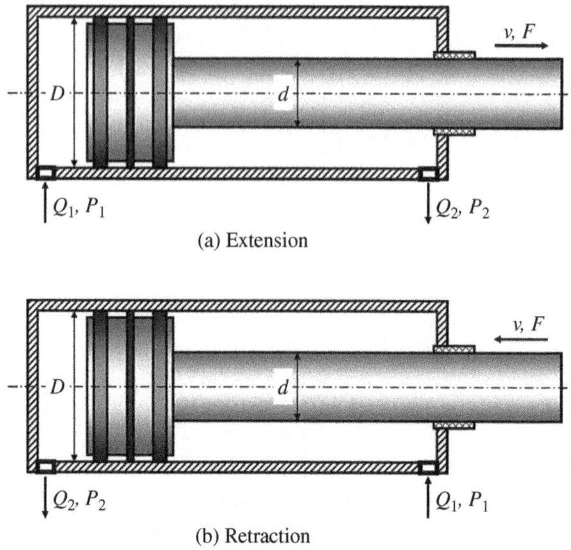


**FIGURE 4.1**
Operating principle of a typical single-rod double-acting hydraulic cylinder. (a) Extension and (b) retraction strokes.

Similarly, in a **retraction operation**, the pressure flow is imported to the rod end of the cylinder. With the same inlet fluid of flow rate $Q_1$ under the pressure $p_1$, the converted velocity and force on the rod can be calculated using the following two equations:

$$v_2 = \frac{4}{\pi\left(D^2 - d^2\right)} Q_1 \tag{4.3}$$

$$F_2 = \frac{\pi\left(D^2 - d^2\right)}{4} p_1 - \frac{\pi D^2}{4} p_2 \tag{4.4}$$

The inlet pressure is actually determined by the total load, including the friction, to be driven by the cylinder. While a hydraulic power unit supplies pressurized fluid into a cylinder, the **incompressibility** feature of hydraulic fluid raises the fluid power rapidly to overcome all the resistance applied on the movable piston to enlarge the space that will allow more flow to enter the cylinder until the maximum space is reached. Because of the imperfect incompressibility of hydraulic fluids, the actuation of a hydraulic cylinder is always softer than a mechanical actuator can achieve. Such **softness** plays an important role in the dynamic performance of hydraulic actuation, and its effect on actuation dynamics can be quantitatively described using a system parameter of **hydraulic stiffness**. The hydraulic stiffness for a hydraulic cylinder is defined as a function of fluid bulk modulus, piston areas, cylinder chamber volume, and the volume of hydraulic hoses connected to both chambers. For a typical single-rod double-acting cylinder, the total stiffness of the cylinder can be determined using the following equation:

$$k_h = \beta\left(\frac{A_1^2}{V_{L1} + V_1} + \frac{A_2^2}{V_{L2} + V_2}\right) \tag{4.5}$$

where $k_h$ is the hydraulic stiffness of a cylinder, $\beta$ is the fluid bulk modulus; $A_1$ and $A_2$ are the cap-end and rod-end piston areas; $V_1$ and $V_2$ are the cap-end and rod-end cylinder chamber volume; and $V_{L1}$ and $V_{L2}$ are the volume of hydraulic hoses connected to cap-end and rod-end cylinder chambers, respectively.

Another important system parameter often used to represent the operating status is the **natural frequency** of a hydraulic cylinder, which can be determined in terms of the combined mass and hydraulic stiffness of the cylinder using the following equation.

$$\omega_n = \sqrt{\frac{k_h}{m}} \tag{4.6}$$

$\omega_n$ is the natural frequency of a hydraulic cylinder and $m$ is the combined mass of the cylinder and the external load applied to the cylinder.

Even though there are numerous types of hydraulic cylinders, different in design, function, or usage, the operation of all those cylinders should follow the basic operating principles of hydraulic cylinders introduced in this section.

Like linear actuators, rotary actuators deploy hydraulic power to do work by converting the potential energy carried by the pressurized flow into mechanical power in the forms of torque and rotational velocity. However, rotary actuators are often mounted at the equipment joint and rotate the load to do designated work. While some of the rotary actuators are designed to perform limited turns and are often called **oscillating motors**, the most commonly used ones are designed to perform continuous rotating work and are simply called **hydraulic motors**. Normally, oscillating motors are used in applications where there is a need to drive a load in a finite turning angle in both directions and requires high instantaneous torque, whereas the hydraulic motors are used in applications doing continuous rotating work.

One commonality between the oscillating motors and the hydraulic motors is that both motors generate torque and rotating motions as the product of energy conversion. This means that regardless of the configuration differences among different types of rotary hydraulic actuators, they all operate following the same basic principle that the rotating velocity is determined by the flow rate supplied to the actuators and the output torque can be calculated based on the pressure drop between the inlet and outlet ports of the actuators. Much the same as for the linear actuator discussed earlier in this section, the inlet pressure to a motor is also determined by the total load to be moved by the motor, including the friction. In general, the motor output rotating velocity and torque can be calculated using the following equations:

$$n = \frac{Q}{D_v}\eta_v \tag{4.7}$$

$$T = \frac{D_v \Delta p}{2\pi}\eta_m \tag{4.8}$$

where $n$ is the motor output rotational velocity; $T$ is the motor output torque; $D_v$ is the motor volumetric displacement; $Q$ is the actual inlet flow rate; $\Delta p$ is the pressure drop between the motor inlet and outlet ports; and $\eta_v$ and $\eta_m$ are the volumetric and mechanic efficiencies of the motor, respectively.

The **volumetric efficiency** of a motor is defined as the ratio of theoretical and actual inlet flow rates to the motor, and the **mechanical efficiency** is the ratio of theoretical and actual output torques from the motor, expressed as follows:

$$\eta_v = \frac{Q_t}{Q} \tag{4.9}$$

$$\eta_m = \frac{T}{T_t} \tag{4.10}$$

where $Q$ and $Q_t$ are actual and theoretical inlet flow rates supplied to the motor, and $T_t$ and $T$ are theoretical and actual output torques from the motor.

Hydraulic motors have the unique **load-limit function** of stall to protect the system from damage when overloaded. Theoretically, a motor is stalled if it is completely stopped by an excessive load. In practical analysis, a **motor stall** occurs when the motor output speed is less than 1 rpm (one revolution per minute). **Stall torque efficiency** is often used to indicate the load-handling capability of a motor and is defined as follows:

$$\eta_s = \frac{T_s}{T_t} \tag{4.11}$$

where $\eta_s$ is the stall efficiency of a motor and $T_s$ is the measured stall torque on the motor output shaft.

The level of volumetric efficiency has a direct effect on the breaking performance of a hydraulic motor, and a high volumetric efficiency normally means a low **internal leakage** within a motor. Such a low internal leakage allows the fluid in different chambers of a motor to provide an effective break on the motor due to the incompressibility of the hydraulic fluids. Different from the volumetric efficiency, the mechanical efficiency will directly affect motor starting performance because a low mechanical efficiency often implies that a low torque is available to start the motor. To provide a convenient way to evaluate both the starting and the breaking performance jointly, an **overall efficiency**, defined as the product of the volumetric and the mechanical efficiencies as follows, is often used.

$$\eta_o = \eta_v \eta_m \tag{4.12}$$

where $\eta_o$ is the overall efficiency.

Taking the overall efficiency into consideration, we can calculate the actual output power from a hydraulic motor using the following equation:

$$P_h = \Delta p Q \eta_o \tag{4.13}$$

where $P_h$ is the actual output power from a hydraulic motor and $\Delta p$ is the pressure drop between the motor inlet and outlet ports.

**Example 4.1: Hydraulic Actuator Capacity**

A single-rod double-action hydraulic cylinder as shown in Figure 4.1 has a 100 mm diameter bore and a 70 mm diameter rod, respectively. When the system can deliver a maximum flow rate of 300 $L \cdot min^{-1}$, with a line relief valve preset at 21 MPa and the return line total resistance of 1 MPa, what is the maximum load the cylinder can push if the cylinder friction can be ignored, and under what speed? If we did want to achieve the maximum speed by using cylinder retraction to push the load, what will be the maximum load driving capability of the cylinder?

a. The maximum load-driving capacity and associated speed can be calculated using Eqs. (4.1) and (4.2):

$$\begin{aligned} F_1 &= \frac{\pi}{4}\left[D^2 p_1 - \left(D^2 - d^2\right)p_2\right] \\ &= \frac{\pi}{4}\times\left[0.1^2\times 21\times 10^6 - \left(0.1^2 - 0.07^2\right)\times 1\times 10^6\right] \\ &= 1.61\times 10^5 (N) = 161(kN) \end{aligned}$$

$$\begin{aligned} v_1 &= \frac{4}{\pi D^2} Q_1 \\ &= \frac{4}{\pi\times 0.1^2}\times\frac{0.3}{60} \\ &= 0.637\left(m\cdot s^{-1}\right) \end{aligned}$$

b. The maximum speed and associated load-driving capacity can be calculated by using Eqs. (4.4) and (4.3):

$$\begin{aligned} F_2 &= \frac{\pi}{4}\left[\left(D^2 - d^2\right)p_1 - D^2 p_2\right] \\ &= \frac{\pi}{4}\times\left[\left(0.1^2 - 0.07^2\right)\times 21\times 10^6 - 0.1^2\times 1\times 10^6\right] \\ &= 7.62\times 10^4 (N) = 76.2(kN) \end{aligned}$$

$$\begin{aligned} v_2 &= \frac{4}{\pi\left(D^2 - d^2\right)} Q_1 \\ &= \frac{4}{\pi}\times\frac{0.3/60}{\left(0.1^2 - 0.07^2\right)} \\ &= 1.25\left(m\cdot s^{-1}\right) \end{aligned}$$

DISCUSSION 4.1: A ratio of the rod diameter over the bore diameter of 0.7 roughly gives a 1:2 ratio on effective pressure-acting areas of the piston in its rod- and cap-end sides. The results obtained from this example indicate that the load-pushing capacity of a cylinder is proportional to the effective acting area, and the actuating speed is inversely proportional to the area.

## 4.2 Hydraulic Cylinders

### 4.2.1 Classification of Hydraulic Cylinders

From the basic operating principle, we know that a hydraulic cylinder utilizes the flow and pressure of the inlet fluid to drive the load performing a linear motion. A typical hydraulic cylinder consists of a **cylinder body**, a **piston**, a **rod**, and **seals**

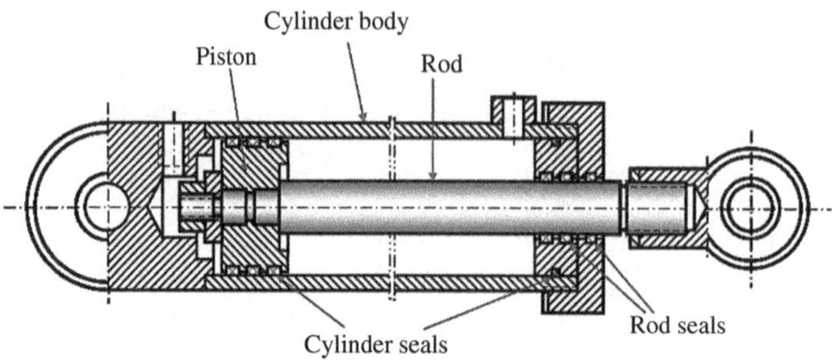


**FIGURE 4.2**
Configuration illustration of typical piston-type single-rod double-acting hydraulic cylinder for (a) extension, (b) retraction, and (c) differential extension operations.

(Figure 4.2). To meet the special needs for a wide range of applications, cylinders have evolved into an almost endless array of configurations, sizes, and special designs. The structural features, rod designs, actuation methods and usage are commonly used criteria for categorizing those cylinders. The most commonly used cylinder classification is based on the actuation, which sorts all hydraulic cylinders into two categories: single acting and double acting. The cylinders can also be sorted into single rod and double rod in terms of rod designs. According to their configuration features, they can be either piston cylinders, ram cylinders, or telescopic cylinders. Another way of classifying hydraulic cylinders is by their usage, normally only for some specialty cylinders. For this category, there are tandem cylinders, duplex cylinders, and many more.

The most common type of cylinder in mobile hydraulic systems is probably the piston-type, **single-rod double-acting** cylinder as depicted in Figure 4.2. This type of cylinder uses a piston–rod assembly as the actuating element to transfer the hydraulic potential energy into mechanical power by converting the pressure acting on piston ends to generate a force and using rod protrusion from the cylinder to transmit the generated force to the load. Because of the structural constraints, the cap-end area of a single-rod piston is always bigger than the rod-end area and results in different operational characteristics during extension and retraction as expressed by Eqs. (4.1) through (4.4):

### 4.2.2 Operating Parameters of Hydraulic Cylinders

Since a single-rod double-acting cylinder holds a basic relationship of $A_1 > A_2$, it carries two basic operational characteristics of $F_1 > F_2$ and $v_1 < v_2$. That is, when the pressurized fluid enters the cap-end chamber, the cylinder will generate a larger pushing force with a slower extending speed, or it will generate a smaller pulling force with a higher retracting speed when the fluid enters the rod-end chamber. Because of this feature, it is common to use the **extension cycle** as the actuating stroke to push the load and the **retraction cycle** as the returning stroke to move the rod back to its original position. An **area ratio**, defined as the ratio of the piston area of cap- and rod-end sides, is often used to quantify the relationship between the rod protrusion and retraction speeds. The following equation reveals that the rod protrusion and retraction speeds ratio is inversely proportional to piston cap-end and rod-end areas.

$$\varphi = \frac{A_1}{A_2} = \frac{D^2}{D^2 - d^2} = \frac{1}{1 - \left(\dfrac{d}{D}\right)^2} = \frac{v_2}{v_1} \tag{4.14}$$

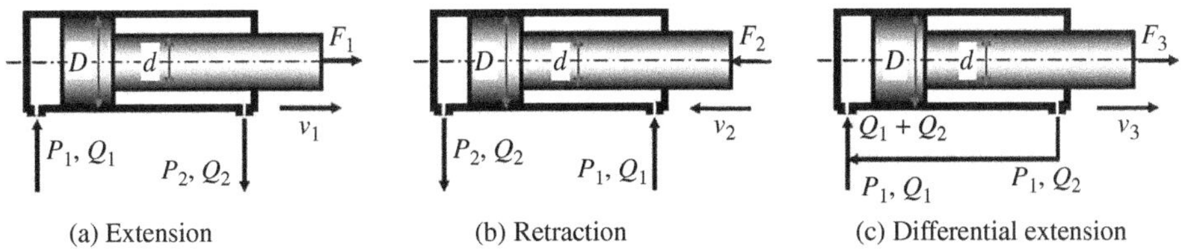


**FIGURE 4.3**
Illustration of the principle of a typical single-rod double-acting cylinder functionalities.

where φ is the piston area ratio of a single-rod cylinder, and $D$ and $d$ are diameters of the cylinder bore and rod.

Normally, the area ratio can range from 1.06 to 5.00 for a typical single-rod cylinder. It represents a diameter ratio between the rod and the piston from 0.25 to 0.90.

A unique function of a single-rod double-acting cylinder is its capability of implementing a differential extension operation. As illustrated in Figure 4.3, the normal operational functions of extension and retraction are implemented by routing pressurized fluid through either the cap-end or the rod-end chamber and releasing the normally nonpressurized fluid in the other side of the chamber (Figure 4.3(a) and (b)). By redirecting the returning fluid from the rod-end chamber (also called the **recycled fluid**), along with the supplied pressurized fluid, back to the cap-end chamber, a single-rod double-acting cylinder can achieve a so-called **differential extension** function. As depicted in Figure 4.3(c), by utilizing the feature of area difference between the piston cap end and the rod end, the cylinder could generate enough force under a condition of equilibrant pressure in both sides of the piston to extend the rod. Because the recycled fluid is extruded by the piston motion, the following equation can be used to determine the rate of this recycling flow.

$$Q_2 = \frac{\pi\left(D^2 - d^2\right)}{4} v_3 \tag{4.15}$$

The extending speed and pushing force in this state can be calculated using the following equations.

$$v_3 = \frac{4}{\pi D^2}\left(Q_1 + Q_2\right) = \frac{4}{\pi d^2} Q_1 \tag{4.16}$$

$$F_3 = \frac{\pi D^2}{4} p_1 - \frac{\pi\left(D^2 - d^2\right)}{4} p_1 = \frac{\pi d^2}{4} p_1 \tag{4.17}$$

where $v_3$ is the rod-extending velocity and $F_1$ is the exerting force during the differential extension operation; $D$ and $d$ are cylinder bore and rod diameters; $p_1$ and $p_2$ are inlet and outlet pressures; and $Q_1$ and $Q_2$ are inlet and outlet flow rates of the fluids, respectively.

From the above equations, one may find that the operating characteristics of **differential extension** are determined by the size of the rod rather than the piston. As a result, this operation mode can achieve a higher protruding speed by paying the price of a reduced pushing force. When the area ratio of such a cylinder is 2.0, that is, the ratio of the rod diameter over the piston diameter satisfies $d = D/\sqrt{2}$, the single-rod double-acting cylinder can achieve an equal extension and retraction speed.

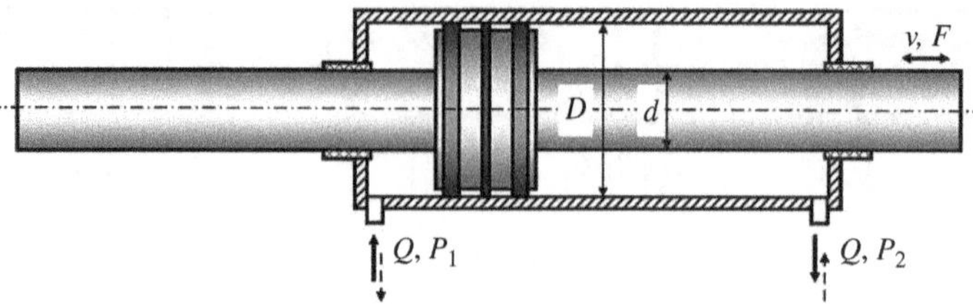


**FIGURE 4.4**
Configuration illustration of typical piston-type double-rod hydraulic cylinder.

The equal extension and retraction speed can also be achieved by using a **double-rod cylinder**. As depicted in Figure 4.4, a typical double-rod cylinder has a rod attached to both sides of the piston, with each rod extending through a rod cap, which eliminates the differential area between both sides of a piston. With the equal areas on both sides of the piston, a given flow produces the same extension and retraction speeds and generates the same force to move a load in both directions. With an inlet flow rate $Q$ at a pressure $P_1$, the flow-generated speed and force on the rod can be calculated using the following two equations:

$$v = \frac{4}{\pi\left(D - d^2\right)} Q \tag{4.18}$$

$$F = \frac{\pi\left(D^2 - d^2\right)}{4}(p_1 - p_2) \tag{4.19}$$

where $v$ is the rod speed and $F$ is the exerting force for both extension and retraction; $D$ and $d$ are cylinder bore and rod diameters; $p_1$ and $p_2$ are inlet and outlet pressures; and $Q$ is the flow rate routed in and out the cylinder.

The **single-acting cylinders** find many applications in mobile hydraulic systems. The basic feature of this type of cylinder is that they only use the pressurized fluid to drive the cylinders in one direction and rely on nonhydraulic forces, such as the gravity force and spring force, to push the pistons to return to their original state. To do so, the pressurized fluid is supplied to only one side of the piston; the chamber on the other side of the piston is normally vented to the atmosphere or connected to the tank. Depending on whether the pressurized fluid is routed to the cap end or the rod end of a cylinder, the pressurized fluid will either extend or retract the cylinder. In many factory automation applications, **spring-return single-acting cylinders** are commonly used. Typically, the pressurized fluid enters the cap end of the cylinder to extend the piston rod. In retraction, the return spring exerts a force to push the piston, returning to its original position, and consequently it pushes the fluid in the cap-end chamber back into the tank. In comparison, the gravity-driven retraction finds more applications in mobile hydraulic systems. A few commonly used single-acting cylinders in mobile hydraulic systems include, but are not limited to, ram cylinders, piston cylinders, and telescopic cylinders. While a double-rod cylinder can theoretically be used as a single-acting cylinder, in practice it is seldom used.

Unlike a piston cylinder, a **ram cylinder** (Figure 4.5) uses a ram as the sole moving element of actuating. As depicted in the figure, the ram is coupled only with the ram cap; therefore, the bore of the cylinder chamber does not need to be finely machined. Such a feature makes fabrication of this type of cylinder easier, especially for those with long

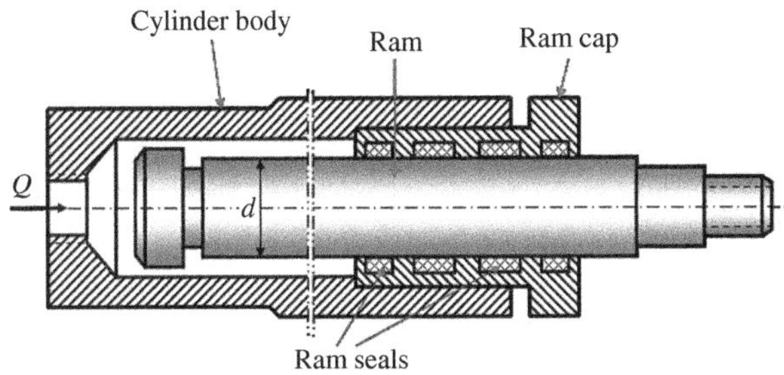


**FIGURE 4.5**
Configuration illustration of a typical single-acting ram cylinder.

barrels. As a single-acting cylinder, this type of cylinder can only perform one direction actuation and requires external forces to complete the traction motion. The output velocity and the actuating force from this type of cylinder can be calculated using the following equations:

$$v = \frac{4}{\pi d^2} Q \tag{4.20}$$

$$F = \frac{\pi d^2}{4} p \tag{4.21}$$

where $v$ is the ram-extending velocity; $F$ is the exerting force; $d$ is the ram diameter; and $Q$ and $p$ are the flow rate and pressure of supplied pressure fluid, respectively.

A few common examples of applying **ram cylinders** in mobile hydraulic systems are hydraulic jacks, which are normally constructed using a single-acting gravity-return cylinder, and hydraulic parking brake actuators, which are usually built using a spring-applied and hydraulic-pressure-released, single-acting spring-extending cylinder. Ram cylinders are normally vertically mounted and used.

In some mobile applications, such as the crane arms, a **telescopic cylinder**, as depicted in Figure 4.6, is often used to satisfy the special requirement of using a very compact retractable arm to extend an excessive distance to perform the designated work. Normally, telescope-type cylinders are single acting. The typical structure of telescope cylinders consists of sets of tubing nesting inside one another. Because of such a structural design, the collapsed length of a telescope cylinder is often a small fraction, commonly ranging from

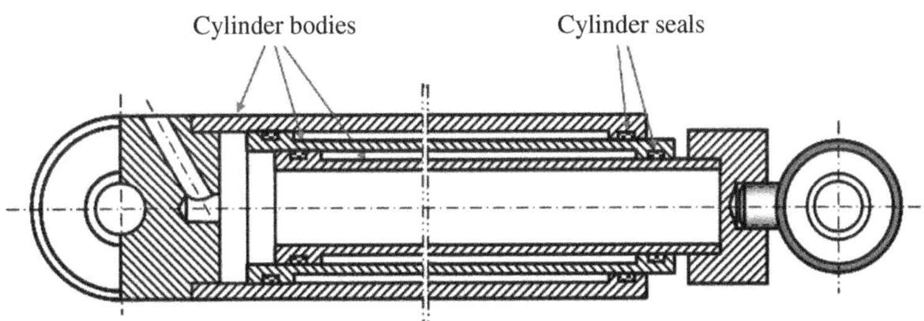


**FIGURE 4.6**
Configuration illustration of a typical telescopic type hydraulic cylinder.

one-half to one-fifth of its extended length. However, the cost is often several times higher than that of a standard cylinder capable of producing equivalent force. Corresponding to the difference in effective pressure-bearing areas, the cylinder will then extend from the largest cylinder to the smallest cylinder as the pressurized fluid is routed into the pressure chamber of the cylinder. The largest cylinder extends first because it requires the lowest pressure to extend. Then the next largest cylinder will follow after the largest cylinder is completely extended. The innermost cylinder will be the last one to extend. During retraction, the opposite order of action can be expected; that is, the innermost cylinder will be the first to retract followed by the next larger cylinder, until all cylinders are collapsed.

Another category of double-acting cylinders are **tandem cylinders** and **duplex cylinders** (Figure 4.7). Both types consist of at least two sets of single-rod pistons. The main structural difference in distinguishing these types of cylinders is whether the piston assemblies are physically connected to a common rod (tandem cylinder, Figure 4.7(a)) or not (duplex cylinder, Figure 4.7(b)). As illustrated in the figure, a tandem cylinder is designed for applications where high force must be generated within a narrow radial space where substantial axial length is available. The velocity and force functions of a tandem cylinder in extension and retraction operations need to be modified according to the structural feature of the cylinder and can be expressed as follows.

In extension:

$$v_1 = \frac{4}{\pi\left(2D^2 - d^2\right)} Q_1 \tag{4.22}$$

$$F_1 = \frac{\pi\left(2D^2 - d^2\right)}{4} p_1 - \frac{\pi\left(D^2 - d^2\right)}{2} p_2 \tag{4.23}$$

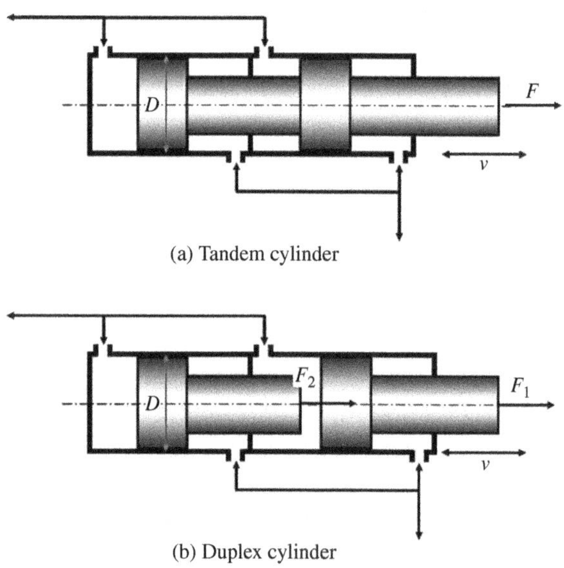


**FIGURE 4.7**
Illustration of the configuration of (a) a typical tandem cylinder and (b) a typical duplex cylinder.

In retraction:

$$v_2 = \frac{2}{\pi\left(D^2 - d^2\right)} Q_1 \tag{4.24}$$

$$F_2 = \frac{\pi\left(D^2 - d^2\right)}{2} p_1 - \frac{\pi\left(2D^2 - d^2\right)}{4} p_2 \tag{4.25}$$

In comparison, the pistons within a duplex cylinder are not physically connected; the rod of one cylinder protrudes into the nonrod end of the second cylinder if they are connected. In addition, the duplex cylinder has another unique feature. It allows the individual composing cylinders to have their own different stroke lengths, with the longer stroke cylinder normally placed at the rod end of the cylinder. It will create a condition that the back cylinder rod will be unable to protrude into the front cylinder. Such a feature makes a duplex cylinder operate either under a **tandem state** or a **high-speed state**. The operation functions of the tandem state of a duplex cylinder are the same as those of a tandem cylinder, and the operation functions of a single-rod cylinder as defined by Eqs. (4.1) to (4.4) can be used for the high-speed state.

As all the cylinders introduced so far are used to actuate a load doing translational work, they are also called **actuating cylinders**. Another type of special cylinder often seen on some mobile hydraulic systems is a **pressure intensifier**. As depicted in Figure 4.8, the core element of a typical pressure intensifier is a **free-piston** assembly, consisting of a large and a small piston. The principal function of a pressure intensifier is to create a higher pressure to push a heavy load in one branch and maintain the rest of the system operations under normal system pressure. In some cases, it can also be inversely used to generate a larger flow with lower pressure to achieve a high-speed operation in a branch with a light load. Different from the actuating type cylinders, the operational functions of a pressure intensifier can be expressed using a flow function and a pressure function, as follows:

$$Q_2 = \frac{d^2}{D^2} Q_1 = \frac{Q_1}{\varphi_A} \tag{4.26}$$

$$p_2 = \varphi_A p_1 \tag{4.27}$$

where $\varphi_A$ is the area ratio of the larger piston over the smaller one in a pressure intensifier, and $D$ and $d$ are the diameters of the larger and the smaller pistons, respectively.

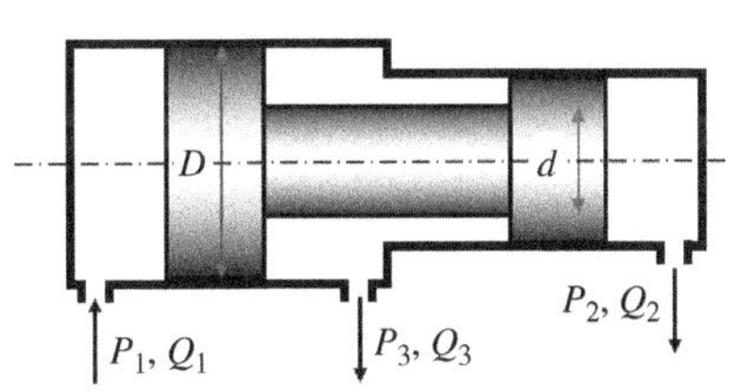


**FIGURE 4.8**
Illustration of the configuration and operation principle of a free-piston type pressure intensifier.

### 4.2.3 Hydraulic Cylinder Cushions

Hydraulic cylinders are often used to drive a heavy load performing fast translational moves; an abrupt stop at the end of a stroke will often induce a large inertia force acting on the cylinder, which will not only cause an excessive impact, but also generate pressure spikes and hydraulic noises. This will be attributed to unexpected cylinder damage, especially when such an inertia force is frequently applied to the cylinder. An effective way to reduce such an inertia force is to add a cushion, a device that will increase the fluid-bleeding resistance to reduce the piston velocity of motion near its end of stroke. As depicted in Figure 4.9, a typical **cylinder cushion** is commonly installed on end caps, and a cushion plunger is often used to form a damping effect either using a carefully designed flow-restriction clearance, a flow-bleeding check valve, or some forms of orifices, or a combination of these means. While the structure of cushion devices may vary, the basic principle is the same: a flow restriction is created to bleed the fluid out of the cylinder chamber when the piston is moved very close to its end-of-stroke. Such a flow restriction will raise the fluid pressure in the cylinder chamber, which will consequently slow down the piston speed to achieve the desirable **cushion effect**. To properly design a cushion device, the length of the cushion stroke, the maximum cushion pressure in terms of the full piston speed, and the mass being driven need to be determined. Quantitatively describing the cushion design principle, without loss of generality, we can assume that the cushion chamber has a cross-sectional area $A_C$, and a generic orifice can be used to represent the flow restriction. Therefore, a set of cushion state equations can be created as follows:

$$F - p_C A_C = ma \tag{4.28}$$

$$Q_C = vA_C = C_d A_o \sqrt{\frac{2}{\rho}(p_C - p_2)} \tag{4.29}$$

where $F$ is the protruding force of the piston; $A_C$ is the area of cushion chamber; $p_C$ is the fluid pressure in the cushion chamber; $m$ is the total mass driven by the cylinder and $a$ is the acceleration of the piston moving; $Q_C$ is the flow rate bled from the cylinder chamber;

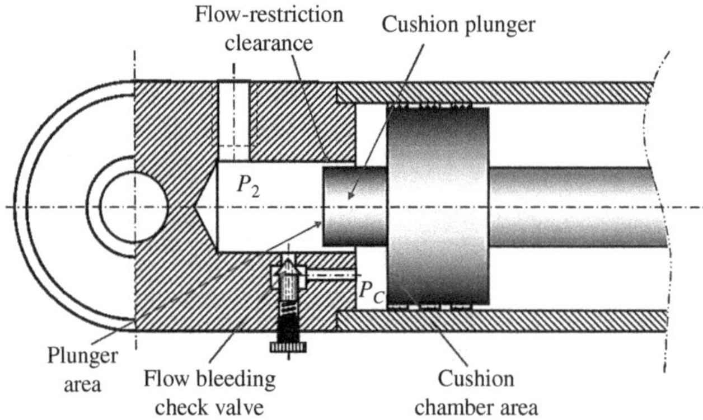


**FIGURE 4.9**
Illustration of the configuration and operation principle of a typical hydraulic cylinder cushion.

$v$ is the piston cushion speed; $C_d$ and $A_o$ are an orifice coefficient and the orifice area of the cushion device; $\rho$ is fluid density; and $p_2$ is the fluid pressure in the return line.

The above equations provided a base to derive a piston deceleration equation during cushion processes, as follows:

$$\frac{dv}{dt} = \frac{1}{m}\left[F - \left(\frac{\rho}{2C_d^2}\varphi_C^2 v^2 - p_2\right)A_C\right] \tag{4.30}$$

where $\varphi_C$ is the area ratio of cushion chamber area over the orifice area ($\varphi_C = A_C/A_o$).

The return line pressure is often close to the ambient pressure in many hydraulic systems and can be ignored without a significant loss of accuracy of piston velocity calculations. Solving the above equation by ignoring the return line pressure results in the decay functions of **cushion velocity** and **cushion pressure** in terms of cushion stroke, $x$:

$$v = \left[\frac{F}{K^2\varphi_C^2 A_C} - \left(\frac{F}{K^2\varphi_C^2 A_C} - v_0^2\right)e^{-\frac{2}{m}K^2\varphi_C^2 A_C x}\right]^{\frac{1}{2}} \tag{4.31}$$

$$p_C = \frac{F}{A_C}\left[1 + \left(\frac{v_0^2}{v_\infty^2} - 1\right)e^{-\frac{2}{v_\infty^2}\frac{F}{m}x}\right] \tag{4.32}$$

In Eqs. (4.31) and (4.32), $K$ is a fluid-specific orifice constant, defined as $K = \sqrt{\rho/2C_d^2}$, and $v_\infty$ is the cushion velocity of the piston at the end of the cushion.

Equations (4.31) and (4.32) reveal that the maximum cushion velocity and pressure are occurring at $x = 0$, namely, the beginning of the cushion. This means that a cushion device can effectively halt a hard stop at the end of a piston stroke by forming a high cushion pressure when the piston travels close to the end stroke, quickly slowing the piston velocity.

### 4.2.4 Hydraulic Cylinder Power Transmission

As a type of commonly used linear actuators, the main function of hydraulic cylinders is to deploy the hydraulic power to move the load. Depending on the potential of the load motion, often driven by the gravity force acting on the load, a hydraulic cylinder may do the work by **pushing** or **pulling** the load. In other words, a hydraulic cylinder may be operated under either a **resistive operation** to push a load or an **overrunning operation** to pull a load to prevent it from moving too fast. As illustrated in Figure 4.10, the cylinder is pushing a resistive load while extending to lift the load and pulling an overrunning load during retraction. Normally, a hydraulic system needs to transmit a large amount of hydraulic energy to the cylinder in a resistive operation and demands only a small amount of energy, mainly to overcome system resistances, during an overrunning operation. The pulling power is actually the amount of kinematic energy added to the hydraulic system by the overrunning load, which is converted into heat energy at the returning line during the controlled load movement.

When a load is driven indirectly by the cylinder via a linkage structure, a resistive load state may be switched to an overrunning state during a stroke of cylinder operation, or vice versa, and result in a **transient operating state**. Such a transient operating state can often be seen in many mobile hydraulic systems.

The amount of power transmitted to move a load is determined by the force applied to the load and the time spent to move the load a certain distance. Based on the magnitude

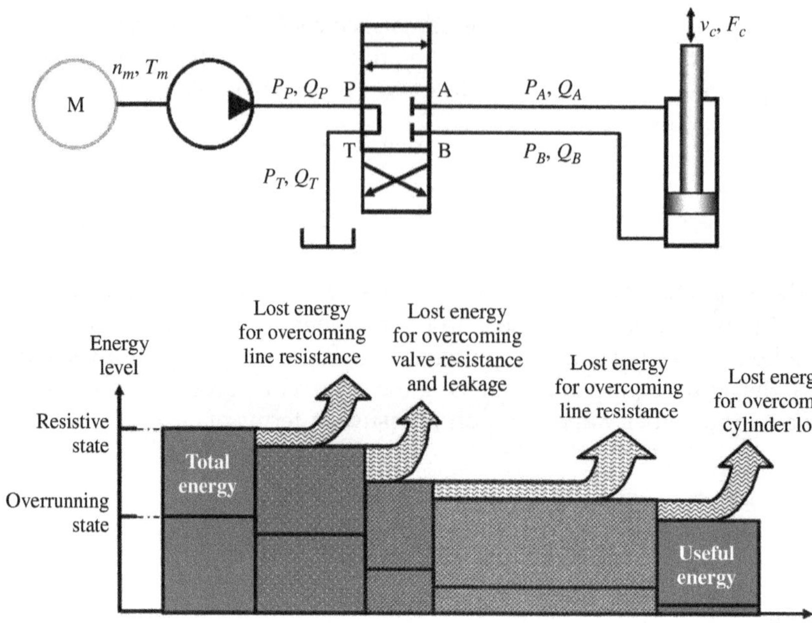


**FIGURE 4.10**
Illustration of the principle of energy distribution during typical cylinder-operating states.

of force and the duration of time involved, a driving force can often be divided into categories of breakaway, inertial, and constant velocity. The **breakaway force** is the critical force needed to overcome the static friction to start moving a load from a state of rest; the **inertial force** refers to the force required to accelerate a load; and the **constant velocity force** is the force that must be provided to overcome the dynamic friction of the load moving along a surface to maintain a constant motion. All three forces form the operation load of a hydraulic cylinder defined as follows:

$$F = F_b + F_i + F_v \tag{4.33}$$

where $F$ is the overall load; $F_b$ is the breakaway force; $F_i$ is the inertial force; and $F_v$ is the constant velocity force acting on a hydraulic cylinder during a typical operation.

These forces play different roles at different stages in a typical operation cycle. The breakaway force dominates the cylinder load when starting a stroke and therefore bears a very large acceleration. The inertial force is mainly formed to overcome the resistance to raise the load-moving speed to a desirable level, which is an accelerated operation state. In comparison, the constant velocity force is used to maintain the motion status and can be treated as a zero acceleration state.

In the load pattern analyzed earlier, a cylinder bears the heaviest load when starting to move. In many cases, the pressure and flow ratings of a hydraulic system are two of the predetermined design parameters, subject to numerous design constraints. To figure out the amount of power transmitted to a cylinder to drive the load, the normal place to start is with determination of the maximum overall load the cylinder needs to drive. One critical calculation in power transmission is to determine the cylinder size, mainly the bore and the rod diameters. The sizing criterion is that the cylinder can generate a sufficient driving force to push or pull the overall load under the limitation of available operating pressure of the fluid entering the cylinder. The general form of driving force can be expressed as follows:

$$F_d = \frac{F}{\eta_m} = \frac{(F_b + F_i + F_v)}{\eta_m} \tag{4.34}$$

where $F_d$ is the driving force of the cylinder available for driving the external load, and $\eta_m$ is the mechanical efficiency of a cylinder (normally $\eta_m = 0.95$ for most single-rod cylinders).

For a single-rod cylinder, the bore and rod sizing equations can be defined based on Eqs. (4.2), (4.4), and (4.14), and are expressed as follows:

$$D = \max\left\{\sqrt{\frac{4\varphi}{\pi(\varphi p_1 - p_2)}F_{d1}},\ \sqrt{\frac{4\varphi}{\pi(p_1 - \varphi p_2)}F_{d2}}\right\} \tag{4.35}$$

$$d = D\sqrt{\frac{\varphi - 1}{\varphi}} \tag{4.36}$$

where $D$ is the bore diameter; $d$ is the rod diameter; $\varphi$ is the area ratio of a single-rod cylinder; and $F_{d1}$ and $F_{d2}$ are the driving forces of the cylinder in extension and retraction, respectively.

The maximum extension and retraction velocities can therefore be determined according to the supplied flow rate and cylinder size using Eqs. (4.1) and (4.3). After knowing the cylinder velocities, the mechanical power $P_m$ used to drive the load can be calculated using the following equation.

$$P_m = F_{d,\max} v \tag{4.37}$$

where $F_{d,\max}$ is the larger driving force of the cylinder in extension and retraction.

### 4.2.5 Hydraulic Cylinder Applications

As discussed earlier, hydraulic cylinders are designed to deploy the potential energy carried by the pressurized fluid in driving loads during linear motions, often using some kind of linkages. Based on Newton's law of reaction, a cylinder must have a reacting support to get the protrusion or retraction force from the pressurized fluid. While various methods of cylinder mounting are available for different applications, they all follow the basic principle that a linear cylinder should not be subject to any rotating torque, but only a linear force in the direction of the cylinder axis in a form of extraction or compression.

One common approach to satisfying the basic requirement of cylinder installation on mobile hydraulic systems is to mount cylinders using **pivot joints** on their centerline. Such pivot joints allow cylinders to be self-aligned and permit only a linear extracting or compression force to be applied on the cylinders. However, in many applications, some forms of noncenterline-type cylinder mountings are unavoidable. Such mountings, often featured by firmly fixing one or both ends of the cylinder on a rigid frame, will often result in unnecessary constraints and make it very difficult to have a perfect cylinder alignment to achieve a bend-free installation of a cylinder, especially when the cylinder is operating

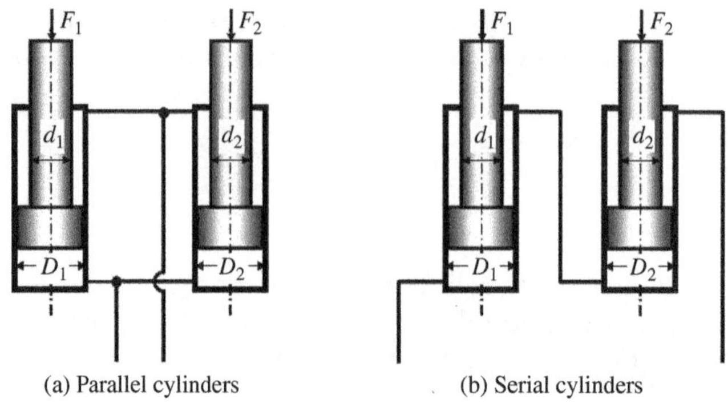


**FIGURE 4.11**
Configuration illustration of typical (a) parallel and (b) serial cylinder arrangements.

under varying pressure and temperature. For example, when a firmly fixed cylinder is operating under a maximum pressure of 40 MPa, it could be elongated about 0.1~0.2%. Fixed, noncenterline-mounted cylinders add another strength problem because mounting bolts will be subjected to increased tension in combination with shear forces. If such a mounting is unavoidable, to ease the adverse effects of noncenterline mountings, it is recommended that a stronger cylinder body be designed to resist bending. When a misalignment between the cylinder and its load occurs, the mounting style may have to be altered to accommodate the skewing movement. If multiple-plane misalignment is encountered, a universal alignment mounting could be used to reduce cylinder bending and side loading.

A hydraulic cylinder is a type of versatile actuator for many applications. While the single cylinder can, in many cases, satisfactorily actuate the load, sometimes there are applications requiring multiple cylinders installed at different locations to actuate collaboratively, either in parallel or in series, to perform the required function (Figure 4.11). The operation characteristics of those cylinders are determined in terms of the role of each cylinder in a particular application. In a **parallel system**, the cylinder with the lowest operating pressure requirement will first operate, and the cylinder requiring a higher operating pressure will not start to work until the one operating at the lower pressure completes its operation. In a **serial system**, all cylinders are operating at the same time, which means that the system should provide sufficient power to drive an accumulated load at the same time.

#### Example 4.2: Cylinders in Parallel

As illustrated in Figure 4.11(a), two cylinders are connected to form a parallel cylinder-actuating system. Assume the bore and rod diameters of the left cylinder are 120 mm and 60 mm, respectively, and that those of the right cylinder are 100 mm and 50 mm. If the mass of the external loads to be driven by the left and right cylinders are 750 and 500 kg, respectively, and the back pressure in both rod-end chambers is 200 kPa, calculate the operating pressure of each cylinder for driving the loads. (Assume that the left cylinder requires 550 kPa for no-load extension and the right one requires 500 kPa.) If the supplying flow is 150 L·min$^{-1}$, what are the extending velocities of those cylinders?

a. The operating pressure of two cylinders:

$$\because F_{drive} = F_{load} + F_{friction} + F_{back\_pressure}'$$
$$\therefore p_1 A_1 = F_{load} + p_f A_1 + p_2 A_2$$

On the left cylinder:

$$\begin{aligned} p_{l1} &= p_{lf} + \frac{F_{lload}}{A_{l1}} + p_{l2}\frac{A_{l2}}{A_{l1}} \\ &= 550 + \frac{750 \times 9.8}{\frac{\pi}{4} \times \left(\frac{120}{1000}\right)^2 \times 1000} + 200 \times \frac{\frac{\pi}{4} \times \left[\left(\frac{120}{1000}\right)^2 - \left(\frac{60}{1000}\right)^2\right]}{\frac{\pi}{4} \times \left(\frac{120}{1000}\right)^2} \\ &= 1,350(kPa) \end{aligned}$$

On the right cylinder:

$$\begin{aligned} p_{r1} &= p_{rf} + \frac{F_{rload}}{A_{r1}} + p_{r2}\frac{A_{r2}}{A_{r1}} \\ &= 500 + \frac{500 \times 9.8}{\frac{\pi}{4} \times \left(\frac{100}{1000}\right)^2 \times 1000} + 200 \times \frac{\frac{\pi}{4} \times \left[\left(\frac{100}{1000}\right)^2 - \left(\frac{50}{1000}\right)^2\right]}{\frac{\pi}{4} \times \left(\frac{100}{1000}\right)^2} \\ &= 1,274(kPa) \end{aligned}$$

b. The extension velocities of those two cylinders:

The left cylinder:

$$\begin{aligned} v_{l1} &= \frac{Q}{A_{l1}} \\ &= \frac{\frac{150}{60 \times 1000}}{\frac{\pi}{4} \times \left(\frac{120}{1000}\right)^2} \\ &= 0.22\left(m \cdot s^{-1}\right) \end{aligned}$$

The right cylinder:

$$\begin{aligned} v_{r1} &= \frac{Q}{A_{r1}} \\ &= \frac{\frac{150}{60 \times 1000}}{\frac{\pi}{4} \times \left(\frac{100}{1000}\right)^2} \\ &= 0.32\left(m \cdot s^{-1}\right) \end{aligned}$$

**DISCUSSION 4.2:** Will those two cylinders extend simultaneously under the stated condition? The answer is no, because the left cylinder requires a lower operating pressure to drive the load. When the system pressure builds up and the supplying flow reaches the required pressure for the left cylinder, the pressure flow will be kept at that level and the supplying flow will drive the left cylinder to push its load, moving until fully extended. After that, the system pressure will resume rising until it reaches the required operating pressure for the right cylinder, pushing that cylinder to extend. After both cylinders extend fully, the system pressure will keep rising until it reaches the maximum system pressure setting.

## 4.3 Hydraulic Motors

### 4.3.1 Classification of Hydraulic Motors

Offering the same functionality as hydraulic cylinders, hydraulic motors are also a kind of hydraulic power deployment device, but they convert hydraulic potential energy into mechanical rotary power. To drive a load via drive shafts, all types of hydraulic motors share some common design features: a driving surface area subject to a pressure differential, a way of timing the porting of pressurized fluid to the pressure surface to achieve continuous rotation, and a mechanical connection between the surface area and an output shaft.

Hydraulic motors also have a large array of configurations, sizes, and special designs. The rotating formats, structural features, and usage are the most commonly used criteria for categorizing motors. In terms of the rotating formats, hydraulic motors can be classified as limited rotation and continuous rotation motors. A noticeable feature of a **limited rotation motor** is that it turns less than a complete revolution and keeps oscillating between clockwise and counterclockwise motion to perform work; therefore, it is also called an **oscillating motor** or a **rotary cylinder**. In contrast, a **continuous rotation motor** always turns complete circles, often endlessly until the supplying flow is stopped. This category of motors can be thought of as an inverse version of a hydraulic pump, redesigned to withstand the different forces involved in motor applications. As a result, this category of motor can be classified as gear, vane, and piston motors in terms of their structural features. In terms of usage, they can be sorted into bidirectional and unidirectional motors, both having fixed and variable displacement options. The **bidirectional motors** can be further separated into high-speed (or high-torque) and low-speed groups. To avoid unnecessary confusion, this textbook reserves the term **motor** solely to continuous rotation motors and uses the term **oscillating motor** to rename the limited rotation motors.

### 4.3.2 Operating Parameters of Hydraulic Motors

Hydraulic motors are rated by their torque and rotating velocity capabilities, which are usually determined by their displacement. The term **motor displacement** refers to the volume of fluid required to turn the motor output shaft one complete revolution. Generally, **cubic centimeters (cc) per revolution** (often **cubic inches per revolution** in the United States) is used to describe motor displacement in engineering practices. For a fixed-displacement motor, this value is a constant after the motor is built, while for a variable-displacement motor, it is possible to change the displacement during an operation. A fixed-displacement motor keeps an unchangeable driving surface area to withstand the pressure drop, which will generate a constant torque under a steady pressure drop. However, the output speed can be varied by controlling the amount of input flow entering the motor. As the driving surface area can be adjusted in a variable-displacement motor, it can result in a regulated output torque in proportion to the displacement of the pump under a steady pressure drop. A variable-displacement motor can also achieve speed control when the input flow is constant.

The **theoretical output torque** from a hydraulic motor is determined by its displacement and the pressure drop across the motor as defined using the following equation:

$$T = \frac{D_v}{2\pi}\Delta p \tag{4.38}$$

where $T$ is the theoretical output torque, $D_v$ is the displacement of the motor, and $\Delta p$ is the pressure drop across the motor.

The theoretical output torque is defined as the torque available at the motor shaft assuming no mechanical losses. In practice, it is impossible to get the theoretical output torque from a motor to drive the load. The **actual output torque** must overcome the mechanical resistances to drive the motor. Often, the mechanical resistance can be determined by figuring out the torque required to drive a motor without carrying any external load. A **breakaway torque** is often used to quantify the amount of torque required to get a stationary motor to start turning, and a **running torque** is commonly used to measure the amount of torque needed to keep a motor turning consistently. The breakaway torque refers to the amount of torque needed to overcome the **static resistance,** and the running torque refers to the amount of torque needed to overcome the **dynamic resistance** of a motor. More torque is always required to start a motor than to keep it turning; therefore, the breakaway torque is always larger than the running torque in value. Importantly, both the breakaway and running torque are **resistance-overcoming torques,** and they are formed by converting hydraulic potential energy to drive the load in different stages of motor operation. When the motor needs to drive an external load, the **total torque** (namely, the theoretical torque) required to drive the motor properly is the sum of the resistance-overcoming torque. We can use a **starting torque** to indicate the motor capacity to generate an output torque to start to turn a load, and an **operating torque** as the motor output torque necessary to keep a load turning. Normally, the starting torque for many hydraulic motors ranges between 70 and 80%, and the running torque of common hydraulic motors is approximately 90% of the theoretical torque. Most of the time, a motor parameter of **mechanical efficiency** is used to quantitatively define the ratio of actual torque deliverable to drive a load to the theoretical torque:

$$\eta_m = \frac{T_A}{T_T} \times 100\% \tag{4.39}$$

where $\eta_m$ is the mechanical efficiency, and $T_A$ and $T_T$ are actual and theoretical output torques from a hydraulic motor.

It is often difficult to calculate the actual output torque in terms of potential energy carried by the pressurized fluid using theoretically derived equations. In practice, the mechanical torque, which can be measured at the motor output shaft or calculated in terms of the load being driven, is often used as an indication of the actual output torque. It is also a common engineering practice to use the difference between the theoretical torque calculated for a loaded condition and the no-load torque measured at the same speed as the actual output torque for the situation.

$$T_A = T_T - T_{NL} \tag{4.40}$$

where $T_{NL}$ is the actually measured torque when the motor bears no external load.

One more motor-operating parameter, the **torque ripple**—defined as the difference between minimum and maximum torques delivered at a given pressure during one revolution of the motor—is commonly used to evaluate the smoothness of the motor operation.

Different from the output torque, the **theoretical output speed** from a hydraulic motor is determined by motor displacement and supplying flow rate to the motor as expressed in the following equation:

$$n = \frac{Q}{D_v} \tag{4.41}$$

where $n$ is the theoretical output speed and $D_v$ is the volumetric displacement of the motor.

The theoretical output speed of a motor is defined as the speed generated under the assumption of no flow loss during the hydraulic power-deploying process. Similar to the output torque, it is impossible for a motor to operate at the theoretical output speed because of the unavoidable fluid leakage within a hydraulic motor, especially when it is operating under high-pressure conditions. The **volumetric efficiency**, defined as the ratio of theoretical flow over the actual flow required to produce a certain speed at the motor output shift, is often used to indicate the capability of a hydraulic motor to convert input flow to output speed:

$$\eta_v = \frac{Q_T}{Q_A} \times 100\% \tag{4.42}$$

where $\eta_v$ is the volumetric efficiency, and $Q_A$ and $Q_T$ are actual and theoretical inlet flow rates to a hydraulic motor.

When comparing this volumetric efficiency equation to Eq. (2.5), it can be easily established that the definition of volumetric efficiency for a hydraulic motor is an inverse of that for a hydraulic pump. That is because a pump cannot produce as much flow as it should theoretically and a motor cannot utilize all the flow being supplied to turn the motor. It is common to call the leakage through a motor the **slippage** of the motor. Therefore, the actual output speed is a slip-free speed on a motor output shaft and can be defined as follows:

$$n_A = n\eta_v \tag{4.43}$$

where $n_A$ is the actual speed and $n$ is the theoretical speed obtainable at the output shaft of a hydraulic motor.

A few operating parameters often used to indicate the performance of a hydraulic motor include the **maximum motor speed**, defined as the speed that the motor operating under a specific inlet pressure can sustain for a limited time without damage, and the **minimum motor speed**, defined as the slowest, uninterrupted, and consistent speed available from the motor output shaft.

Similar to that for a hydraulic pump, the **overall efficiency** of a hydraulic motor is defined as the product of the mechanical efficiency and the volumetric efficiency of the motor.

$$\eta_o = \eta_m \eta_v \tag{4.44}$$

where $\eta_o$ is the overall efficiency of a hydraulic motor.

However, because of the difference in the direction of energy flow in a motor and in a pump, the overall efficiency of a motor is actually an inverse of that of the pump in terms of hydraulic potential energy and mechanical power, as expressed in the following equation.

$$\eta_o = \frac{n_A T}{p Q_A} \tag{4.45}$$

### 4.3.3 High-Speed Hydraulic Motors

Based on their normal ranges of operation speed, hydraulic motors are often classified into two categories: high-speed and low-speed/high-torque motors. **High-speed motors**, in general, share a lot of similar structural features with hydraulic pumps as introduced in Chapter 2. Similar to gear pumps, **gear motors** also consist of a pair of meshed gears enclosed in housing. At a turning speed in proportion to the supplied flow rate being delivered, the motor generates a driving torque in terms of the hydraulic pressure acting on the exposed teeth surface. It should be remembered that as a reacting force, the hydraulic pressure of the supplied fluid is determined by the load the motor is driving. In terms of the types of gears being used, the gear motor can be classified as the **external gear motor** (Figure 4.12(a)) and the **internal gear motor** (Figure 4.12(b)). Normally, both gears in a typical external gear motor have the same number of teeth. The internal gear motor often uses an inner gear of one tooth less than that of the outer gear. The only difference between a typical gear pump and a typical gear motor is that the driving gear in a pump is driven by an external prime mover and is used to drive the idler gear. The gears mesh to form a

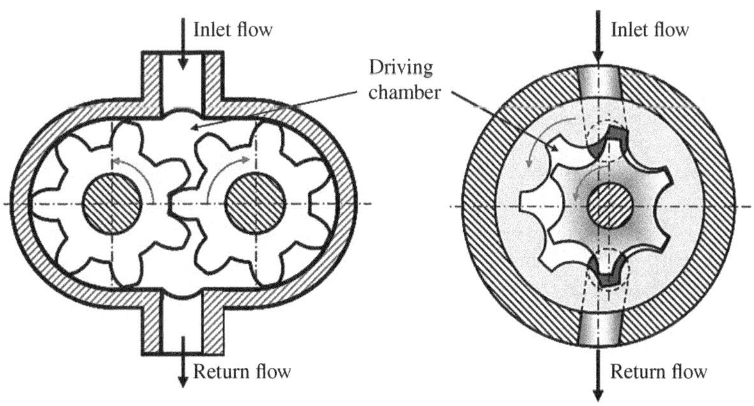


**FIGURE 4.12**
Illustration of the configuration and operation principle of typical (a) external and (b) internal hydraulic gear motors.

suction chamber by creating a vacuum through an increase in the volume at the inlet side and a compression chamber by decreasing the volume at the output side of the pump to create a pumping process.

A gear motor can have exactly the same structural design; the driving gear shaft would connect to a load device. When pressurized fluid enters the housing, it will push both gears to rotate in the direction of least resistance around the periphery of the housing and form a mesh to enlarge the space to hold the inlet flow. Such a mesh will drive the shaft to turn and therefore drive the load to do the work. Precise tolerances between gears and housing help to control the fluid leakage and increase the volumetric efficiency. In many high-performance gear motors, wear plates are often installed on the sides of the gears to keep the gears from moving axially and compensate for the worn-off tolerance to help control internal leakage. The output torque from a gear motor is a function of pressure on one tooth because pressure on other teeth is in hydraulic balance. An attribute of the structural similarity between gear motors and gear pumps is that they are often interchangeable in many applications.

A gerotor, a special type of gear, is often used in internal gear motors for smoother and quieter operations. There are two categories of internal gerotor motors: **direct-drive gerotor motor** (Figure 4.13(a)) and **orbiting gerotor motor** (Figure 4.13(b)), commonly used in many mobile hydraulic systems. A typical direct-drive gerotor motor consists of an internal-external gear pair, with the inner external gear always installed on the output shaft. The external gear always has one less tooth than the internal counterpart, and all its teeth are in contact with some teeth of the internal gear at all times. Both gears rotate in the same direction, with their centers separated by a fixed eccentricity pushed by the inlet-pressurized flow. The center of the inner gear coincides with the center of the output shaft. As can be seen in Figure 4.13(a), a fluid pocket is formed between external teeth 1 and 2 on the inner gear and their meshed internal teeth on the outer gear. As the pressurized fluid enters the motor through the inlet port (the dashed kidney-shaped inlet port in Figure 4.13(a)), it causes the pocket to enlarge. This pocket-enlarging process consequently forces both gears to turn. The gears will rotate teeth 6 and 1 into the position to form a new pocket, as formed by teeth 1 and 2 just a moment ago. Then teeth 5 and 6 follow, and so on. This continuous change in teeth positions results in the gear turning continuously and generates a smooth output rotation on the shaft to drive the load.

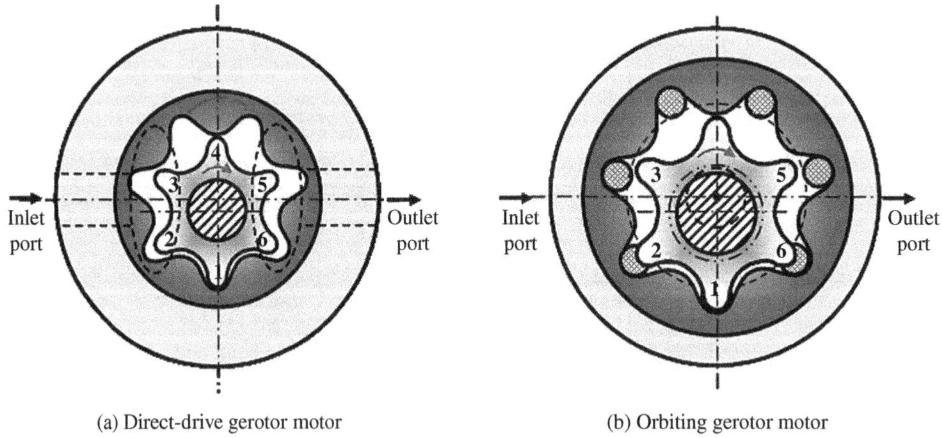


**FIGURE 4.13**
Illustration of the configuration and operation principle of typical (a) direct-drive and (b) orbiting type of gerotor motors.

By comparison, typical orbiting gerotor motors consist of a rotating inner external gear meshing with a stationary outer internal gear. As in the direct-drive motor, the inner gear in an orbiting motor also has one less tooth than the outer one. While the inner external gear of an orbiting motor is also installed on the output shaft, the axial line of the output shaft has a fixed eccentricity to the center of the inner gear and normally coincides with the center of the outer stationary gear. During rotary actuation, the orbiting gerotor motor also utilizes the pockets formed between the teeth of the inner gear and their meshed outer gear to generate continuous turning on the output shaft, like a direct-drive motor does. One unique structural feature of orbiting gerotor motors is their flow-distributing system, which uses a **commutator** (often called the **valve plate**) to lead the pressurized fluid to appropriate working pockets of the motor and to create a flow path to bleed the returning fluid from the motor back to the tank. To offer such functions, a commutator has the same number of both fluid inlet and outlet openings as the number of inner gear teeth arranged alternately around the commutator. During operation, the commutator turns synchronously with the inner gear and alternately connects each fluid pocket, formed by all the teeth between the inner and outer gears via a fixed fluid path normally on the tip of the outer internal gear teeth, to either the inlet or the outlet ports. For example, as the external teeth 1 and 2 on the inner gear and their meshed internal teeth on the outer gear form an **actuating pocket** as illustrated in Figure 4.13(b), the commutator connects this pocket to the inlet port via the corresponding fluid inlet opening. Meanwhile, it blocks the corresponding outlet opening to prevent the inlet pressure fluid from bleeding to the outlet port. This inlet pressure fluid pushes the inner gear, rotating on the meshed teeth of the stationary outer gear. At the same time, a pathway to the tank is provided for the opposite side of the pocket (called the **exhaust pocket**) to bleed the fluid from this pocket as it diminishes as the result of the inner gear rotating. The motor will continuously form new actuating pockets and diminish old exhausting pockets as the inner gear rotates. Gear motors are in general fixed-displacement and bidirectional motors.

Another type of commonly used hydraulic motors is the **vane motor**. Similar to gear motors, vane motors generate driving torque by converting the hydraulic pressure, a reacting force in fluid medium based on the system load, to act on the exposed surfaces of the vanes. Like the pumps of the same type, vane motors can be classified as unbalanced and balanced designs. Figure 4.14 depicts the configuration and operation principle of a

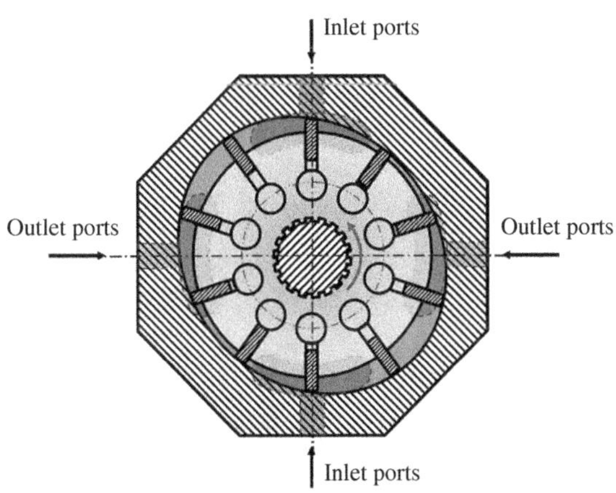


**FIGURE 4.14**
Illustration of the configuration and operation principle of a typical balanced vane motor.

**balanced vane motor**, which consists of a rotating rotor, a stationary cam or cam ring, and a number of sliding vanes. A balanced vane motor always has two actuating and exhausting zones, both of which are arranged on the opposite side of the rotor.

A **circular cylinder rotor**, which carries several sliding vanes, is always pushed by a back pressure on the radial directions to separate the actuating and exhaust zones in between the rotor and cam walls to form contained pockets for carrying either pressure or exhaust fluids in an operation. To form two actuating and exhausting zones, the cam or the cam ring should have two major and two minor radial sections joined by transitional sections or ramps, as graphically illustrated in Figure 4.14. Radial grooves and holes through the vanes equalize radial hydraulic forces on the vanes at all times. In some designs, light springs are used to force the vanes radially against the cam to ensure sealing at zero speed to help the motor develop a starting torque. During the operation, pressurized fluid enters and leaves the motor through openings in the side plates of the ramp. In a balanced vane pump as illustrated in Figure 4.14, pressurized fluid entering at two inlet ports pushes the rotor turning counterclockwise. The fluid pockets transport the fluid to the outlet ports returning to the tank. In many vane motors, their structures are symmetrically designed, and their inlet and outlet ports are switchable. This type of vane motor is often bidirectional; their operational direction can be easily switched by simply supplying the pressurized fluid to the outlet ports and connecting the inlet ports to the returning line. However, not all vane motors are bidirectionally workable. If the fluid inlet/outlet ports of a unidirectional motor were switched, the opposite direction pressure flow could damage such a motor.

As depicted in Figure 4.15, the driving torque on a vane motor is created by the differential force acting on the surfaces of vanes forming a fluid pocket. The area of a vane extending from the rotor is the effective surface, the vane exposed to the pressurized fluid. The hydraulic force acting on this vane is the product of the fluid pressure and the vane effective area. Because two vanes are involved in forming a fluid pocket, the torque-generating

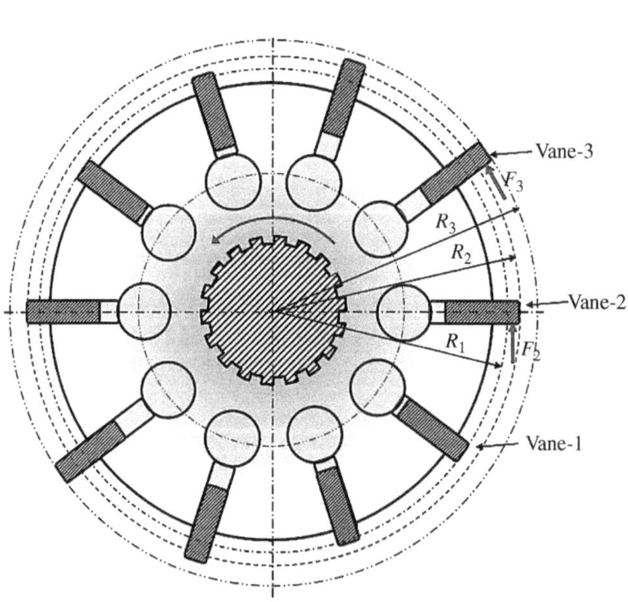


**FIGURE 4.15**
Illustration of a differential force acting on a vane motor in creating the driving torque.

surface of a pocket is the area difference between two pocket-composing vanes and is always located on the top portion of the longer extended vane as shown in Figure 4.15. The torque generating zone in a vane motor is only in the pressurized portion of the motor. The following equation expresses how much torque can be theoretically generated in a vane motor.

$$T = n\sum_{i} \frac{(R_{i+1} - R_i)p}{2} L \tag{4.46}$$

where $T$ is the total theoretical torque generated on a vane motor; $R_i$ and $R_{i+1}$ are the radiuses of the shorter and the longer composing vanes of fluid pocket $i$ in the pressurized portion(s), respectively; and $n$ is a motor structure coefficient, $n = 1$ for unbalanced motors and $n = 2$ for balanced ones.

The rotor in a vane motor is usually separated axially from the surface of side plates using a thin fluid film. Normally, the front side plate is clamped against the cam ring by pressure, which makes it possible to compensate for temperature and pressure variations, as well as plates or rotor wear, and to maintain an optimum clearance for the best efficiency. This feature also helps to extend the service life of vane motors.

Another category is **piston-type hydraulic motors**. Similar to piston-type pumps, piston motors are composed of two major types, including radial-piston motors and axial-piston motors. Among them, **radial-piston motors** have broad applications in mobile hydraulic systems, especially in hydraulic transmission systems. As illustrated in Figure 4.16, the main components in a typical radial-piston motor include a rotary motion cylinder block consisting of several (often an odd number of 5, 7, 9 or more) radial cylinders evenly arranged in the block, with a sliding piston that reciprocally moves within each cylinder. The output shaft is always attached on the cylinder block, and the fluid inlet and outlet ports are normally placed in a **pintle** located in the center of the cylinder block. As the pressurized fluid is supplied to the motor through the inlet port, the fluid pushes the

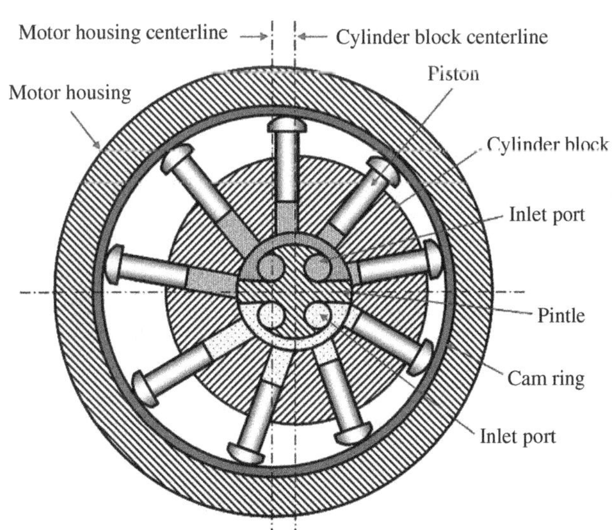


**FIGURE 4.16**
Illustration of the configuration and operation principle of a typical radial-piston motor.

piston outward, moving against the cam ring which creates a reacting force on the piston shoe installed on the outer end of the piston to rotate the cylinder block. Like vane-type counterparts, radial piston motors can be designed in fixed- and variable-displacement models. While there are numerous actual designs for implementing the variable-displacement actuation, they all follow the same principle of changing the piston stroke by shifting the cylinder block laterally in relation to the motor housing to adjust the eccentricity between the centerlines of the cylinder block and the motor housing. When the eccentricity is zero, namely, when the centerlines of the cylinder block and motor housing overlap, there will be no piston stroke induced by the pressurized fluid, and therefore the motor will not turn. Many variable-displacement radial motors are designed in such a way that the eccentricity of the cylinder block can go from positive to negative. That is, the centerline of the cylinder block can go from one side of the motor housing centerline to the other side. Such an over centerline eccentricity change will cause a reverse in the rotating direction on the output shaft of the motor, thus making the motor bidirectional.

Another major type of piston motor is an **axial-piston motor**. Like an axial-piston pump, this type of hydraulic motor can be sorted further into two styles of **inline axial-piston motors** and **bent-axis piston motors** in terms of structural features as depicted in Figure 4.17. Similar to radial-piston motors, both types of axial-piston motors also rotate the output shaft by converting the reciprocating piston motion, axial direction in this case, to the cylinder block rotational motion. As depicted in Figure 4.17, a typical axial-piston motor, inline or bent, creates a torque to generate the rotating motion of the cylinder block and then the output shaft, by applying pressure on one end of the pistons to drive them to perform an outward reciprocating motion in the cylinder block. Such an outward reciprocating motion transmits the pressure-generated force to the other end of the pistons and creates a reaction against a tilted drive plate, which causes the cylinder block and motor shaft to rotate. The amount of torque being generated in this type of motor is proportional to the total area of all pistons and is a function of the angle at which the drive plate is positioned. In an inline design, the motor output shaft and the cylinder block are centered on the same axis, and a swash plate is used to convert the reciprocating motion of pistons into a rotating motion of the cylinder block (Figure 4.17(a)). In a bent design, the cylinder block and motor output shaft are mounted at an angle to

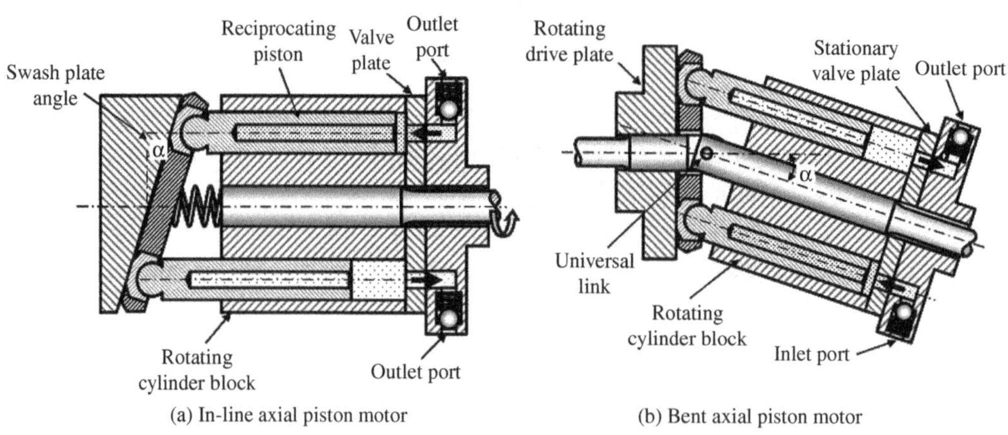


**FIGURE 4.17**
Illustration of the configuration and operation principle of typical (a) in-line and (b) bent axial piston hydraulic motors.

each other, and the reaction is against a drive plate flange. A universal link is used to connect the output shaft and the cylinder block shaft to transmit the rotating motion of the cylinder block, converted from the reciprocating motion of pistons, into that of the output shaft (Figure 4.17(b)).

Like its radial version, axial-piston motors are also available in fixed- and variable- displacement models. The displacement-adjusting mechanism of **variable-displacement axial-piston motors** are the same as their pump counterparts, as described in Section 2.1.5. In general, increasing the angle of the swashplate or the drive plate in a piston motor will increase the torque capacity but reduce the output shaft speed of a motor, and vice versa. Similar to load-sensing functions in piston pumps, a pressure and/or torque compensator can be used to regulate the motor displacement in response to load changes for maximum performance under all load conditions. Typically, the adjusting range of the swashplate/drive plate in many axial-piston motors is between 7.5° and 30°. A motor will operate under its maximum speed capacity with its minimum displacement and torque when the angle is adjusted to its low boundary and work at its maximum displacement and torque with the minimum speed when the angle is adjusted to the maximum value.

### 4.3.4 Low-Speed High-Torque Motors

In many applications, the rotary-actuating devices of a hydraulic system are often operating at a low speed with a high torque. Specially designed **low-speed, high-torque** (LSHT) hydraulic motors are commonly used in such applications. The use of low-speed motors eliminates the need for gearboxes for many applications, which will not only reduce system complicity and lower the initial cost, but often, more importantly, reduce the requirement for maintenance and improve the system reliability, both contributing to reduced operation costs. Among them, the elimination of gearboxes in a power transmission system is the most attractive feature because of its operational requirement to transmit a large amount of power within a comparatively small-space envelope. The elimination of a gearbox also removes a significant amount of inertia in the power transmission system, which allows the rotary actuator to rapidly reverse its operation direction by simply reversing the direction of fluid supply to the motor. All these indicate that the use of a low-speed motor is a cost-effective, highly reliable design for many rotary-actuating hydraulic systems. Typically, the output speed for LSHT motors ranges from 10 to 1000 rpm. With suitable closed-loop electronic control, some specially designed LSHT motors can operate smoothly at a very low speed of 0.1 rpm level. All LSHT motors generally exhibit good starting efficiencies, with fairly constant torque over their entire speed range.

Like high-speed counterparts, LSHT motors also come with gear, vane, and radial- and axial-piston designs. One common feature of LSHT gear and vane motors is their use of wider gears or vanes to form large fluid-carrying pockets and larger areas of effective pressure-bearing surface to slow the motor down and meanwhile create a larger torque. Because of the relatively poor sealing capacity between the vanes and the cam ring, the volumetric efficiency of classic LSHT vane motors is normally low. To solve this problem, a type of specially designed **rolling-vane motor** can offer a nearly constant volumetric efficiency at the entire speed range. As illustrated in Figure 4.18, the fluid is supplied to the high-pressure chambers through two flow supply slots located on one side of the rotor, and two flow discharge slots located on the other side of the rotor. Four timed rolling vanes act as two groups of flow control valves (formed by the opposite vanes) to separate the high- and low-pressure chambers through ensuring high pressure against the trailing

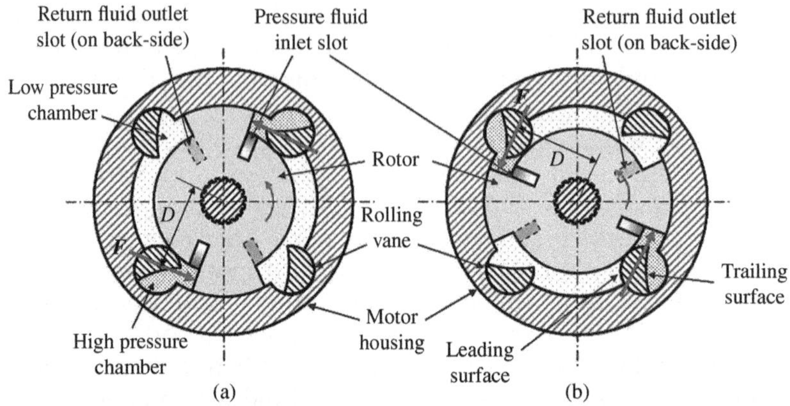


**FIGURE 4.18**
Illustration of the configuration and operation principle of typical rolling-vane hydraulic motors.

surfaces and low pressure on the leading surfaces of the vanes. The symmetric structure of such a rolling-vane motor makes it radially balanced, as pressure fluid always acts on equal and opposite areas. The inlet fluid is always supplied to two pressure chambers of equal cross-sectional area, resulting in a smooth, pulseless rotating speed on the rotor, consequently on the output shaft of the motor. This type of LSHT can be used in applications requiring smooth and precise motion controls, such as on precise hydraulic presses.

Due to their high-torque characteristics, LSHT motors are usually used in high-pressure applications. Since vane motors are normally good for low-pressure applications, it is much more common to use piston-type LSHT motors in high-pressure applications. Generally, radial-type piston motors have large piston sizes and often have a larger displacement range than axial types. This type of motor can always create larger torque under the same operating pressure. In addition, these types of motors in general have excellent leaking-resistance characteristics that result in good volumetric efficiency through the entire speed range. All these features make radial-piston motors the most popular designs of LSHT motors.

One commonly used radial-piston motor is the **cam-type radial-piston motor**. As illustrated in Figure 4.19, a typical cam-type radial-piston motor consists of a cam ring, a cylinder carrying rotor, a fluid distribution shaft, and a few piston and roller sets. The cam ring is normally made of a few internal cam-lobe curves (four for the illustrated example). When the rotor turns one cycle, this cam ring will push (jointly with the pressure fluid) a piston, performing the same number of reciprocating motions (four in the example case). The fluid inlet and outlet ports are designed on the shaft to deliver pressurized fluid to the extending pistons to create a turning torque and discharge return fluid as the pistons retract to allow the rotor to continue the operation. The roller attached on the top of the pistons can greatly reduce the friction formed by the relative motion between the pistons and the cam ring, and more importantly transmit only the axial force to pistons, which can also effectively reduce the friction between the side surface of pistons and the cylinder bore, resulting in a much longer service life. The concentric configuration of this type of motor offers a balanced radial force, which makes the motor easier to start and can operate smoothly under very low speeds. However, the concentric configuration often results in a slightly less efficient configuration than the eccentric configuration.

Another commonly used LSHT radial-piston motor is the so-called **static balanced radial-piston motor**. As depicted in Figure 4.20, a unique configuration feature of a typical

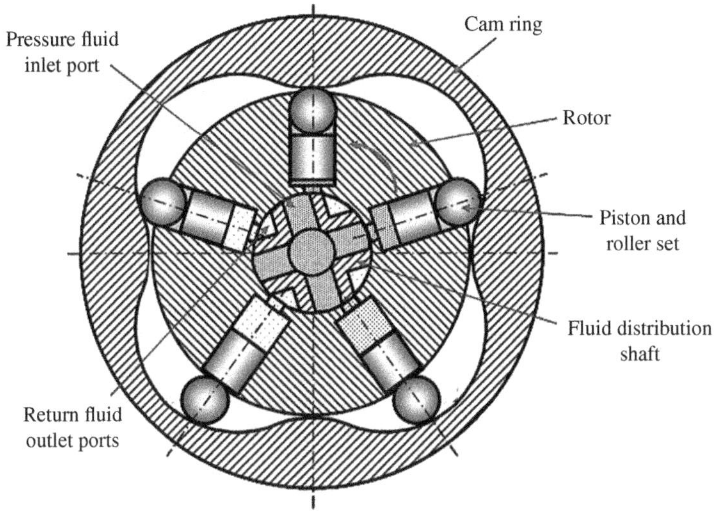


**FIGURE 4.19**
Illustration of the configuration and operation principle of a typical cam-type radial piston motor.

static balanced type of LSHT motor is the multilateral slider (in the illustrated example, it is a pentagon). This pentagon slider is installed on a crankshaft, which also serves as a fluid distributor as the fluid inlet and outlet ports are placed on it. Driven by the pressurized fluid, this pentagon slider makes a translational move to push the crankshaft, causing a rotational move. A pressure ring is used for each piston to secure the sealing between pistons and sliders during high-pressure operations. Consequently, the rotating of the crankshaft changes the cylinders connected to either the fluid inlet or outlet port alternatively to sustain the operation. Arranging all cylinders centrically, this motor can achieve a static balance between the pistons, pressure rings, and the pentagon slider.

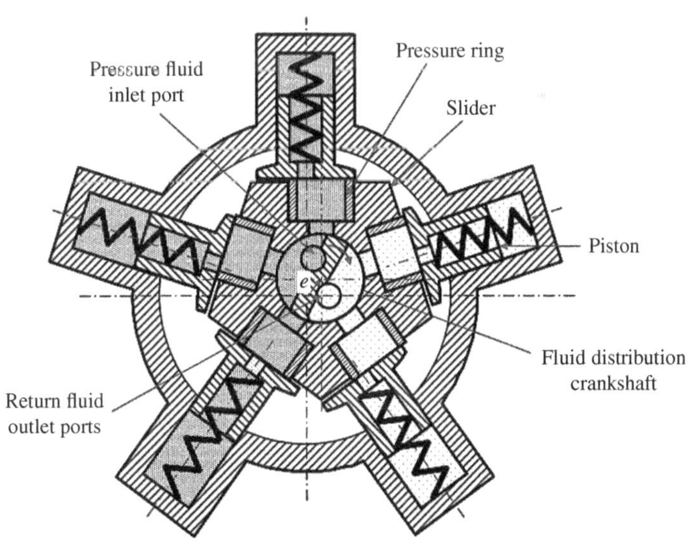


**FIGURE 4.20**
Illustration of the configuration and operation principle of a typical static-balanced radial piston motor.

Normally, radial-piston motors require a high degree of manufacturing precision to ensure reliable functioning. While such high-precision manufacturing always increases the initial costs, radial-piston motors generally have a high volumetric efficiency and a long-expected service life. In selecting this type of motor for different applications, it is strongly recommended that the specified operational speed ranges provided by the manufacturers be followed closely because operating the motor at too low a speed may often cause torque ripples or speed flutters and at a too high-speed may result in unexpected leakage in between the moving pairs in these motors.

Axial-piston motors have excellent high-speed performance in general. When used in low speed applications, this type of motor often requires operation within a certain speed range in order to function efficiently. Typically, inline-type axial-piston motors will operate smoothly only at a relative high speed of 100 rpm or above, and the bent-axis type can smoothly operate at a very low speed of 10 rpm or below. Axial-piston-type LSHT motors are often used in applications requiring good starting torque characteristics and good volumetric efficiencies, especially at lower pressure conditions. The configuration features of LSHT axial-piston motors are very similar to their high-speed counterparts. Similarly, axial-piston LSHT motors also require high-precision manufacturing, and their efficiency characteristics and expected service life are similar to those of radial-piston motors. Initially, axial-piston motors always cost more than vane or gear motors of comparable horsepower. However, their high efficiency and long service life can effectively recover the higher initial cost during the life of machinery systems.

### 4.3.5 Oscillating Rotary Actuators

Some applications, such as swing operations of a hydraulic excavator and pick-and-place operations of a robot arm, require limited rotating actuations rather than continuous ones. **Oscillating rotary actuators** (also called oscillating motors or rotary cylinders), operating in limited rotations, are designed specially for such applications. A typical oscillating motor is actually a **vane actuator**, in which the output shaft is driven by the oscillatory rotating vane. A **single-vane actuator** as depicted in Figure 4.21(a) has an expendable

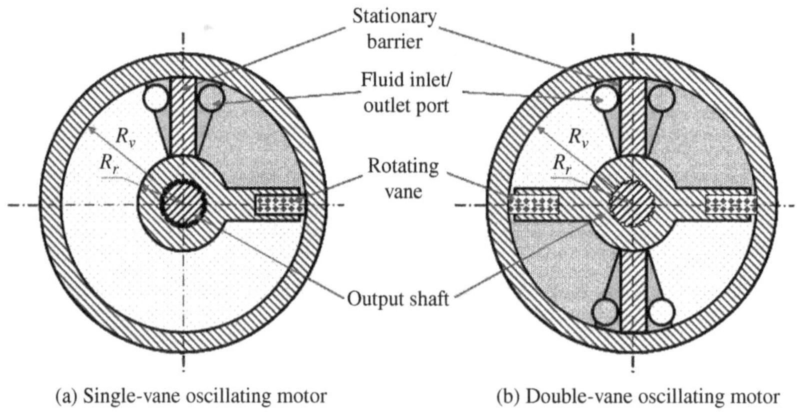


**FIGURE 4.21**
Illustration of the configuration and operation principle of typical (a) single-vane and (b) double-vane type oscillating rotary motors.

cylindrical chamber in which a vane connected to the output shaft rotates through an arc up to 280°. Two ports are separated by a stationary barrier. When pressurized fluid enters the chamber on one side of the vane (the right side of the stationary barrier in the depicted illustration in the figure), the differential pressure applied across the vane rotates the output shaft until the vane meets the barrier. The limited rotation is reversed by reversing the pressurized fluid at the inlet and outlet ports.

Figure 4.21(b) depicts a **double-vane actuator**, which has two diametrically opposed vanes and barriers. At the price of a reduced rotation range of generally less than 100°, this configuration provides twice the torque in the same space as a single-vane oscillating motor. To control the flow alternation entering different chambers, it is common to use a four-way directional control valve to regulate the fluid entering into or discharging from those chambers. Because of configuration characteristics, oscillating rotary motors are generally insensitive to contamination and are capable of realizing zero internal leakage, and therefore can achieve highly accurate position control. Other than the illustrated oscillating motors, there are a few other designs of rotary motors, such as **rotary bladder motors** and **rotary abutment motors.**

Because of the noticeable difference in configuration between the continuous and oscillating motors, the determination of driving torque created on an oscillating motor is usually somewhat different from that of a continuous motor as expressed in Section 4.3.2. In general, the hydraulic force acting on the vane surface equals the pressure applied on this surface area, and the created torque can be calculated in terms of the mean radius, expressed as follows:

$$F = p(R_v - R_r)L \tag{4.47}$$

$$T = \frac{R_v + R_r}{2}F = \frac{pL}{2}\left(R_v^2 - R_r^2\right) \tag{4.48}$$

where $F$ is the hydraulic force acting on a vane; $T$ is the total theoretical torque created on the motor; $p$ is the pressure of the supplied flow; $R_v$ and $R_r$ are the vane and the rotor radiuses; and $L$ is the width of the motor.

The volumetric displacement of a single-vane oscillating motor can be approximately calculated using the following equation:

$$D_v = \pi\left(R_v^2 - R_r^2\right)L \tag{4.49}$$

where $D_v$ is the volumetric displacement of a motor.

The torque equation can, therefore, be expressed as:

$$T = \frac{D_v p}{2\pi} \tag{4.50}$$

**Example 4.3: Operating Parameters of an Oscillating Motor**

Figure 4.21(a) illustrates a single-vane oscillating motor. Assuming that a vane radius of the motor is 200 mm, the rotor radius is 50 mm, and the width of the motor is 100 mm, try to estimate the operating pressure on the motor when it is used to drive a 6000 N·m rotating load. (Assume that it requires an operating pressure of 1.3 MPa to drive the motor with no external load.)

a. Calculate motor volumetric displacement according to Eq. (4.49):

$$\begin{aligned} D_v &= \pi\left(R_v^2 - R_r^2\right)L \\ &= 3.14\times\left(0.20^2 - 0.05^2\right)\times 0.1 \\ &= 1.18\times 10^{-2}\left(m^3\right) = 11.8(L) \end{aligned}$$

b. Solve the pressure needed to drive the load in terms of Eq. (4.50):

$$\begin{aligned} p &= 2\pi\frac{T}{D_v} \\ &= 2\pi\times\frac{6000}{1.18\times 10^{-2}} \\ &= 3.2\times 10^{6}\left(N\cdot m^{-2}\right) = 3.20(MPa) \end{aligned}$$

c. The required operating pressure to drive the motor to do the work:

$$\begin{aligned} p &= p_L + p_R \\ &= 3.2 + 1.3 \\ &= 4.5(MPa) \end{aligned}$$

DISCUSSION 4.3: As driving any other mechanical devices, a hydraulic motor also consumes a certain amount of energy over its turning resistance to maintain normal operation. Ideally, such an operational resistance is a constant value after a device is manufactured.

### 4.3.6 Speed Control and Power Transmission of Hydraulic Motors

The application of hydraulic motors determines the required power and speed, although the actual speed and torque required for some applications may vary while maintaining a required output power. As defined by Eq. (4.7), the operating speed of a hydraulic motor is independent of the operating pressure but is determined jointly by motor displacement and supplying flow rate. Therefore, control of a motor speed can be achieved by either controlling the inlet flow to the motor or adjusting motor displacement if it is a variable displacement model. The inlet flow control is commonly accomplished by either using a variable-displacement pump to generate just the required amount of flow or using a proportional control valve to regulate the amount of flow distributed to the motor. The former is an energy-efficient approach and is often configured as a hydrostatic transmission design (which is discussed in detail in a following section). The latter is a highly responsive approach and often requires that it be powered by a constant-pressure source, such as a pressure-compensated pump.

A typical valve-regulated motor speed control system is schematically depicted in Figure 4.22(a). This approach to designing a motor speed control system is normally accomplished by two design tasks of sizing components capable of delivering sufficient power for driving the required load and configuring a feedback control scheme capable of

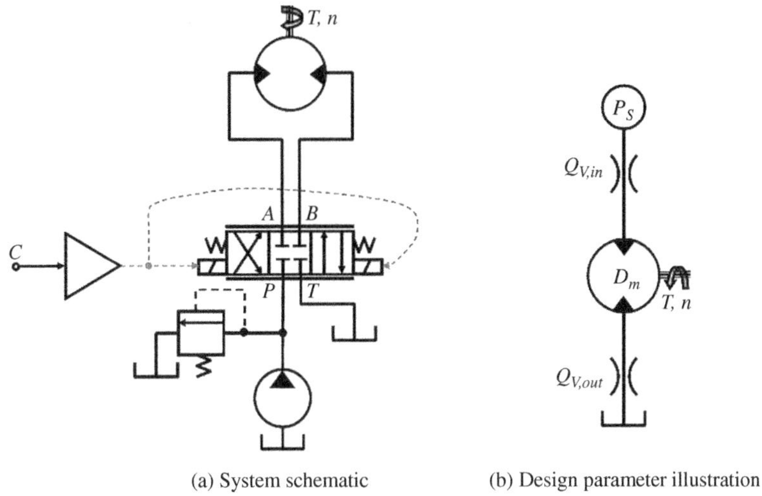


**FIGURE 4.22**
Illustration of the system schematics and design parameter of a typical valve-regulated motor speed control. (a) System schematic and (b) design parameter illustration.

maintaining a constant speed under a changing load within entire motor-operating speed ranges. This design method requires selecting system parameters at the worst-case operating point, namely, at the maximum power point, as the design parameters. Optimally sizing the components normally involves selecting three key hydraulic parameters of supply pressure ($p_S$), motor volumetric displacement ($D_v$), and the valve-controlled flow ($Q_v$, often represented as orifice-regulated flow) to define the maximum capacity of power delivery under a worst-case scenario, as illustrated in Figure 4.22(b). In many cases, one can easily create two equations to support this design optimization process. Since this optimization involves three hydraulic parameters, the design engineer needs first to specify one parameter out of the three, based on the design scenario, to calculate the other two. For example, in many design practices, the supply pressure for the maximum power scenario is always specified as a design constraint since a hydraulic system is often specified by its rating of the allowable maximum operating pressure. In such a case, one needs to calculate the volumetric displacement of the motor and the maximum valved flow to optimally size the major components. As depicted in Figure 4.22(b), a design equation incorporates both the system flow capacity limited by the control valve and the motor volumetric capacity, often called the **valve control of motor motion** (VCMM) equation in terms of the system supplying pressure and pressure drops across the valves and the motor. From the orifice equation defined by Eq. (3.2), and the motor speed and torque equations defined by Eqs. (4.7) and (4.8), this VCMM equation can be expressed as follows.

$$\begin{aligned} p_S &= \sum \Delta p_V + \Delta p_m \\ &= \left(\frac{nD_v}{\eta_v}\right)^2 \left(\frac{\rho}{2C_d^2 A_{v,in}^2} + \frac{\rho}{2C_d^2 A_{v,out}^2}\right) + \frac{2\pi T}{\eta_m D_v} \\ &= \frac{n^2 D_v^2}{\eta_v^2 k_v^2} + \frac{2\pi T}{\eta_m D_v} \end{aligned} \tag{4.51}$$

where $D_v$ is motor volumetric displacement; $n$ is the rotating speed of the motor in a unit time; $T$ is the motor output torque, $\eta_v$ and $\eta_m$ are volumetric and mechanic efficiencies of the motor; $\Delta p_V$ and $\Delta p_m$ are the total pressure drop across both valves and across the motor; $C_d$ is the orifice coefficient; ρ is the density of hydraulic flow; $k_v$ is the total valve coefficient determined by the effective valve opening area and fluid density; and $A_{v,in}$ and $A_{v,out}$ are the flow-passage areas of the inlet and outlet flow control valve, respectively.

This VCMM equation expresses the relationships between supply pressure, valved flow rate, motor displacement, plus load torque and motor speed. In most design practices, the torque and speed are often among the required performance parameters the designed system must meet. In those cases, the design problem is often simplified by using the VCMM equation to solve for motor displacement. Theoretically, after the displacement is selected, the speed is determined solely by the inlet flow rate as expressed in the following equation:

$$n = \frac{Q}{D_v} \tag{4.53}$$

where $D_v$ is the motor displacement and $Q$ is the inlet flow rate to the motor.

In practice, the obtained motor speed will always be lower than the theoretical value, mainly due to internal fluid leakages. When the volumetric efficiency of a motor is known, the actual speed of the motor can be calculated using Eq. (4.7), which is equal to multiplying a volumetric efficiency to Eq. (4.53). However, it is difficult to determine the volumetric efficiency accurately because it is often affected by many factors. Research indicated that motor operation pressure has a great effect on the volumetric efficiency, mainly through the pressure-induced leakage. Often, the volumetric efficiency can be defined using the following equation:

$$\eta_v = \frac{nD_v - \Delta Q}{nD_v} = 1 - \frac{\Delta Q}{nD_v} \tag{4.54}$$

where $\Delta Q$ is fluid leakage within a motor.

Because the motor leakage is mainly attributed to operating pressure, not speed, one can easily find from the equation that the volumetric efficiency of a motor will decrease as its speed reduces if the operating pressure is maintained at the same level. Motor manufacturers often specify both the maximum and minimum operating speeds, namely, the motor speed range, as part of the core motor performance parameters. If the speed surpasses this maximum level, the friction of fluid motion within the motor will be greatly increased, which will induce a significant decline in motor efficiency or even result in motor malfunction. If the speed is too low, the motor rotation will be unstable induced by the friction, fluid leakage, and flow pulsation.

One drawback of valve-regulated motor speed control is that this approach is often associated with noticeable pressure and/or flow loss, which often results in low energy efficiency. An alternative motor speed control can be accomplished by adjusting motor displacement. Because motor displacement adjustment can regulate shaft speed if the inlet flow rate is constant or regulate the output torque if the pressure drop across the motor remains the same, this approach can achieve a **dual adjustment**. In practice, one can set a motor operating at a small displacement to satisfy the requirement for high-speed low-torque operations, or at a large displacement to meet the need for high-torque low-speed

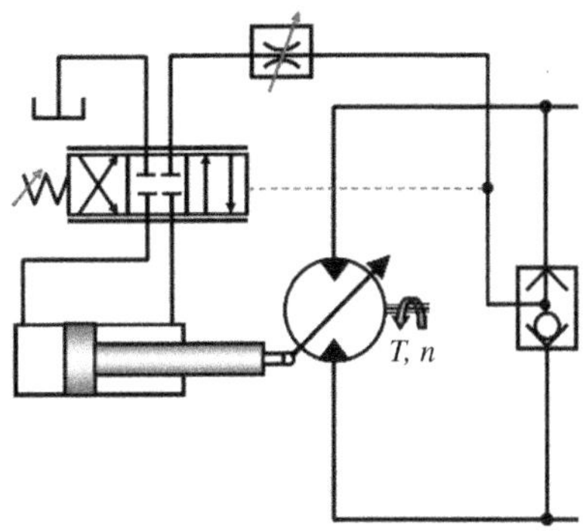


**FIGURE 4.23**
Illustration of the principle of typical constant-power approach of motor speed control.

operations. In such a way, it allows use of a small pump to supply a small rate of high-pressure flow to the motor for performing a wide range of operations. This approach will not only decrease the initial cost for building the system, but more importantly can improve the overall energy efficiency by keeping the pump operating at its heavy load condition. Depending on applications, the displacement adjustment of motor speed can be realized by either a constant power approach or a constant-torque approach.

By a **constant power approach** of speed control, the inlet flow to the motor is normally constant, and the speed control is achieved by adjusting the motor displacement in proportion to the pressure of inlet flow as depicted in Figure 4.23. If the load on the motor changes during an operation, it will induce a pressure increase since the inlet flow is constant. This increased pressure will then push the pressure control valve to move left against the spring force, which will make the motor displacement actuating cylinder move left to increase the motor displacement, and consequently reduce the pressure of inlet flow to the motor. When the pressure is balanced, the spring force acting on the left end of the pressure control valve spool will push the spool back to its neutral position to shut off the pressure delivery to lock the motor-displacement actuating cylinder at the current position. This speed control approach maintains the motor operation at a constant power by keeping both the inlet flow and pressure constant regardless of load variations. It is suitable for applications required to alternatively operate under either high-speed low-torque or low-speed high-torque conditions.

Another method of motor speed control via displacement adjustment is the **constant-torque approach,** which controls the product of the motor displacement and the pressure drop across the motor to be constant. This function can be achieved using either a mechanical or an electrical control. As illustrated in Figure 4.24, an electrically implemented constant-torque motor speed controller uses a displacement sensor to get motor displacement information in terms of piston position and a pressure sensor to obtain the inlet flow pressure reading required to calculate the generated torque. Then the speed controller compares the calculated data to the set torque for the operation and adjusts the motor displacement if those two do not agree to control the speed in maintaining a constant torque. The main advantage of using an electrically implemented approach is that it can easily adopt feedback control strategies to achieve better system characteristics in both dynamic and steady states.

As is true of linear actuators, the main function of hydraulic motors is to deploy hydraulic power to move the load, but in a rotating form. It is also useful to repeat the fundamental concept that the operating pressure of a motor is determined by the load torque applied

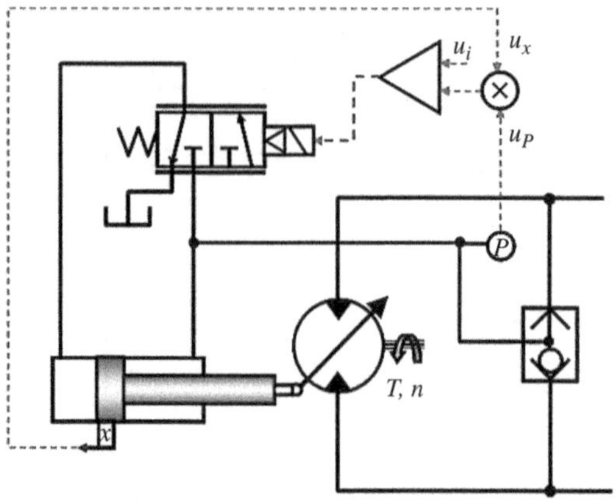


**FIGURE 4.24**
Illustration of the principle of typical constant-torque approach of motor speed control.

on this motor. The motor transmits its maximum power capacity when it is operating at the maximum allowable system pressure and shaft speed. In an ideal condition without considering the energy losses during the power transmission, the theoretical torque being generated on a hydraulic motor is determined by the displacement of the motor and the pressure drop across the motor as expressed in the following equation:

$$T_T = \frac{D_v \Delta p}{2\pi} \tag{4.55}$$

where $D_v$ is the motor displacement, $\Delta p$ is the pressure drop between the motor inlet and outlet ports, and $T_T$ is the theoretical torque generated by the motor.

As discussed in Section 4.1.2, the actual output torque from a motor is always smaller than its theoretical value, and a mechanical efficiency $\eta_m$ has to be considered for counting all mechanical losses. Equation 4.8 is commonly used in engineering design to figure out the actual output torque from a motor. Because the mechanical efficiency of a motor is normally affected by many factors, it is difficult to measure it accurately. In most engineering practices, it is common to consider all power losses, except fluid leakage, as mechanical losses.

A higher torque is always required to start moving a stationary load than to keep the load in motion. This higher torque is necessary to overcome the static friction. Extra torque is also needed to move the load after it has been stalled. Because the maximum torque being created in a motor is determined by the motor displacement and the system pressure, the extra demand on torque to restart the motor will result in less torque available for driving the external load. This loss of output torque during the motion-breaking stage is often called the **stall torque** and the **stall torque efficiency**, defined by the following equation.

$$\eta_s = \frac{T_s}{T_T} \tag{4.56}$$

where $\eta_s$ is the stall torque efficiency and $T_s$ is the stall torque.

It is common practice to select motor displacement based on the stall torque rather than the maximum load torque to ensure sufficient torque to drive a stationary load. For most mobile hydraulic systems, the stall torque must be at least 1.5 times higher than the maximum running torque. This theoretically calculated motor displacement needs to be further inflated to compensate for the motor mechanical and volumetric efficiencies. Before knowing anything about the motor, it is reasonable to approximate the mechanical efficiency at stall varying from 75 to 95%, and the maximum volumetric efficiency between 80 and 95% for most motors. In engineering practices, a motor of exact calculated displacement is often not commercially available. Therefore, it is recommended that the nearest larger displacement be selected to ensure the suitability of a selected motor for the designed applications. It needs to be pointed out that motor efficiencies at the full load condition may be quite different from the efficiencies at stall.

The power generated in a hydraulic motor can be calculated after the operating speed and torque of the motor are specified. Theoretically, the power generated in a motor is purely determined by the motor output speed and torque, which is exactly the same as the hydraulic power delivered to the motor by the pressure fluid, as expressed by the following equation:

$$P_T = T_T \omega = \Delta p Q \tag{4.57}$$

where $P_T$ is the theoretical output power from the motor, $T_T$ is the theoretical motor output torque; $\omega$ is the motor shaft angular velocity; $\Delta p$ is the pressure drop across the motor; and $Q$ is the inlet pressure flow to the motor, respectively.

Because of both fluid leakages and all other mechanical losses, the actual output power from a motor is always less than its theoretical value. The actual output power can, therefore, be calculated using the following equations:

$$P_A = T_A \omega = P_T \eta_o \tag{4.58}$$

$$\eta_o = \frac{T_A \omega}{\Delta p Q} \tag{4.59}$$

where $P_A$ is the actual output power from the motor, $T_A$ is the actual motor output torque, and $\eta_o$ is the overall efficiency of the motor.

Note that the actual power delivered to the motor by pressure fluid, also called the **hydraulic power**, is the maximum power the motor can theoretically generate. The actual output power available to drive the load is a mechanical form of power and is often called the **break power**.

**Example 4.4: Sizing a Hydraulic Motor**

An application requires driving a maximum torque load of 500 N·m when rotating at 3000 rpm. Assume the maximum allowable system pressure of 20 MPa and a return line pressure of 0.7 MPa. If the volumetric and overall efficiencies are 93 and 82%, respectively, try to size the motor and the required supplying flow rate.

a. Calculate motor mechanical efficiency using Eq. (4.12):

$$\begin{aligned}\eta_m &= \frac{\eta_o}{\eta_v} \\ &= \frac{82.0\%}{93.0\%} \times 100\% \\ &= 88.2\%\end{aligned}$$

b. Calculate motor volumetric displacement using Eq. (4.8):

$$\begin{aligned} D_v &= \frac{2\pi T}{\Delta p \eta_m} \\ &= \frac{2 \times 3.14 \times 500}{(20-0.7)\times 10^6 \times 0.882} \\ &= 1.84 \times 10^{-4}\left(m^3\right) = 0.184(L) \end{aligned}$$

c. Calculate required supplying flow rate using Eq. (4.17):

$$\begin{aligned} Q &= \frac{nD_v}{\eta_v} \\ &= \frac{3000/60 \times 1.84 \times 10^{-4}}{0.93} \\ &= 0.01\left(m^3 \cdot s^{-1}\right) = 600\left(L \cdot \min^{-1}\right) \end{aligned}$$

DISCUSSION 4.4: Both the volumetric displacement and the output torque of a hydraulic motor are inverse proportional to the pressure drop between inlet and outlet ports of the motor. This indicates that any back pressure at the outlet port will reduce the torque output from the motor. The efficiency factor for most motors is fairly constant when operating from half- to full-rated pressure and over the middle portion of the rated speed range. Reducing displacement from maximum in variable-displacement motors will also reduce the overall efficiency.

## 4.4 Hydrostatic Transmission

### 4.4.1 Overview of Hydrostatic Transmission

A basic **hydrostatic transmission** (HST) is actually a coupled pump-motor system that transmits power from the prime mover, often an internal combustion engine, to the final drive, or directly to the wheels using static hydraulic media. HST has found wide applications on many off-road vehicles such as tractors, forklifts, and lawn mowers, mainly because of the following advantages:

1. Good maneuverability supported by continuous variable speed, torque, and power control in either direction and over the entire speed and torque ranges;
2. Good controllability provided its low rotating inertia permits very fast and smooth acceleration and deceleration;
3. High system safety backed by the capability of being stalled and undamaged under excessive load; and
4. High application flexibility carried by extremely high power-to-weight ratio and the simplicity in reconfiguration for suiting diverse requirements to different applications.

HSTs often outperform mechanical and electrical counterparts when a variable output speed is required mainly because of their ability to offer either **variable-power and variable-torque transmissions** using variable-displacement pump and motor, or **constant-torque and variable-power transmissions** using a variable-displacement pump and a fixed-displacement motor, or **constant-power and variable-torque transmissions** using a torque-compensated variable-displacement pump with a fixed-displacement motor. However, like any other systems, HSTs provide such superior characteristics at a cost of lower efficiency than its mechanical counterparts. Compared to the overall efficiency of 92% or higher for a typical mechanical transmission, a typical HST for the same application often has an overall efficiency of around 80%. While some specially designed HSTs can reach an overall efficiency of 85% or higher, none has achieved the same efficiency level as a mechanical transmission.

Being constructed by a pair of hydraulic pump motors, an HST is normally sized in terms of system corner power. As defined in Chapter 2, the corner power of a system is determined by the maximum force and maximum speed required to perform the designed functions, even though these two conditions rarely occur simultaneously. The corner power required to overcome the traction force to push a vehicle to move can be determined using the following equation.

$$P_c = \frac{F_T v}{\eta_{of}} \tag{4.60}$$

where $P_c$ is the corner power required for driving a vehicle; $F_T$ is the maximum vehicle traction force; $v$ is the maximum vehicle speed; and $\eta_{of}$ is the overall efficiency of the final drive.

To convert this mechanical corner power into a hydraulic form, the transmission corner power is often defined as the product of maximum pressure drop across the hydraulic motor and the maximum supplied flow rate as follows:

$$P_h = \frac{\Delta p Q}{\eta_o} \tag{4.61}$$

where $P_h$ is the HST corner power; $\Delta p$ is the maximum pressure drop across the motor; $Q$ is the maximum flow rate supplied to the motor; and $\eta_o$ is the overall efficiency of the motor.

The transmission corner power calculation gives a base for initial transmission selection. Selection is normally refined by considering the effects of duty cycle, final-drive ratio, rolling radius, primer-mover speed, and service life. Early HSTs were intended primarily for light-duty and low-cost applications such as agricultural and garden tractors. With the continuous improvement in designs, and more importantly on performance, HSTs have now been adopted to a broader range of applications. For example, light-duty units (15 KW or less) are continuously being used on small agricultural and consumer equipment, such as lawn tractors and golf-course maintenance equipment; medium-duty transmissions (15 to 45 KW) are commonly used on many types of middle-size off-road equipment, such as loaders, excavators and harvesters; and heavy-duty systems (45 KW or larger) are used on large agricultural and construction equipment. One reason for the increasing attractiveness of HSTs attributes is the improved design of pumps and motors, particularly higher flow and pressure ratings in a more compact package.

### 4.4.2 Configurations of Hydrostatic Transmission

Conceptually, HST can be constructed by pairing many different designs of hydraulic pumps and motors. In terms of the ways of arranging pumps, motors, and control valves, HSTs can be easily configured into six different designs, as depicted in Figure 4.25. The simplest form of HST uses a fixed-displacement pump to drive a fixed-displacement motor (**fixed pump-fixed motor** (FP-FM; Figure 4.25(a)). The motor speed of this type of HST is adjusted by varying the speed of the prime mover to control the pump output flow rate. Although the FP-FM type of HST is simple and inexpensive, it has limited applications primarily due to low energy efficiency. The main factor contributing to this low-energy efficiency is that the pump must be sized to produce enough flow to drive the motor at a fixed speed under full load. Because the pump displacement is fixed, when full speed is not required at the motor, fluid from the pump outlet passes over the relief valve, which converts the energy carried by the released portion of pressure flow into heat and is wasted.

One common way to improve HST efficiency in creating a constant torque transmission is to use a variable-displacement pump instead of a fixed-displacement one as shown in Figure 4.25(b). In a typical **variable pump-fixed motor** (VP-FM) HST, the motor speed is controlled by varying pump displacement to adjust the flow rate supplied to the motor, with the torque remaining constant over the entire speed range because the torque depends only on fluid pressure and motor displacement. Since the motor speed changes in proportion to pump displacement, the delivered power from this type of HST varies with the pump displacement (Figure 4.26).

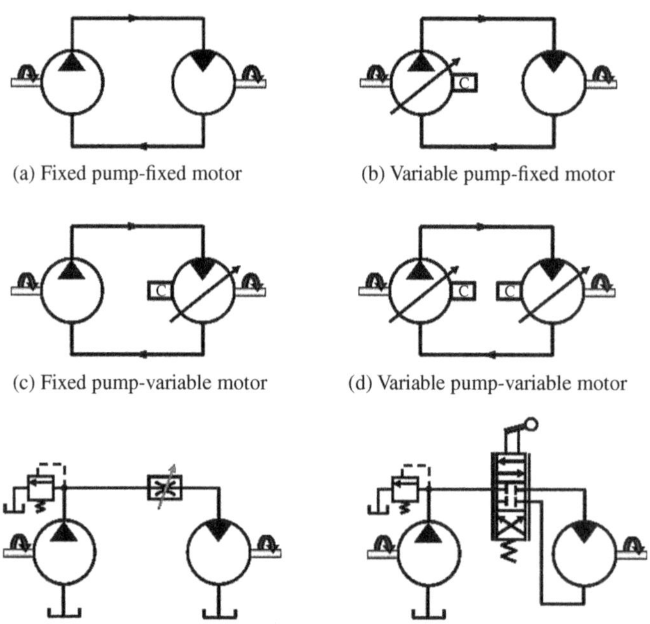
(a) Fixed pump-fixed motor (b) Variable pump-fixed motor
(c) Fixed pump-variable motor (d) Variable pump-variable motor
(e) Needle valve controlled fixed pump-fixed motor (f) Four-way valve controlled fixed pump-fixed motor

**FIGURE 4.25**
Conceptual illustration of six typical configuration arrangements of hydrostatic transmission, (a) fixed pump-fixed motor, (b) variable pump-fixed motor, (c) fixed pump-variable motor, (d) variable pump-variable motor, (e) needle valve controlled fixed pump-fixed motor and (f) spool valve controlled fixed pump-fixed motor.

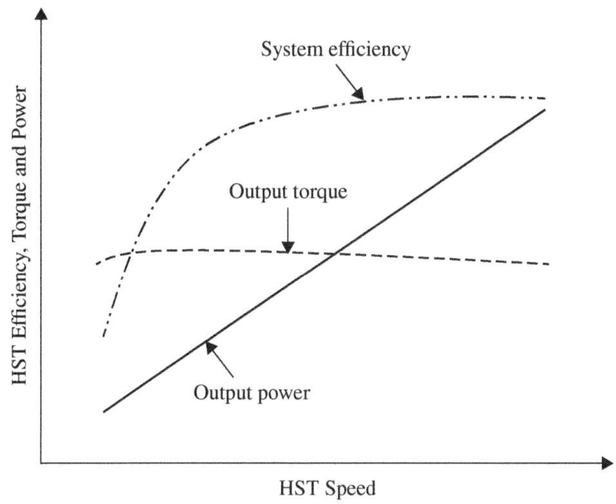


**FIGURE 4.26**
Conceptual illustration of typical performance curves of a VP-FM hydrostatic transmission.

While a VP-FM constant torque HST can significantly improve the power delivering efficiency over that of a FP-FM HST, as it delivers only needed flow to drive the load, it is not very economical for applications with a wide range of load variations. For example, if an application requires driving a load varying from 10 to 50 N·m, the motor has to be sized in order to be capable of driving the heaviest load, which in this example is 50 N·m. Meanwhile, the pump should be sized to be capable of supplying sufficient flow to drive the motor when running at its highest desirable speed, for example, 1800 rpm, and when driving the lightest load, which is 10 N·m in this example. Therefore, a VP-FM constant-torque HST should be sized to satisfy both motor- and pump-sizing requirements, 50 N·m and 1800 rpm, respectively. That is, the HST has to be sized to produce 9.5 kW power. Since a heavy load is always required to be driven at a reduced speed in proportion to the load, a **variable-torque** configuration could be much more attractive than the constant-torque counterpart in those applications.

A simple way of obtaining variable torque in HST is to use a variable-displacement motor to form a **fixed pump-variable motor** (FP-VM) HST (Figure 4.25(c)). Such a modification makes the HST capable of delivering a constant power limited by the constant flow supplied to the motor. During operation, the motor displacement can be varied to maintain the product of speed and torque constant, namely, a constant power (Figure 4.27). Such an HST is often used to increase the motor displacement for raising the torque by decreasing the motor speed, and vice versa. Using a FP-VM HST to drive the variable load described in the previous case, we find that the motor is sized to be capable of driving the heaviest load, which is 50 N·m, and the pump is sized to be capable of driving the lightest load when operating at the highest speeds, which are 10 N·m and 1800 rpm, respectively. By adjusting the motor displacement to drive different loads, 1.9 kW is consumed to drive a 10 N·m load at a full speed of 1800 rpm or a 50 N·m load at a reduced speed of 360 rpm. One limitation of the FP-VM configuration is that the motor speed is coincidently changed as motor displacement is adjusted to generate the required torque without flexibility.

The most versatile HST configuration is a variable-displacement pump and a variable- displacement motor (**variable pump-variable motor**, or VP-VM) as shown in

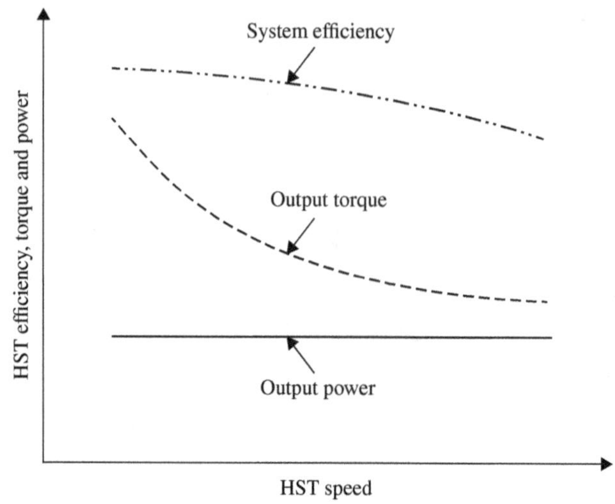


**FIGURE 4.27**
Conceptual illustration of typical performance curves of a FP-VM hydrostatic transmission.

Figure 4.25(d). Theoretically, a VP-VM configuration provides infinite ratios of torque and speed to deliver the power. With the motor at maximum displacement, the HST speed and power output can be adjusted by varying the pump output while keeping the output torque constant. Adjusting the motor displacement to full pump flow can control the motor speed at a desired level, and with the torque varying inversely with the speed delivered, the power remains constant. As illustrated in Figure 4.28, the performance curves of a typical VP-VM HST are composed of two ranges of adjustment: the motor displacement is normally fixed at maximum, and the pump displacement is increased from zero

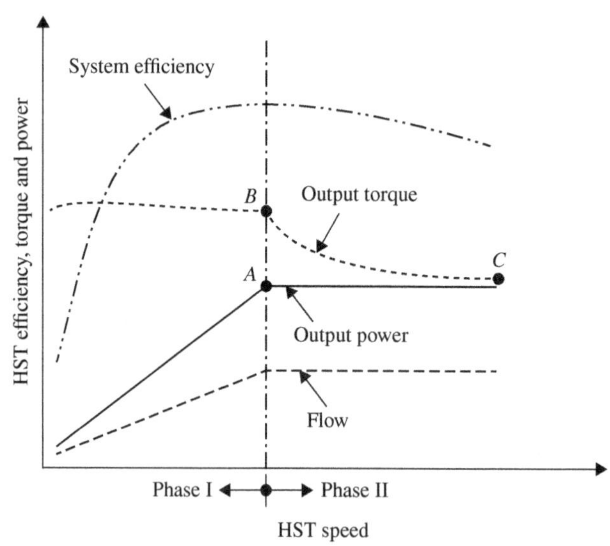


**FIGURE 4.28**
Conceptual illustration of typical performance curves of a VP-VM hydrostatic transmission.

to maximum like a VP-FM in Phase I; and the motor displacement varies to get the highest overall performance after the pump reaches its maximum displacement as in a FP-VM in Phase II. Therefore, a VP-VM HST can be treated as a combination of a VP-FM and a FP-VM HST. The power transmission capability of this HST is determined by the lowest output speed at which the constant power must be transmitted. Theoretically, the maximum power can be transmitted in an HST as a function of flow and pressure, and is limited by the lowest output speed because of the constraint of supplying flow.

In reality, the actual output power in a constant power transmission over the entire speed range can be determined by dividing the theoretical power by a **torque-to-speed** ratio, defined by the ratio of the maximum to minimum speeds for constant power transmission. For example, if the minimum speed (also called **critical speed**) for constant power transmission in a VP-VM HST, represented by point *A* on the power curve in Figure 4.28, is one-half the maximum speed, a torque-to-speed ratio of 2:1 is represented and indicates that the maximum power that can be transmitted by this HST is about one-half of its theoretical maximum. The critical speed of an HST is always determined by the dynamics of its composing components. Corresponding to the critical speed (point *A*) on the power curve there is a point *B* on the torque curve. At any speed above this critical speed, the torque decreases as the speed increases, with the output torque dropping to its minimum level (point *C* on the torque curve) when the HST reaches its maximum output speed. When the HST output speed is less than the critical value, the output torque remains nearly constant at its maximum level, but the output power decreases in proportion to speed.

Figure 4.25(e) and (f) illustrate the control of a special type of HST, the valve-controlled HST, with the supplying flow in the former being controlled using a needle valve and the latter using a four-way proportional valve. Comparing this type of HST with the nonvalve-controlled ones, we find that the most important difference is that the valve-controlled HSTs can supply only a portion of pump outlet flow to drive the motor, while the nonvalve-controlled ones have to utilize all pump outlet flow. Such flow control characteristics in these designs make it possible to realize FP-VM functionality using a fixed-displacement pump and a fixed-displacement motor, which not only significantly reduces the cost for building the HST but also achieves more accurate control of flow supply, and consequently motor output speed. The main drawback of this type of HST is the relatively low overall efficiency because the flow supply is controlled by throttling. However, the simplicity and flexibility in design still makes this type of HST very attractive in many applications, especially on mobile hydraulic systems. More discussions will be provided in section 4.4.4 (*Applications of Hydrostatic Transmission*).

The schematics of typical HSTs illustrated in Figure 4.25 show two different configurations of **closed circuits** (Figure 4.25(a), (b), (c) and (d)) and **open circuits** (Figure 4.25(e) and (f)). As illustrated in Figure 4.25, in an open-circuit HST, the pump delivers the fluid from the reservoir to the motor, with the returning flow being discharged directly back to the reservoir; in a closed-circuit HST, the fluid is supplied from the pump to the motor and is then returned to the pump inlet port as the charging flow.

Due mainly to unavoidable internal fluid leakage within both the pump and motor, a small portion of flow will be involuntarily removed from the circuit and result in a deficiency in charging flow to the pump, which, in turn, will produce insufficient discharge flow from the pump and result in the HST having difficulty in building up adequate pressure to drive the load. Therefore, one important component in a closed-circuit HST is a charge pump (Figure 4.29). The charge pump is commonly designed either as an integral part of the main pump of the HST or as an independent pump installed separately. Regardless of its arrangement, a charge pump performs two main functions: (1) it

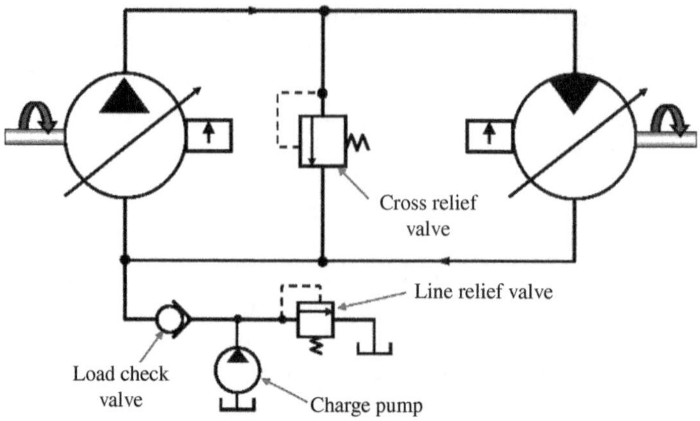


**FIGURE 4.29**
Typical configuration of a unidirectional VP-VM HST with a charging pump and a load-release valve.

prevents cavitation in the main pump by refilling the fluid lost in the closed circuit, and (2) it provides pressurized fluid for actuating variable-displacement control on either or both the main pump and the motor in the HST. Another important component in a typical closed-circuit HST is the **cross-relief valve** as depicted in Figure 4.29, designed to prevent excessive operating pressure from building up in the supply line. By bleeding the extra fluid to the low-pressure line, this cross-relief valve can provide two very useful features, remaining stalled without damage and recirculating fluid to achieve the best possible functionalities. The cross-relief valve is integrated into the motor package in many HSTs.

In many applications, it is required that the HST be able to reverse operating directions. To offer this capability, a bidirectional HST often uses an integrated load-releasing and charging circuit to prevent either excessive pressure from being built up in the supply line due to overload or cavitation in the inlet line because of insufficient flow. As illustrated in Figure 4.30, integrated circuits commonly consist of two pairs of check valves, one for releasing high pressure from the supply line and the other for recharging additional flow to the inlet line as needed. The released flow can be either recharged to the inlet line or dumped back to the tank according to the operating condition. For design simplicity, in

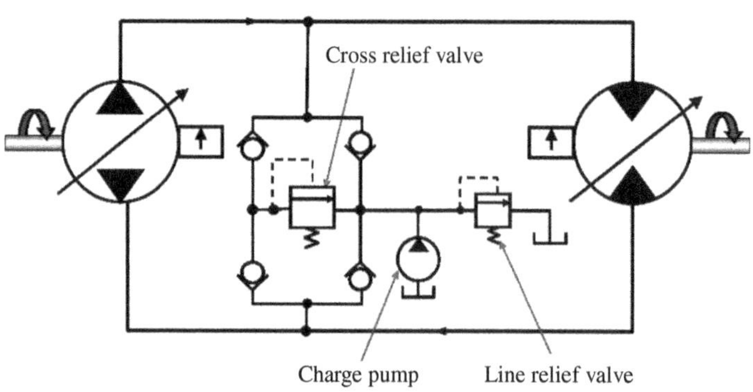


**FIGURE 4.30**
Typical configuration of a bidirectional VP-VM HST with an integrated load releasing and fluid charging circuit.

many cases, a shuttle valve instead of a pair of check valves is used to release excessive high pressure. This shuttle valve is always shifted by high-pressure fluid to connect the high-pressure line to the cross-relief valve to offer the excessive high-pressure releasing function. Similar to the unidirectional HST, a charge pump is also commonly designed in an integrated releasing–recharging circuit to provide the two main functions of preventing cavitation in the main pump by refilling the fluid lost in the closed circuit and providing pressurized fluid for actuating variable-displacement control on either or both the pump and motor in an HST.

As illustrated in Figure 4.25 (e) and (f), the open-circuit HST circuits are more like regular valve-controlled hydraulic circuits than the closed-circuit ones. Indeed, the valve-controlled open-circuit HST systems in general have superior performance over closed-circuit HSTs mainly attributed to broader bandwidth controllability carried by the valves to offer the necessary system response rate and overall performance to achieve sensitive and accurate HST control. However, the closed-circuit HSTs do offer a higher efficiency in general than their open-circuit counterparts because no power will be generated by the pump until the load needs it.

### 4.4.3 Control of Hydrostatic Transmission

To offer enhanced energy efficiency and power variability, most HST systems use a pair of coupled pumps and motors to transmit the power. To estimate the performance of an HST, it is important to study the dynamics of such systems. Among all the pertinent dynamic parameters, the **response time**, defined as the time required for the output motor speed to reach a new set point of operation status in responding to an input control command, is often the most critical one to be evaluated.

Analyzing a basic HST system without unnecessary details and without loss of generality, an **open-loop closed-circuit** HST system is used to serve as the basis in this book. As illustrated in Figure 4.31, a typical open-loop closed-circuit HST normally uses a fixed-displacement pump and motor. Because of its fixed-displacement feature, such an HST is designed to deliver the needed power to drive the load at a constant operating speed up to the capacity limitation of the system. Theoretically, the maximum power an open-loop closed-circuit HST system can transmit is limited by the flow rate the pump can supply and the maximum operating pressure the system can withstand. The **corner power** defined in Section 2.2.1 can be used to quantify the capacity limitation of the system.

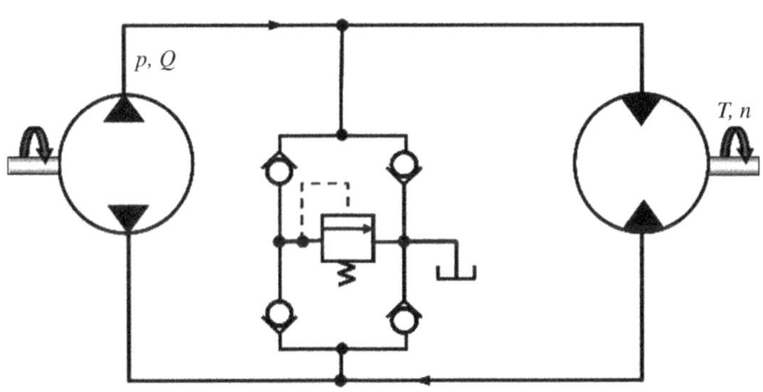


**FIGURE 4.31**
Typical configuration of an open-loop closed-circuit hydrostatic transmission.

As defined in Section 1.3, the hydraulic power delivered by the pump can be determined in terms of the pump-discharge flow and pressure using Eq. (4.62):

$$P_h = pQ \tag{4.62}$$

where $P_h$ is the corner power delivered by the pump, $p$ is the discharge pressure, and $Q$ is the discharge flow from the pump.

The mechanical power, as defined in the following equation, available to drive the external load on the motor output shaft should be the same as the hydraulic power generated at the pump:

$$P_m = 2\pi nT \tag{4.63}$$

where $P_m$ is the output mechanical power on motor shaft, $n$ is the motor output shaft rotating speed, and $T$ is the motor output torque for driving the load.

Assume the overall power transmission efficiency in a hydraulic motor is $\eta_o$ and $\eta_l$ in the connecting lines between pump and motor. Substituting Eqs. (4.38) and (4.41) in Eq. (4.63), we see that the mechanical power available at the output shaft of the motor can be represented in terms of pump-discharging pressure and flow rate, as follows:

$$P_m = pQ\eta_o\eta_l \tag{4.64}$$

Because the pump-discharge flow in a closed-circuit open-loop HST is a constant when the pump is operating at a constant speed, and the discharge pressure is determined by the system load (in this case, the external torque load applied on motor output shaft), such an HST can limit power transmission to the motor in response to the torque requested to drive the external load, thus achieving higher efficiency in power transmission.

In comparison, a typical closed-loop closed-circuit HST system, normally constructed using a variable-displacement pump to drive a fixed-displacement motor as illustrated in Figure 4.32, can also control the operating speed of the motor in terms of the load. To achieve such a function, the variable-displacement pump in a closed-loop closed-circuit

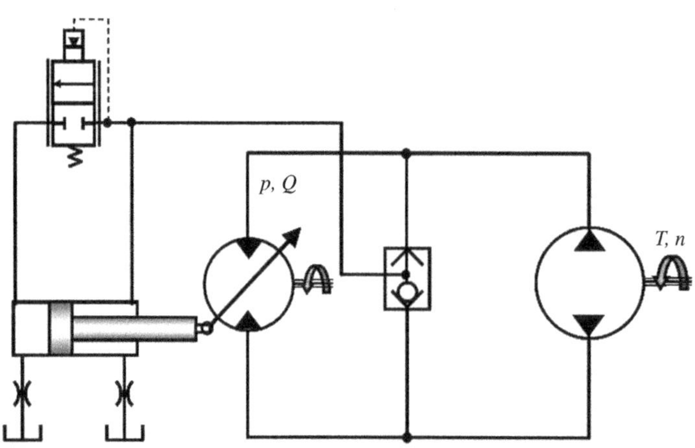


**FIGURE 4.32**
Typical configuration of a closed-loop closed-circuit hydrostatic transmission.

HST system is often equipped with a proportional pressure control valve to sense the system load and adjust the pump displacement in terms of the load. In operation, as the external load applied to the output shaft of the motor increases, it will raise the system pressure accordingly, which in turn increases the power needed to drive the load. To control the amount of power transmitted to the motor, the pressure control valve will be opened in proportion to the load pressure to lead a certain amount of fluid to the head chamber of the pump-displacement control cylinder and reduce the pump-discharge volume to supply less flow. The minimum value of the discharge volume is determined by the position of the bleeding orifice on the cylinder. When the external load is reduced, it will consequently bring down the system pressure, which in turn will close the proportional pressure control valve creating a bigger pressure drop across the valve and allow high system pressure to retract the pump control cylinder, increasing the discharge volume to supply more flow. The maximum discharge volume, however, is restricted by the physical structure of the pump design.

The closed-circuit HST is normally used in applications of driving a certain load, restricted by its power-transmitting capability. In other words, a closed-circuit HST can drive an external load operating at full speed when the load to be driven is below a certain level. Under such a circumstance, the pump can deliver the maximum amount of fluid to drive the load. As the load increases to surpass the critical value, the pump control device will reduce the pump-discharging flow by adjusting its displacement in response to the rising operating pressure. In some HSTs, the motor displacement is also adjustable for increasing the operating speed when the load is light. In both cases, control of the motor speed is achieved by adjusting the displacement of the pump and/or the motor in response to system pressure variation. The speed control **response time** of a closed-circuit HST, defined as the time required for motor output speed to reach a new set point corresponding to a pump or motor displacement control signal, is determined by the natural frequency of the HST. Neglecting the energy dissipation in the system, we can define the **natural frequency** of an HST as follows:

$$f_n = \frac{1}{2\pi}\sqrt{\frac{k_h}{m}} \tag{4.65}$$

where $f_n$ is the natural frequency, $k_h$ is the stiffness of the HST system, mainly determined by the fluid elasticity within the HST, and $m$ is the total mass of the system.

**Stiffness** is one of the dominant factors affecting the response time of the HST system. A stiff system (a system with a large $k_h$ value) means the system will have little deformation when a large load is applied and therefore can respond to an actuating action promptly. Stiffness itself is affected by many factors, such as compressibility of the fluid, expandability of the hydraulic lines, and stiffness of the pump/motor displacement control devices. For example, if an HST has a long hydraulic line, the stiffness level of this HST will be low. It means that a significant deformation will occur when an actuating action is applied to the system, which in turn will result in slow response to the action. Fortunately, the hydraulic lines in typical closed-circuit HSTs are often very short, with many using solid conduits as the fluid-transmitting lines. Therefore, most of the closed-circuit HSTs are normally very stiff.

Always remember that the stiffness of an HST is also dependent on the compressibility of the fluid and the compliance of the system-connecting components, such as tubing and hoses. Normally, hydraulic fluids are treated as incompressible, and when the fluids are fully filled in an HST, high stiff characteristics are presented. However, if a cavitation exists in the system, it will significantly increase the compressibility of the fluid, which will dramatically decrease the system stiffness and in turn noticeably slow down the response to a control action. Increasing inlet flow pressure to the pump is the most effective way to prevent cavitation from being formed in an HST system. A simple, practical way to increase inlet pressure to a pump is to use a charge pump. In cases where the changes in an external load are infrequent and only last a short period of time, an accumulator (which will be discussed in detail in the next chapter) can be added to the circuit to make the system stiffness more stable.

Another key parameter having a significant influence on system natural frequency is the total **mass** to be driven. As expressed in Eq. (4.65), a large mass results in a low natural frequency. Normally, a heavy load often means a large mass to be driven. In practice, a closed-circuit HST changes its output speed by adjusting the displacement of either the pump or the motor, which often requires moving a large mass to adjust the displacement. As result, the bandwidth for a fast-response closed-circuit HST will normally not go beyond 20 to 30 Hz.

In some cases, the low-frequency response characteristics of closed-circuit HSTs cannot provide sufficient bandwidth for applications requesting a fast response, such as closed-loop positioning control. In many of those cases, valve-controlled, open-circuit HSTs can be a practical alternative because they have the capability of offering a high-frequency response in speed control implementation. When a servo valve is used, it is possible to achieve a frequency response up to 150 to 200 Hz, whereas the fastest pump frequency response is only a small fraction of that. Even when a proportional valve is used, it is possible to implement a pump control that is around a 50 Hz response speed.

Similar to its closed-circuit counterparts, open-circuit HSTs can also be categorized as open loop and closed loop in terms of their capability of adjusting motor speed. As illustrated in Figure 4.33, the motor output speed of a typical valve-controlled open-loop open-circuit HST is solely controlled by the supplying flow regulated by the control valve, regardless of the load, as long as it is below the maximum allowable operating pressure set by the line relief valve. In this open-circuit HST, the amount of pressured fluid delivered to the motor is controlled proportionally to the opening of a directional control valve. The speed control of such an HST is achieved by adjusting the valve opening to regulate the amount of pressurized fluid supplied to the motor. The amount of hydraulic power

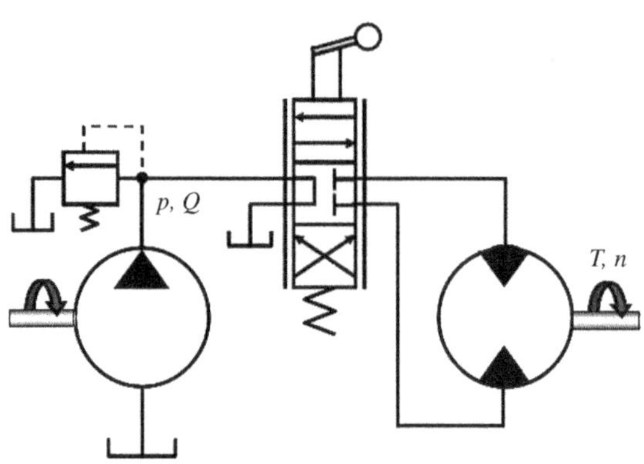


**FIGURE 4.33**
Typical configuration of an open-loop open-circuit hydrostatic transmission.

transmitted to the motor can be determined by the supplied flow rate times the pump-discharge pressure using the following equation:

$$P_h = pQ_v \tag{4.66}$$

where $Q_v$ is the controlled supplying flow passing through the valve.

A comparison of Eq. (4.66) with Eq. (4.62) shows that the only difference between the two is that the flow rate used in Eq. (4.62) is the pump-discharge flow, while in Eq. (4.66) it is the supplying flow controlled by the valve. This difference in the supplying flow completely changes the speed control characteristics of open-circuit HSTs from their closed-circuit counterparts, as a result of the use of a flow control valve in such a circuit. As discussed in Chapter 3, different flow control valves have their specific flow control characteristics, and valve-controlled open-circuit HSTs will certainly inherit those characteristics in their speed control. For example, if a tandem-center proportional valve is used in an open-loop open-circuit HST, the motor can be controlled running at a constant speed as long as the load is unchanged. However, as the motor load changes, the operating pressure will change accordingly, which in turn will vary the resistance ratio between **pump-to-motor** (P-to-M) and **pump-to-tank** (P-to-T) passages in case the tandem-center proportional valve is partially open (see Section 3.1.4 for details). Such a resistance ratio variation will then induce a change in the amount of flow supply to the motor and will eventually cause a motor speed change when corresponding to the load variation. Because the tank pressure keeps an almost constant value, and the motor-operating pressure and the pump-discharging pressure always reflect the load being driven, a variation in the load will induce a greater change in pressure drop across the P-to-T pass than that across the P-to-M pass, which in turn results in a variation in the ratio of flow distribution between those two passes. In general, a lighter load will reduce P-to-T pressure drop due to the decrease in pump-discharge pressure, which will reallocate more flow to the motor and result in a higher operating speed at the motor. Similarly, a heavier load will cause the motor to slow down.

Just as the closed-loop design can control the motor speed at a fairly constant level for closed-circuit HSTs, a closed-loop design for open-circuit HSTs can achieve a similar performance. Figure 4.34 depicts how an electrically controlled proportional control valve,

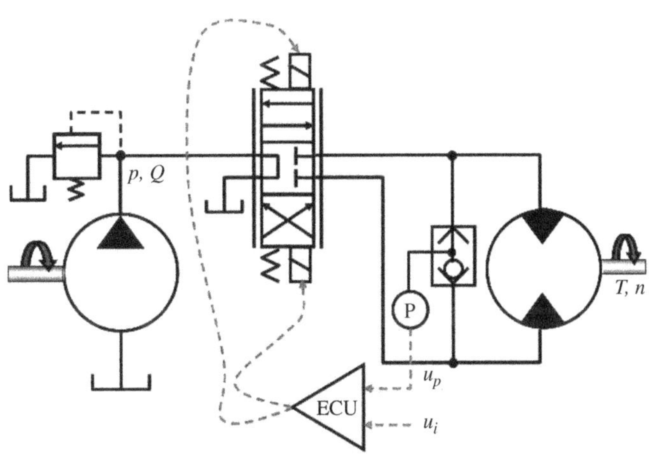


**FIGURE 4.34**
Typical configuration of an electrically controlled closed-loop open-circuit hydrostatic transmission.

supported by a pressure sensor, can be used to realize the load-compensated HST speed control. As illustrated in this figure, an **electronic control unit** (ECU) is used to sense the load pressure on a motor inlet port, compute the needed correction for valve spool position based on the sensed data, and create an appropriate correction signal for adjusting the valve opening to keep the motor operating at a desired speed.

While valve control provides a wider bandwidth in HST control, it has two major disadvantages in comparison to a pump/motor control. The first is a valve-realizing flow control by creating an adjustable restriction in the supply line to limit the amount of flow supplied into the motor, which will induce an **energy loss** in the control implementation. In comparison, a variable-displacement pump in a closed-circuit HST performs its control task by generating only the needed power, and therefore it is an **energy conservation method** of control. The other disadvantage of valve-controlled HST is its high cost, especially when a servo valve is used for a very wide bandwidth in speed control. One practical method for solving the high-cost problem is the use of solenoid-driven electrohydraulic valves. However, they dramatically reduce the bandwidth in speed control and therefore can only be used in applications that do not require highly accurate speed control.

**Example 4.5: Hydrostatic Transmission**

Assume we need to design a fixed-displacement HST for driving a 350 N·m torque load at an operating speed of 1000 rpm. What is the power required to drive this load? If we choose a pump with rated operating pressure of 15 MPa, how much flow should be discharged in order to provide enough hydraulic power to drive the torque load? (Assume that both the pump and motor efficiencies are negligible.)

a. The mechanical power needed to drive the torque load is:

$$\begin{aligned} P_m &= 2\pi n T_m \\ &= 2 \times 3.14 \times \frac{1000}{60} \times 350 \\ &= 36633(W) \approx 37(kW) \end{aligned}$$

b. The flow rate required at the pump-discharge port is:

$$\begin{aligned} Q_p &= \frac{P_h}{p} \\ &= \frac{37 \times 10^3}{15 \times 10^6} \\ &= 2.47 \times 10^{-3}\left(m^3 \cdot s^{-1}\right) = 148\left(L \cdot \min^{-1}\right) \end{aligned}$$

DISCUSSION 4.5: The pump flow calculated here is the theoretical flow needed to drive the load. Other than that one has to take both the pump and motor volumetric and mechanical efficiencies into consideration as discussed in the previous chapters or sections, it is also of interest to point out that when the pump attempts to deliver this quantity of hydraulic fluid to the fixed-displacement motor, the load inertia will always keep the motor from being accelerated instantaneously to full speed. Such delay in speed response is an important characteristic of hydraulic system control.

### 4.4.4 Applications of Hydrostatic Transmission

HSTs have been widely used in various mobile power transmission systems, mainly because of the improved maneuverability over their mechanical counterparts. Many HSTs permit fast starting or stopping and can output speed and torque in either direction over the full speed and torque-variation ranges. However, this HST advantage is achieved at the cost of lower power transmission efficiency in general. Compared to a 92% or higher energy efficiency in a typical mechanical power transmission, it would be considered a highly efficient system if an HST could deliver 85% of the input energy to the load.

The flexibility in design, compactness in size, and high power-to-weight ratio are the other major advantages that make HSTs especially suitable for use in mobile machinery, which often requires delivery of engine output power to actuators at various locations and normally has limited space in between the power source and the power consumers for installing power transmission devices. Figure 4.35 depicts four common configurations of HSTs. The first three configurations (Figure 4.35 (a), (b) and (c)) are suitable for applications with limited installation space, which can often be found on many light- or medium-duty mobile machinery. The design illustrated in Figure 4.35(d) is more commonly used on heavy-duty machinery. The split design also allows the use of one pump to drive multiple motors.

The **inline configuration** can often be found in many light- to medium-duty mobile machines. As illustrated in Figure 4.36, this design allows replacing the mechanical transmission using a HST without changing other components. When a VP-FM HST is used in a wheel-type tractor, the traveling speed of the tractor can be controlled simply by varying the displacement of the pump. Because pump displacement can be infinitively changed in either direction, it not only allows accelerating or decelerating the tractor without interruption of power over the entire speed range, but also permits reversing travel direction smoothly without applying the brake. In addition, the HST can also provide an overload protection by stalling the HST under excessive load and opening the line relief valve in the HST to avoid damage to the tractor.

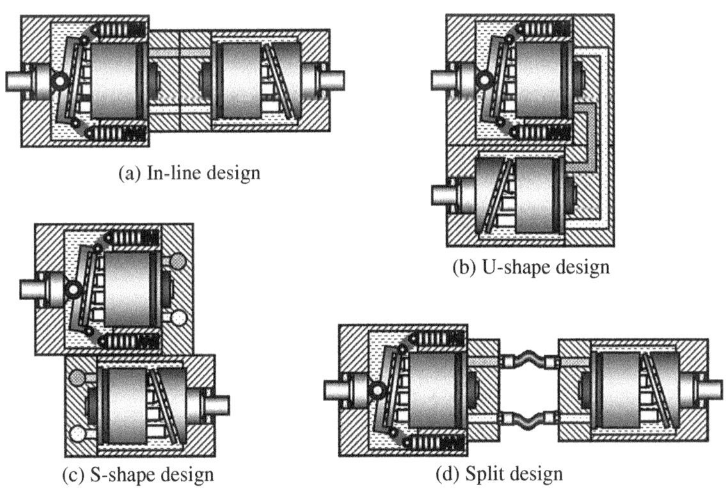


**FIGURE 4.35**
Configuration illustration of four common designs of HSTs. (a) Inline, (b) U-shape, (c) S-shape, and (d) split design.

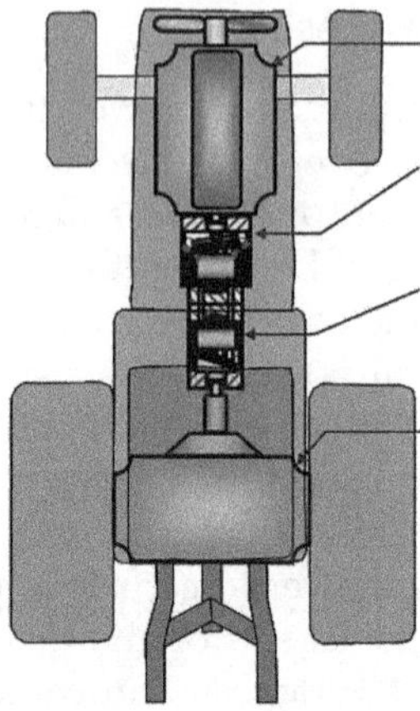


**FIGURE 4.36**
Conceptual illustration of in-line design of HST arrangement in a wheel-type tractor.

The inline design still needs to use a final drive and differential drive train to change the motion. An alternative split design solution has been applied on many off-road vehicles by using one pump to directly drive two wheel-mounted motors to eliminate the differential drive train, as illustrated in Figure 4.37. Due to the low inertia nature of hydraulic power transmissions, a **split configuration** also allows coupling the pump directly to the engine without adding much starting torque to the engine. When both features are put together, the split configuration can considerably simplify the design of a tractor power train and result in a meaningful weight reduction and cost savings.

It is important to remember that when two motors are driven by one pump, each motor only receives one-half of the pump flow and therefore will reduce motor speed by 50% at its output shaft. However, it does not necessarily reduce the vehicle travel speed because of the elimination of the final drive, which often carries a speed reduction function to provide higher driving torques to the wheel. To achieve a high driving torque without using a final drive, the direct-mounting design often requests low-speed high-torque radial motors. There are two types of wheel mountings: shaft mounting and housing mount, the latter of which is commonly used on off-road vehicles. While the housing of the shaft-mounting

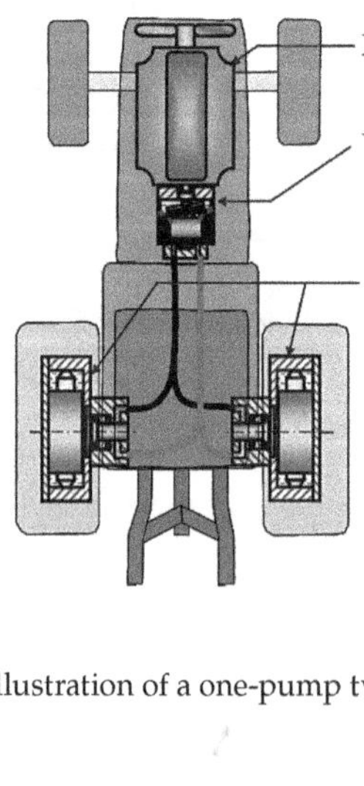


**FIGURE 4.37**
Conceptual illustration of a one-pump two-motor split HST design.

type is normally bolted directly to the vehicle frame with the wheel mounted on the shaft, the housing-mount type often has the wheel bolted on a motor housing with its shaft fixed on a vehicle frame.

It is also common to see final drives being used in some shaft-mounted two-motor split HSTs. Such a design is often selected when the mobile machinery requires two ranges of operating speed: (1) field speed, meeting the high-torque requirement in field operations, and (2) road speed, meeting the high-speed requirement for road transport. Such a design may require the pump to be mounted on a speed-changeable gearbox, allowing the pump to operate either at low speed to supply less high-pressure flow during heavy-load field operations or at high speed to provide more low-pressure flow for road transport.

In some mobile machinery applications, the speed of two motors is controlled solely by the pump-supplying flow, either by load compensation using a load-sensing mechanism or by speed regulation of the prime mover. During vehicle turning, the inner wheels are normally subject to more traction resistance than the outer ones. Such a resistance increase will induce higher operating pressure at the corresponding motor, which will reduce the pressure drop from the pump to the motor and result in less flow being supplied to the inner motor. In turn, it will cause the inner wheel to rotate more slowly than the outer wheel, the same as a differential drive function. To drive mobile machinery effectively on unprepared natural terrain, many times **all-wheel drive** is required. The one-pump two-motor split design can be easily reconfigured into a one-pump four-motor split design to realize all-wheel drive. The front-wheel steered all-wheel drive vehicles require that the front wheels turn at a slightly higher (1 to 2%) speed than the rear ones to enhance the steering performance and improve the tract effort.

In many other mobile machinery applications, the engine is often set to operate at a constant speed, regardless of traveling speed, in order to support other operations designed in parallel to power transmission. Such a problem can be easily solved by adopting the valve-controlled open-circuit HST design on those vehicles. As illustrated in Figure 4.38, by simply adding a proportional control valve to the split open-circuit HST, control flexibility can easily be achieved on either driving the vehicle at an infinitely changeable speed at both forward or reverse motion or stopping the vehicle without affecting the operating status of the engine. Such a design has been commonly used on mobile machinery, with multi-actuator systems supported by one-pump hydraulic systems.

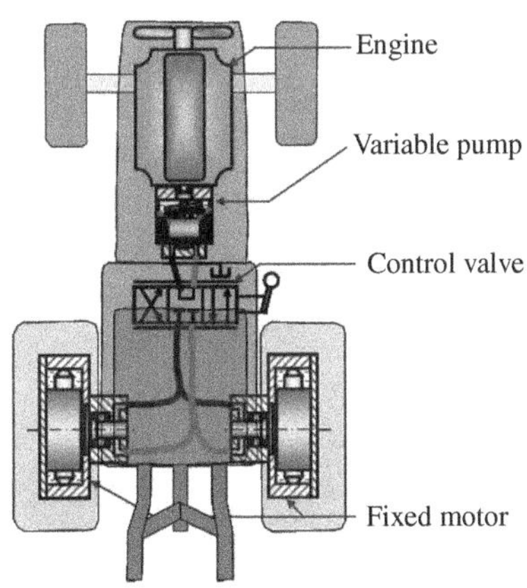


**FIGURE 4.38**
Conceptual illustration of a valve controlled one-pump two-motor split HST design.

A common operation scenario for mobile machinery is that the transportation of the machinery itself normally requires only a small percentage of the power. However, high-powered machinery occasionally needs to move a heavy load. Therefore, it is required that such machinery be capable of delivering sufficient power for all possible applications in the design of the power transmission system. The challenge is that if we design the transmission according to the maximum load, heavy-duty transmission has to be used. Because in most operations only a small portion of its capacity will be used, a very low efficiency of operation will result. If we can design the transmission in terms of the norm load in regular operations, it will allow use of a much smaller transmission to achieve power efficiency. But such a system will be incapable of delivering sufficient power to move the heavy load. One practical engineering solution to this problem is to use a hydromechanical **split-torque power transmission.**

Although there are many different designs in split-torque power transmission, Figure 4.39 illustrates a typical split-torque transmission constructed using traditional planetary gearing power transmission and split hydrostatic power transmission. In principle, this split-torque transmission receives the input power from one shaft and outputs the power on another shaft. However, it splits the total amount of power being delivered from the input to the output through two parallel branches in between. As illustrated in the figure, when the power is delivered to the input shaft of the transmission in a form of input torque and rotating speed, two driving gears installed on the shaft will instantaneously drive both the planet gear of the planetary transmission and the variable pump of the hydrostatic transmission, which split the input torque to both mechanical and hydrostatic paths. The delivered torques will then be merged again at the planetary gears since the hydraulic motor is installed on the planet carrier.

Being physically connected either to the input shaft or the planet carrier, the variable-displacement pump is turning with the input shaft all the time during operation, and the rotating status of the planet carrier is controlled by the rotating speed of the fixed-displacement motor. The relationship between the input and output rotating speed can be determined in terms of both the planet carrier rotating speed and the ratio of the radiuses of planet gears to the sun gear, defined as follows:

$$n_o = \frac{n_i \pm 2(1+a)n_c}{1+2a} \tag{4.67}$$

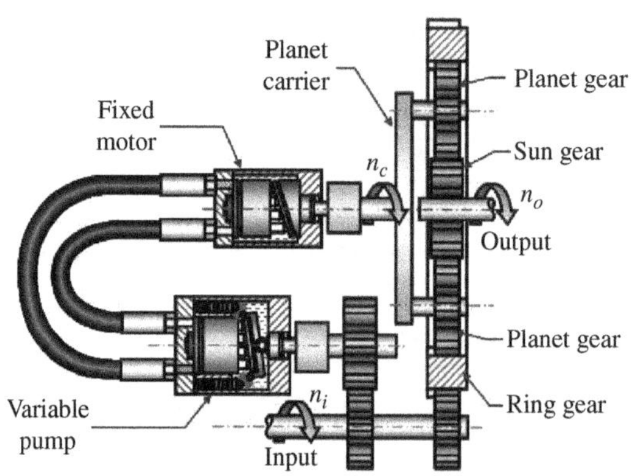


**FIGURE 4.39**
Illustration of the principle of a typical split-torque transmission design.

where $n_i$ and $n_o$ are the input and output rotating speeds at corresponding shafts; $n_c$ is the rotating speed of the planet carrier, and $a$ is the ratio of the radiuses of planet gears to the sun gear.

Under a light load, the displacement of the variable pump will be turned to the minimum. This results in a zero flow being discharged from the pump. The motor is, therefore, kept in standing and delivers zero torque to the planet carrier. The sun gear will then turn at a **base speed** of the split-torque transmission: $n_i/(1+2a)$.

As the system load is increased to surpass a certain threshold level, it will engage the secondary path of torque transmission by increasing the displacement of the variable pump to discharge more flow to drive the motor. Depending on the direction in which the motor turns, the sun gear may turn at either a higher or a lower speed than the base speed, as needed.

A split-torque transmission normally uses the mechanical branch as the primary path and the hydrostatic branch as the secondary path for torque delivery to achieve a high-efficiency power transmission. The size of the hydraulic branch is always smaller in comparison to its mechanical counterpart. The initial cost of a hydromechanical split-torque power transmission is always higher than that of either a hydrostatic transmission or a straight mechanical transmission. It is recommended that a split-torque transmission be selected when the operating economy or the size of the transmission is a primary concern.

## References


1. Akers, A., Gassman, M., Smith, R. *Hydraulic Power System Analysis*. CRC Press, Boca Raton, FL (2006).
2. Anderson, W.R. *Controlling Electrohydraulic Systems*, Marcel Dekker, New York (1988).
3. Dasgupta, K. Analysis of a hydrostatic transmission system using low speed high torque motor. *Mechanism and Machine Theory*, 35: 1481–1499 (2000).
4. Eaton Corp., *Heavy Duty Hydrostatic Transmission Application*. (*Revised Ed.*). Eaton Corporation Hydraulic Division, Eden Prairie, MN (1992).
5. Esposito, A. *Fluid Power with Applications* (*6th Ed.*). Prentice-Hall, Upper Saddle River, NJ (2003).
6. Goering, C.E., Stone, M.L., Smith, D.W., Turnquist, P.K. *Off-road Vehicle Engineering Principles*. ASAE, St. Joseph, MI (2003).
7. Guan, Z. *Hydraulic Power Transmission Systems* (in Chinese). Mechanical Industry Press, Beijing, China (1997).
8. Hansen, D.L., Krutz, G.W. Hydraulic motor speed, torque, and rotational displacement sensing. *Proc. Nat. Conf. on Fluid Power*, 38: 237–246, Chicago (1984).
9. Hedges, C.S. *Industrial Fluid Power* (*3rd Ed.*). Womack Educational Publications, Dallas, TX (1988).
10. Hydraulics & Pneumatics. *Fluid Power Basics*. http://www.hydraulicspneumatics.com/200/FPE/IndexPage.aspx. Accessed on November 20 (2006).
11. Keller, G.R. *Hydraulic System Analysis*. Penton Media Inc., Cleveland, OH (1985).
12. Kugi, A., Schlacher, K., Aitzetmuller, H., Hirmann, G. Modeling and simulation of a hydrostatic transmission with variable-displacement pump. *Mathematics and Computers in Simulation*, 53: 409–414 (2000).
13. Kwaśniewski, J., Piotrowska, A., Raczka, W., Sibielak, M. The mathematical model of a hydrostatic transmission for controller design. In: *Proc. IASTED Int. Conf. on Modelling and Simulation*, pp. 275–280, Palm Springs, CA (2003).

14. Lambeck, R.P. *Hydraulic Pumps and Motors: Selection and Application for Hydraulic Power Control Systems*, Marcel Dekker, New York (1983).
15. Li, Z., Ge, Y., Chen, Y. *Hydraulic Components and Systems* (in Chinese). Mechanical Industry Press, Beijing, China (2000).
16. McClay, D., Martin, H.R. *The Control of Fluid Power.* John Wiley & Sons, New York (1973).
17. Manring, N.D. *Hydraulic Control Systems.* John Wiley & Sons, New York (2005).
18. Merrit, H.E. *Hydraulic Control Systems.* John Wiley & Sons, New York (1967).
19. Murin, J. A controlled diesel drive line with a hydrostatic transmission: Part 1—mathematical model. *Int. J. Vehicle Design*, 38: 109–122 (2005).
20. Murin, J. A controlled diesel drive line with a hydrostatic transmission: Part 2—dynamic properties at periodic loading. *Int. J. Vehicle Design*, 38: 123-138 (2005).
21. Pease, D.A. *Basic Fluid Power.* Prentice-Hall, Englewood Cliffs, NJ (1967).
22. Stringer, J. *Hydraulic Systems Analysis: An Introduction.* John Wiley & Sons, New York (1976).
23. Thoma, J.A. *Hydrostatic Power Transmission.* Trade and Technical Press, Morden, Surrey, UK (1964).
24. Vickers, Inc. *Vickers Mobile Hydraulics Manual (2nd Ed.),* Vickers, Inc., Rochester Hills, MI, (1998).
25. Watton, J. *Fluid Power Systems, Modeling, Simulation, Analog and Microcomputer Control.* Prentice-Hall, New York (1989).
26. Zhang, Q., Goering, C.E. Fluid power system, In: Bishop, R. (ed.), *The Mechatronics Handbook.* CRC Press, Boca Raton, FL, pp. 10–11 ~ 10–14 (2001).

## Exercises

**4.1** What is the primary function of a cylinder in a hydraulic system?

**4.2** How can hydraulic cylinders be classified, and what are they?

**4.3** What is the primary function of a motor in a hydraulic system?

**4.4** How can hydraulic motors be classified, and what are they?

**4.5** How can a differential extension functions be accomplished on a single-rod double-action hydraulic cylinder?

**4.6** When a single-rod double-action hydraulic cylinder satisfies $d = D/\sqrt{2}$, prove that this cylinder can achieve an equal extension and retraction speed during differential extension operations.

**4.7** Name three types of commonly used single-acting cylinders, and use layperson's language to describe their configuration features.

**4.8** In a typical telescopic cylinder extension operation, which tubing will be extended first, and why?

**4.9** Explain how the needle valve control in the open-circuit FP-FM HST depicted in Figure 4.25(e) adjusts the output speed.

**4.10** Figure 4.11(b) illustrates two cylinders being connected together to form a serial cylinder actuating system. Assume that the bore and rod diameters of the left cylinder are 120 mm and 60 mm, respectively, and that those of the right cylinder are 100 mm and 50 mm. If the mass of the external loads to be driven by the left and the right cylinder are 750 and 500 kg, respectively, and the back pressure in the rod-end chamber is 200 kPa, figure out the operating pressure of each cylinder for driving those loads. (Assume that the left cylinder requires 550 kPa for no-load extension and the right one requires 500 kPa.) If the supplying flow is 150 $L \cdot min^{-1}$, what are the extending velocities of those cylinders?

**4.11** A single-rod double-acting cylinder is intended to be used for implementing a differential extension operation. If the bore and rod diameters of the cylinder is 120 mm and 90 mm, respectively, with a flow supply of 40 $L \cdot min^{-1}$, what will be the extending speed and pushing force of the cylinder in a differential extension if the system operating pressure is 8.0 MPa? What will be the extending speed and pushing force of the cylinder in a normal extension under the system operating pressure? (Assume that a back pressure 1.0 MPa).

**4.12** A ram cylinder as depicted in Figure 4.5 uses a ram as the one-direction actuating element. If the diameters of the cylinder bore and the ram are 100 mm and 80 mm, respectively, what will be the actuating velocity when the cylinder supplies a 25 $L \cdot min^{-1}$ flow? When the pressure of the supplying flow is 10 MPa, how much actuating force from this cylinder can be used to drive an external load if the cylinder mechanical efficiency is 0.95?

**4.13** As depicted in Figure 4.8, the diameters of the larger and smaller pistons in a pressure intensifier are 120 mm and 90 mm, respectively. When a 25 $L \cdot min^{-1}$ flow of 20 MPa is supplied to the larger piston chamber, what are the rate and pressure of the output flow from the smaller piston chamber?

**4.14** Assume the hydraulic cylinder shown in Figure 4.9 is vertically installed to lift or lower the load using the extendable rod. If the cylinder has a 75 mm diameter bore with a 40 mm diameter cushion plunger of 20 mm long, and the steady-state velocity for lowering a 100 kg mass is 0.25 $m \cdot s^{-1}$, what will be the fluid pressure in the cushion chamber? (Assume that the back pressure and cylinder friction are negligible in this problem.)

**4.15** A hydraulic motor receives 80 $L \cdot min^{-1}$ flow at a pressure of 20 MPa. If it drives the motor operating at 800 rpm, estimate the maximum torque that the motor can generate for driving a load. (Assume that the motor has 100% efficiency.)

**4.16** A hydraulic motor receives 85 $L \cdot min^{-1}$ flow at a pressure of 21 MPa to drive the motor operating at a constant speed of 850 rpm. If the motor has a power loss of 3 kW, estimate the actual torque output from the motor and the overall efficiency of the pump.

**4.17** Assume that a 100 mm width single-vane oscillating motor, as illustrated in Figure 4.21(a), has a vane of 160 mm radius and a rotor of 40 mm radius. What will be the operating pressure on the motor when it is used to drive a 3600 $N \cdot m$ rotating load? (Assume it requires an operating pressure of 1.0 MPa to drive the motor with no external load.)

**4.18** A radial-piston hydraulic motor has a 50 cc volumetric displacement. If the motor is driven by a 0.001 $m^3 \cdot s^{-1}$ supplying flow of 35 MPa, what will be the theoretical speed, torque, and power output from the motor. If the motor is actually operating at a speed of 1000 rpm to drive a 260 $N \cdot m$ torque, what are the volumetric, mechanical, and overall efficiency of the motor? What is the actual power the motor can generate?

**4.19** One design requires using a FP-FM HST to provide a 500 $N \cdot m$ torque capability to drive a load at 800 rpm. If both the pump and the motor have the same mechanical efficiency of 95% and volumetric efficiency of 85%, what is the power required for driving this load?

**4.20** In designing a new model of self-propelled agricultural machinery, the design team needs to select an HST for its power transmission. Based on the predefined design specifications, this machine will be powered by a diesel engine of 40 kW at a rated speed of 2000 rpm. The rolling radius of the drive wheels to be used is 0.50 m. This type of machinery is designed mainly for infield operation, which requires a maximum driving torque of 1200 $N \cdot m$ to overcome the traction force for driving the machinery traveling in the field at a maximum speed of 2.5 $m \cdot s^{-1}$. Try to select a proper HST for this application if the rate of pressure drop across the motor is 20 MPa. (Assume that the wheels are directly driven by the motor and that the typical HST pump and motor both have a volumetric efficiency of 0.90 and a mechanical efficiency of 0.86.)

# 5

# *Hydraulic Power Regulation*

## 5.1 Overview of Power Regulation

### 5.1.1 Regulating Hydraulic Power

In addition to the principal functions of power generation, distribution, and deployment, a hydraulic system also needs to perform a few supporting functions to ensure that the system is working safely, smoothly, accurately, efficiently, and economically under all conditions. Some examples of those supporting functions are **power absorption**, **power storage**, and **power regeneration**.

Hydraulic devices are always driven by pressurized fluids, either dynamically or statically. When either an external or internal force hits the pressure-bearing surface (often the piston surface) in a hydraulic device during operation, the fluids will always reactively form a resistance to change the motion status of the piston. The impact of such motion changes will always induce some significant pressure spikes due to the flow momentum and fluid compressibility. Those impact-induced pressure spikes will be propagated to the rest of a hydraulic system, resulting in instability in many operations and therefore should be removed if possible. As illustrated in Figure 5.1, a pressure spike is formed as a piston is pushed by an external moving load. This pressure spike will be propagated to the rest of the hydraulic system, often in a wave, and will decrease as the piston is pushed away from its original position.

The physics behind the impact-induced pressure spikes are fluid compressibility and piston deceleration. In previous discussions of hydraulic fluid properties, it has always been assumed that fluids are incompressible. However, this assumption is only valid for ideal fluids. In reality, fluids used in hydraulic systems are non-ideal and actually have a limited compressibility. Because of this limitation, when the front layer of the fluid is stopped by a piston, the rest of the fluid still tries to flow in. Such asynchronous motions between adjacent layers will compress the front layers to induce a rapid rise in pressure in those layers and form pressure waves. While the fluid compressibility provides a sufficient condition to develop pressure spikes, an acceleration (or deceleration, which can also be treated as a negative acceleration) on the piston in a hydraulic device is the necessary condition to make such a development possible.

A few methods can eliminate, or at least reduce, the pressure spike to achieve smoother and more accurate motion control of hydraulic systems. Two of the most common ways are the use of some kinds of hydraulic capacitive or resistive elements. Hydraulic capacitive elements, such as **shock absorbers**, temporarily absorb the kinetic energy carried in pressure spikes and then dissipate the absorbed energy over a period of time. Hydraulic resistive elements, such as **hydraulic springs**, convert the kinetic energy to a potential energy providing additional resistance to slow down the acceleration.

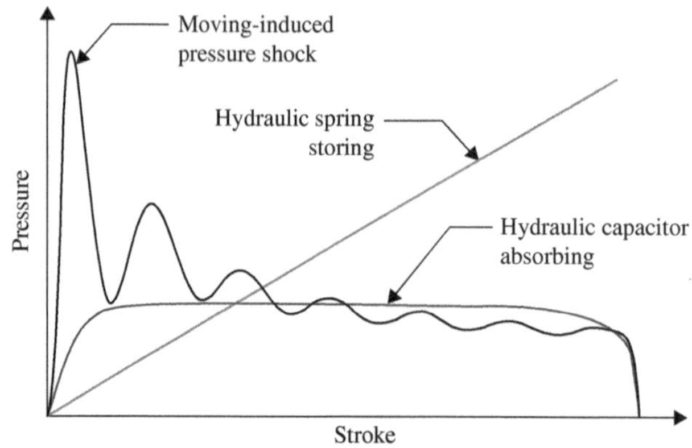


**FIGURE 5.1**
Concept illustration of hydraulic shock formation and methods of hydraulic shock absorbing.

Another common way of handling the extra energy in a hydraulic system is to store it using some specially designed **hydraulic energy storage devices**. The stored energy can either be released after the stroke or kept for later use, depending on the devices being used. To effectively store and retrieve the extra potential energy carried by the high pressure, a special class of energy storage devices, hydraulic accumulators, is a widely used practice.

To meet some special operational requirements, many hydraulic systems, especially mobile ones, occasionally need to work either at an extremely high speed or under an extremely high pressure. In theory, this requires that the hydraulic system be sized capable of supplying the maximum amount flow under the highest possible operating pressure. While such a design can ensure having a functional system all the time, it will result in low efficiency when used at full capacity. An alternative design for such systems is to size the system according to normal operation and to add some types of **power regenerating functions** to supply extra power over a short period of time to meet special needs. As illustrated in Figure 5.2, by integrating proper power regeneration functions, a hydraulic system can be sized according to its regular operating conditions, and also be capable of providing enough flow to realize high-speed operation at a reduced operating pressure level or drive a heavy load with a reduced-flow capacity.

All functions discussed above involve the regulation of energy-absorbing and releasing processes by absorbing or storing hydraulic power in the form of fluid under pressure. We can therefore categorize this set of functions as **hydraulic power regulation** functions.

### 5.1.2 Commonly Used Power-Regulating Devices

To furnish a hydraulic system with the above-defined regulating functions effectively and dependably, a group of specially designed hydraulic power regulation components, including but not limited to hydraulic shock absorbers, liquid springs, hydraulic accumulators, pressure intensifiers, and two-speed cylinders, has been used in hydraulic systems.

A **hydraulic shock absorber** is a hydraulic device that absorbs the kinetic energy carried by the impact-induced pressure waves. A typical shock absorber works in such a way that it first converts the kinetic energy into a heat form by forcing the pressurized flow to flow through an orifice and then dissipates the generated heat from the system. The energy transformation occurs as the fluid is forced through orifices at high velocities.

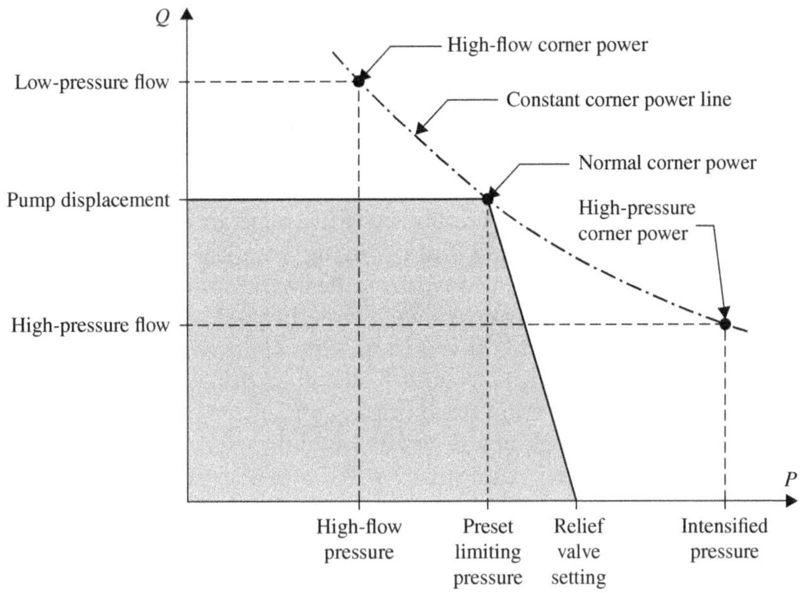


**FIGURE 5.2**
Concept illustration of corner-power point switching for different operating modes when a power regeneration function is furnished to the system.

A **hydraulic fluid spring** can be treated as a special type of shock absorber and is designed to provide a controlled soft stop to the piston to avoid jerky operations caused by sudden stops by applying extra resistance induced during the hydraulic energy absorption. It works in such a way that when an external force is applied to the device, it compresses the contained fluid to absorb and store the energy in the form of higher pressure, which will then develop an extra resistance in proportion to the stroke of the piston to slow down its motion. The amount of stored energy will be gradually released through the carefully designed orifices over time.

**Hydraulic accumulators** are widely installed in many hydraulic systems to smooth out pressure pulsations and to store hydraulic potential energy temporarily. It is very common to utilize the stored energy in hydraulic systems as a supplementary power source to provide extra power that will overcome instantaneous extra load, refill leakage, or even serve as emergency power to achieve higher efficiency and a more reliable operation. Because of these features, it is possible for a hydraulic system with an accumulator to use a smaller pump by using the accumulator capability of storing a certain amount of energy during periods of low demand. In addition, accumulators can also reduce the shocks caused by rapid operation or sudden starting and stopping of actuators in a hydraulic circuit.

A **pressure intensifier** is a special free-piston-type device and is used to increase the fluid pressure supplied to a branch at a level higher than the pump-discharge pressure. It works by using high-volume low-pressure flow to push a smaller portion of flow into a higher pressure, and therefore it is commonly treated as a type of **power regeneration device**. Another commonly used power regeneration device is the **two-speed cylinder**, which uses a higher pump-discharge pressure to recycle the returning flow from the rod-end chamber of the cylinder back to the head-end chamber to gain more flow in order to get a higher actuating speed.

## 5.2 Power-Absorbing Devices

### 5.2.1 Hydraulic Shock Absorbers

As introduced in the previous section, a hydraulic shock absorber is a power-absorbing device commonly used to give a hydraulically driven load a soft stop by absorbing the extra kinetic energy induced by acceleration. As illustrated in Figure 5.3, a typical shock absorber is always filled with hydraulic fluid. During a typical power-absorbing process, the kinetic energy is absorbed by first converting the energy into a thermal form by pushing the hydraulic fluid through a set of orifices. This thermal energy is stored in the contained fluid and then dissipated into the environment over a period of time. One common application of such a design is an automotive shock absorber. The major absorption of energy in the hydraulic device is from the damping effects.

Many types of hydraulic shock absorbers are widely used in different applications. Figure 5.4 depicts the structural configurations of a few examples of shock absorbers commonly used in applications with limited energy absorption. Such variations in the structural design result in some noticeable differences in power-absorbing characteristics. For example, the stiffness of a **simple orifice type** shock absorber (Figure 5.4(a)) is basically determined by piston velocity in compressing the trapped fluid, and that of a **multiple-orifice type** (Figure 5.4(b)) is more controlled by piston position because of a fast reduction in total orifice area as the piston approaches the end of stroke. The stiffness of the two **spear-type** shock absorbers is first determined by the piston velocity before the spears start to regulate the fluid release rate, and they are then controlled by the piston position as the spears form an additional resistance to restrict the fluid release rate from the compressing chamber. The main difference between the **tapered spear type** (Figure 5.4(c)) and **stepped spear type** (Figure 5.4(d)) absorber is that the former presents a linear increase in stiffness and the latter presents a stepped increment. An **annular clearance type** (Figure 5.4(e)) shock absorber is in general harder than an **orivis clearance type** (Figure 5.4(f)) because of the higher resistance to release the fluid from the compressing chamber.

Ideally, the power-absorbing process should keep the output force from the absorber constant throughout the absorbing stroke. However, it is very difficult to attain this feature for a reasonable cost. A more practical method of absorbing shock—using a few square waves instead of a constant absorbance—is often acceptable for many applications. As illustrated

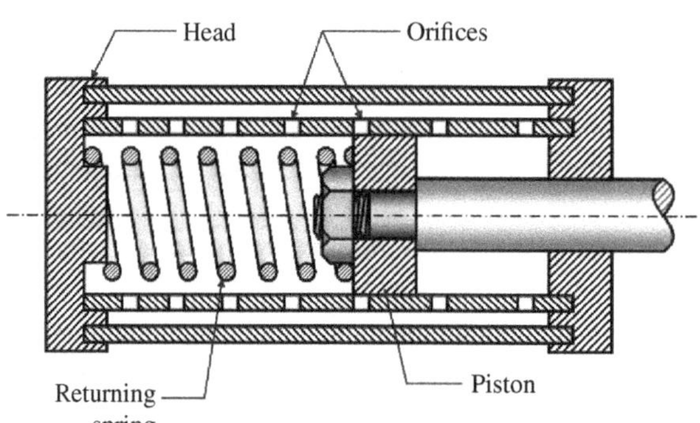


**FIGURE 5.3**
Illustration of the configuration and operation principle of a typical automotive shock absorber.

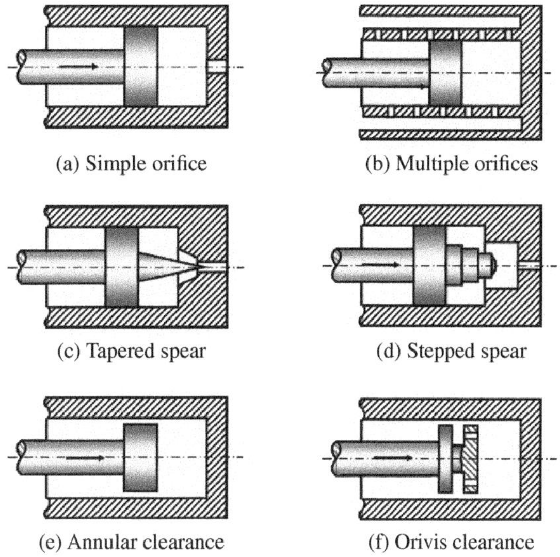


**FIGURE 5.4**
Typical configurations of a few commonly used hydraulic shock absorbers: (a) and (b) orifice types, (c) and (d) spear types, (e) and (d) clearance types.

in Figure 5.5, when a straight spear is used, a shock will always induce a large pressure spike, which is often not an acceptable characteristic.

The simplest alternative is probably the use of a tapered spear as depicted in Figure 5.4(c). Because it has a large flow area at the beginning of the cushion and rapidly reduces as the spear reaches to the end of stroke, such a modification in absorber design could improve the pressure profile by reducing the peak of the pressure spike. It is impractical, in general, to obtain performance close to the ideal absorbing pattern as illustrated in Figure 5.5.

The tapered spear design can be improved by reshaping the spear in a series of stepped spears as depicted in Figure 5.4(d). As shown in Figure 5.5, a stepped spear shock absorber can make the pressure pattern closer to ideal by effectively reducing the maximum pressure level of an extended straight spear in the early stages of cushioning. This modification can greatly improve the shock-absorbing performance without significantly raising cost.

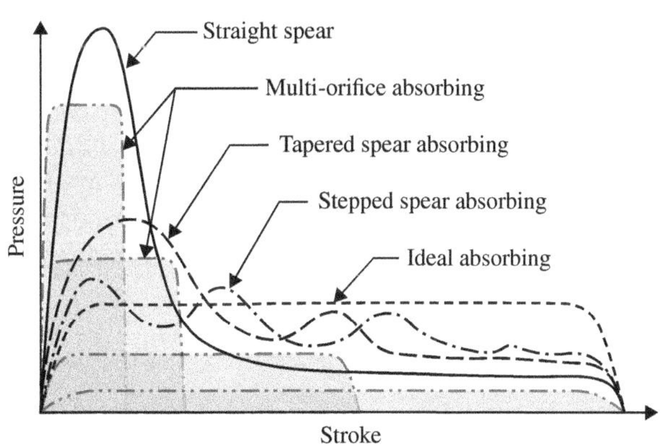


**FIGURE 5.5**
Characteristic illustrations of absorbing compromises using different hydraulic shock absorbers.

A common method for offering a practical compromise to the ideal absorbing is the piecewise constant absorbing by using a multi-orifice type shock absorber (Figure 5.4(b)). As depicted in Figure 5.5, this alternative provides a few close-to-ideal methods of absorption in several stroke ranges. Such a design can offer a series of pseudo-ideal absorbing patterns for different ranges of piston stroking.

Theoretically, the power-absorbing process can be described as an orifice-controlled steady acceleration (or a negative acceleration in case of deceleration) process to maintain a constant piston velocity. Four governing equations commonly used to mathematically represent this process are the basic equations of:

1. Newton's second law of motion

$$F = ma \tag{5.1}$$

2. Burnoulli's equation

$$v_o = \sqrt{\frac{2}{\rho}(p_c - p_e)} \tag{5.2}$$

3. Fluid continuity law

$$A_p v_p = C_d A_o v_o \tag{5.3}$$

4. Energy conservation law

$$E_a = E_k - E_p \tag{5.4}$$

In these four equations, $F$ is the force acting on the piston; $m$ is the total mass of the piston and the driven load; $v_p$ and $a$ are the piston velocity and acceleration; $A_p$ is the piston area; $A_o$ is the total orifice area; $v_o$ is the flow velocity at the orifice; $p_c$ and $p_e$ are the fluid pressure in cushion and back chambers; $C_d$ is the orifice discharge coefficient; $\rho$ is the fluid density; $E_k$, $E_p$, and $E_a$ are kinetic, potential, and absorbed energy, respectively.

These four basic equations provide the needed theoretical basis to design hydraulic shock absorbers. In practice, absorber performance is analyzed by using some **device-specific empirical equations**, formulated based on experimental data. The shock-absorbing time and the piston displacement equations are among the most commonly used absorber performance empirical equations. The following empirical equations are examples of shock-absorbing time and the piston displacement equations applicable to simple orifice type shock absorbers as illustrated in Figure 5.4(a). The shock-absorbing time is normally defined as the time interval in which the piston speed decreases from its initial velocity to 10% of the initial value, and it is expressed as follows:

$$t_a = \frac{m - v_a C_2}{v_a C_1} \tag{5.5}$$

where $t_a$ is the shock-absorbing time and $v_a$ is the final velocity of piston, which is commonly defined as 10% of the initial value.

The two constants of $C_1$ and $C_2$ in Eq. (5.5) are device-specific and impact force dependent constants. The following equations reveal the main attributing factors to those two constants:

$$C_1 = -\frac{m}{v_p^2}\frac{dv_p}{dt} \tag{5.6}$$

$$C_2 = \frac{m}{v_p} \tag{5.7}$$

The distance of the piston stroke while the shock absorbs can also be determined in terms of those two constants using the following equation:

$$x = \frac{m}{C_1}\ln(C_2 + C_1 t_a) - \frac{m}{C_1}\ln(C_2) \tag{5.8}$$

Both the device-specific and impact force dependent constants, $C_1$ and $C_2$, are normally obtained experimentally because the differences in structure may result in huge variations of the values of those constants.

### 5.2.2 Hydraulic Fluid Springs

Hydraulic fluid springs (also called **hydraulic springs**) absorb energy by utilizing the physical phenomena that hydraulic fluids are compressible with very high stiffness in general. Hydraulic springs are designed based on the same principle as that for hydraulic shock absorbers, but with some noticeable differences in their structures, and therefore performances, to meet the requirements for different applications.

Figure 5.6 illustrates the structural features of a widely used **simple orifice type** hydraulic spring. This type of hydraulic spring works simply by compressing the contained fluids to absorb the load-impacting energy to achieve a smoother operation. When an external force is acting on the piston, it pushes the piston leftward to compress the fluid and results in increased pressure in the fluid, which in turn forces a certain amount of fluid to flow through the orifices to the back chamber of the piston. Because the leftward movement of the piston reduces the total effective volume of the contained space formed by two connected chambers in the device in proportion to the piston position, it results in a pressure increase in the contained fluid in reverse proportion to the total effective volume and forms a hydraulic spring to store the potential energy. While the external force is removed from the piston, the hydraulic spring will release stored energy by pushing the piston rightward to its initial position, restoring the force formed by area differences on both sides of the piston.

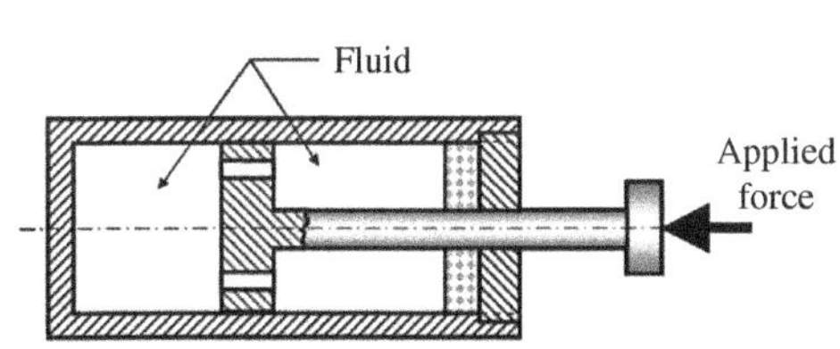


**FIGURE 5.6**
Illustration of the configuration and operation principle of a typical simple orifice type hydraulic spring.

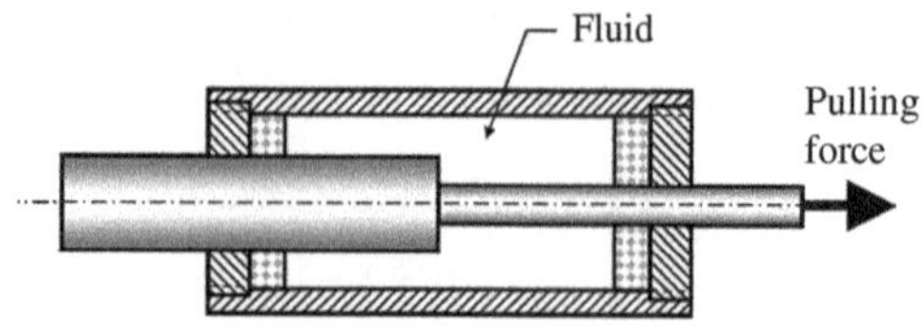


**FIGURE 5.7**
Illustration of the configuration and operation principle of a typical tension type hydraulic spring.

Another commonly used type is the **tension-type** hydraulic spring. Similar in principle, but completing the energy storage process without utilizing orifices, this type of hydraulic spring is only effective when subjecting a tension load as illustrated in Figure 5.7. In operation, as the large-diameter rod moves rightward, it compresses the contained fluid by replacing a portion of effective fluid storage volume using the extra volume of the bigger rod to increase the fluid pressure. After the external tension load is released, the high-pressured fluid will push the bigger rod leftward to regain the fluid-containing volume to reset the fluid spring to its unloaded condition.

It is practically achievable to design a **compound hydraulic spring** by integrating a few simple designs in one device. As illustrated in Figure 5.8, a simple compound type hydraulic spring can be constructed using two sets of simple orifice type hydraulic springs, one embedded in the other. Due to its structural feature, such a compound type hydraulic spring offers a dual spring rate. At the zero load condition, the primary rod is always resting at the rightmost position and the secondary rod at the leftmost position, actuated by the static force generated from the pressure of the contained fluid. When an external pushing load is applied, the primary rod retracts and compresses the fluid in the primary chamber to form the first spring rate to store the kinetic energy. When the fluid pressure in the primary chamber surpasses a threshold value, it will then push the secondary rod to retract, which will increase the stiffness of the fluid spring to form the second spring rate. After the external load is released, the pressure differences in both the primary and secondary chambers will push both rods, returning them to their original positions.

This discussion of the three primary types of hydraulic springs shows that they all follow the same fundamental operating principle of elastic actions: when an external load is applied, the kinetic energy induced by the impact is absorbed and stored in pressurized fluid. The stored energy will be fully released as the external load is removed. A single

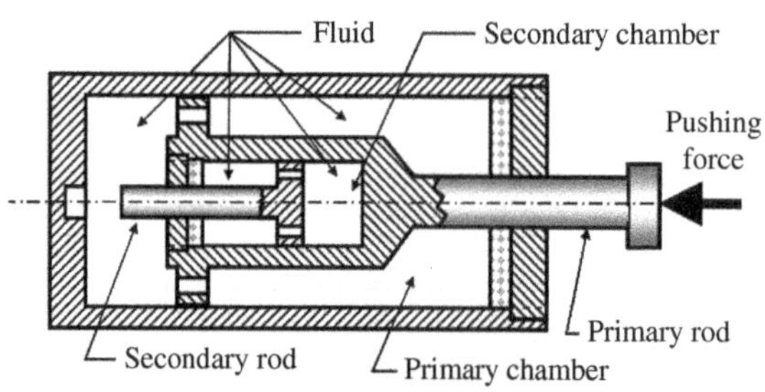


**FIGURE 5.8**
Illustration of the configuration and operation principle of a typical compound type hydraulic spring.

hydraulic spring can provide very high stiffness. By restricting the fluid flow through the orifices in the piston, a hydraulic spring can provide controlled shock-absorbing and damping functions. The major drawback of a fluid spring is its high cost.

**Example 5.1: Hydraulic Shock Absorber Operations**

A simple orifice type shock absorber as illustrated in Figure 5.4(a) takes in power generated by a 1200 kg mass at an initial velocity of 0.6 $m \cdot s^{-1}$. If the initial deceleration of the absorber piston is 3.6 $m \cdot s^{-2}$, how long will it take the absorber to completely take in the power? What will be the piston-stroking distance during the shock absorbing?

a. The shock-absorbing time can be determined by using Eq. (5.5). However, we need to first obtain $C_1$ and $C_2$ according to Eqs. (5.6) and (5.7), as follows:

$$C_1 = -\frac{m}{v_p^2}\frac{dv_p}{dt}$$
$$= -\frac{1200}{0.6^2}\times(-3.6)$$
$$= 12000\left(kg \cdot m^{-1}\right)$$

$$C_2 = \frac{m}{v_p}$$
$$= \frac{1200}{0.6}$$
$$= 2000\left(kg \cdot s \cdot m^{-1}\right)$$

$$t_a = \frac{m - v_a C_2}{v_a C_1}$$
$$= \frac{1200 - 0.06 \times 2000}{0.06 \times 12000}$$
$$= 1.5(s)$$

b. The piston-stroking distance according to Eq. (5.8):

$$x = \frac{m}{C_1}\left[\ln(C_2 + C_1 t_a) - \ln(C_2)\right]$$
$$= \frac{1200}{12000}\times\left[\ln(2000 + 12000 \times 1.5) - \ln(2000)\right]$$
$$= 0.23(m)$$

DISCUSSION 5.1: This exercise reveals the relationships of the shock-absorbing time and stroke to the impact force, initial velocity, and acceleration. One should always remember that both constants ($C_1$ and $C_2$) used in this estimation are strongly device-specific and impact force dependent, which may result in huge variations in the values of those constants using the approach demonstrated in this example. In practice, those constants are normally obtained experimentally.

## 5.3 Power Storage Devices

### 5.3.1 Functions of Hydraulic Accumulators

A commonly used type of power storage device in many hydraulic systems is the **hydraulic accumulator**. Designed to reserve a certain amount of fluid at the system pressure, the main function of this device is to provide or absorb momentary flow, namely, a temporary storage tank of fluid potential energy, for reducing pressure pulsation, ensuring system functionality, lowering heat generation, and improving energy efficiency in a hydraulic system. To effectively perform an energy storage function, an adequately designed hydraulic accumulator should carry a sufficient volume of fluid with very low inertia while releasing and absorbing energy.

As illustrated in Figure 5.9, one of the most typical applications of hydraulic accumulators is to provide an adjustable volume for temporarily bleeding off or returning extra fluid from or to the hydraulic line to maintain a constant system pressure. In such a design, the accumulator can serve as an auxiliary power source, a pulsation absorber, a shock damper, a leakage makeup source, and a thermal expansion compensator.

By utilizing its primary functions, a hydraulic accumulator is commonly used as an **auxiliary power source** to provide supplemental power and pressurized fluid in cases where the main power source can only provide little fluid within a limited time interval. Therefore, with a carefully sized accumulator, it is possible to select the pump in terms of the average flow rate, instead of the maximum rate required for an application, especially when there is a noticeable variation in the flow demanded during its operating cycle. In the example illustrated in Figure 5.10, when an accumulator is used as an auxiliary power source, it is often installed in the supplying line close to the primary power source. Such a location allows the accumulator to absorb and store the fluid energy when there is a surplus and to discharge it when there is a temporary deficiency.

In almost all hydraulic systems, there exists some degree of pressure pulsation, mainly induced by fluctuating discharge flow from the pump, especially when a gear pump is used. If a hydraulic accumulator is used as a **pulsation absorber**, it is often installed near the pump before the main control valve, as illustrated in Figure 5.10, to efficiently absorb fluctuating pump discharge-induced pressure pulsations.

Most impact-induced hydraulic shocks in a system are caused by sudden cylinder (or motor) stops, such as fast valve shifts to stop or to redirect the flow supply, sudden impacts of excessive external loads, and unexpected pump stops. Such sudden stops will create a pressure shock wave that travels back through the system. This shock wave can develop

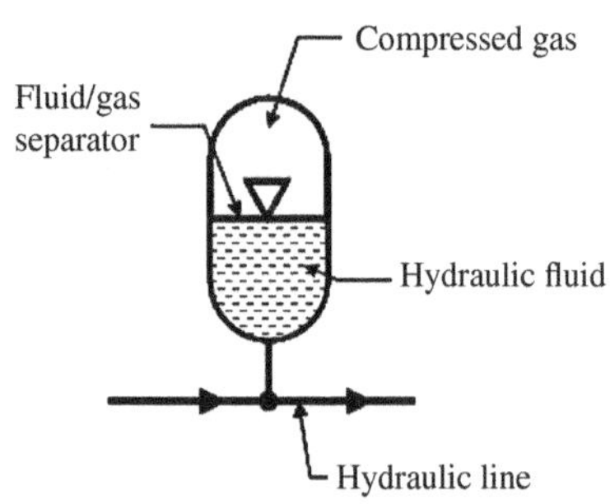


**FIGURE 5.9**
Concept illustration of the operation of a typical gas-loaded type hydraulic accumulator.

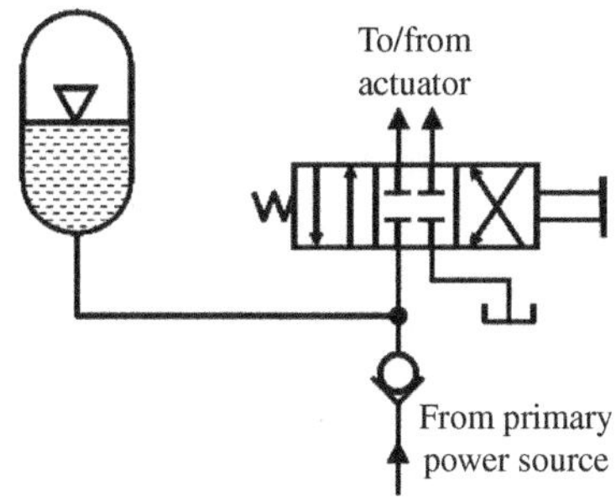


**FIGURE 5.10**
Concept illustration of the operation of an accumulator used as an auxiliary power source or pulsation absorber.

peak pressures several times greater than the normal working pressure, and line-relief valves often cannot respond quickly enough to release those peak pressures, causing noticeable noise or even system failure. An accumulator, often installed close to the actuator, as illustrated in Figure 5.11, can function as a **shock damper** to effectively diminish such hydraulic shocks. An example of this application is the absorption of shock caused by suddenly stopping the bucket on a wheel-type loader. Without using a hydraulic accumulator, a large inertia force induced by the shock wave will cause the bucket to continue to move, which could completely lift the rear wheels of the loader off the ground and transfer the severe shock to the tractor frame and axle. By adding an adequate accumulator to the hydraulic system, the amount of shock acting on the machinery can be effectively reduced.

Another common application of accumulators is to serve as a **leakage makeup source** for pressure maintenance for cases when the primary power supply is discontinued. An example of such an application scenario is when there is a need to hold a load for a long period of time without a power supply from the pump. However, it is unavoidable for any hydraulic system to have some small leakage, either internal or external. As illustrated in Figure 5.11, the accumulator would act as a leakage makeup source to provide a small amount of pressure fluid to maintain the system pressure needed to hold the load for a reasonably long period of time.

Because pressure changes often occur in hydraulic systems when the fluid is subjected to rising or falling temperatures, an accumulator can be used as a **thermal expansion compensator** to receive or deliver a small amount of hydraulic fluid to compensate for the temperature-induced pressure changes in a hydraulic system.

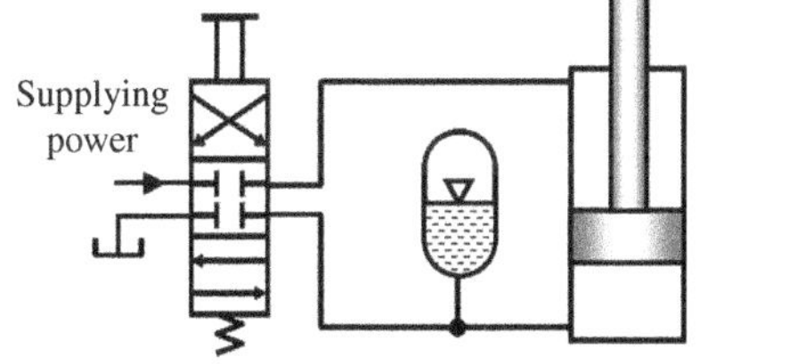


**FIGURE 5.11**
Concept illustration of the operation of an accumulator used as a shock damper and cylinder holder.

### 5.3.2 Operation Principles of Hydraulic Accumulators

Hydraulic accumulators can also be categorized as weight-loaded, spring-loaded, and gas-loaded. They respectively convert the hydraulic energy into the potential energy of a heavy weight, in a loaded spring, or in the compressed gas during the absorbing process, and they convert those forms of energy back to hydraulic energy during discharge.

The **weight-loaded hydraulic accumulator** is the most basic type. Illustrated in Figure 5.12, a typical weight-loaded accumulator has a weight loaded on the top of its rod. The operating pressure, which is equal to the system pressure required to force the fluid to enter the accumulator, is determined by the area of the piston and the total weight loaded on it.

This type of accumulator can hold a constant operating pressure regardless of the amount of fluid being stored. This **fluid volume insensitive** feature makes the operating pressure insensitive to leakage or temperature variations, and no other type of accumulators will have such constant pressure behavior. However, this type of accumulators normally requires a large space to hold the weight, and it often has a slow response when absorbing and discharging fluid energy due to the large inertia because of the weight. In addition, the weight-holding structure requires a vertical installation, making it unsuitable for mobile applications with rolling motion.

Another common type of hydraulic accumulator is the **spring-loaded accumulator** (Figure 5.13). A typical spring-loaded accumulator works in a similar manner to its weight-loaded counterpart but uses a mechanical spring to replace the weight. This modification makes for a very simple design that can be installed in any orientation. In addition, the use of a spring allows for a faster response than its weight-loaded counterparts and offers a better dynamic performance to hydraulic systems. However, the physical law of a spring force acting on the piston in proportion to the spring-compressed length makes this type of accumulator lose the capability of operating at a constant pressure in exchange for the above-mentioned improvements. To design the device compactly, the length of the spring being used is limited, which means a relatively stiff spring has to be

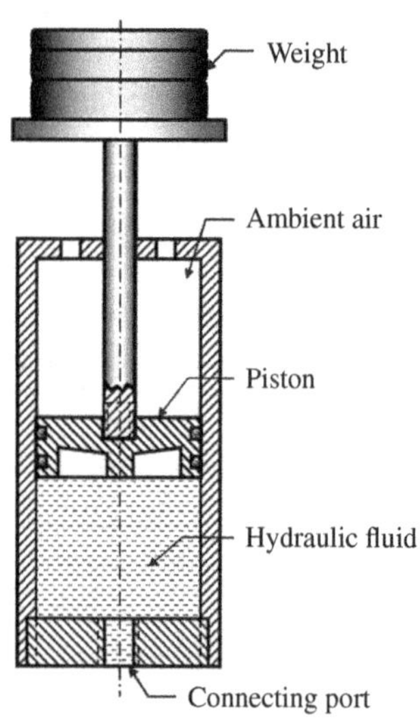


**FIGURE 5.12**
Illustration of the principle of a typical weight-loaded hydraulic accumulator.

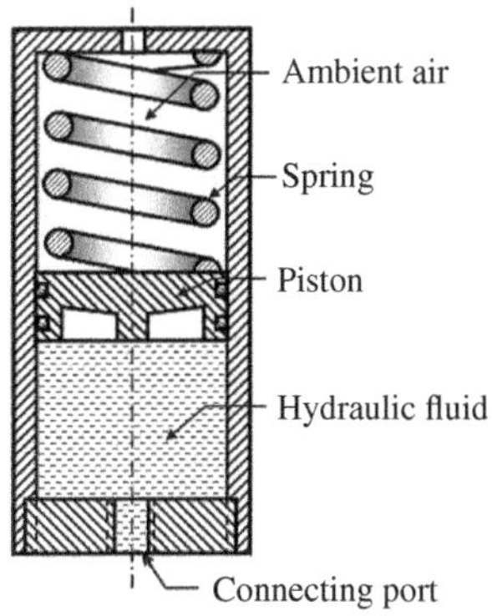


**FIGURE 5.13**
Illustration of the principle of a typical spring-loaded hydraulic accumulator.

used. As a result, a substantial pressure variation can be expected when the accumulator is almost empty and when it is full.

The most commonly used accumulators are **gas-loaded accumulators**, which varies the fluid storage volume by keeping the compressible gases in one isolated chamber and the incompressible liquid in another contained chamber. As shown in Figure 5.14, there are three styles of gas-loaded accumulators: the **piston**, **diaphragm**, and **bladed style**, commonly used in modern hydraulic systems. Regardless of their structural differences, all gas-loaded accumulators work under the same principle. Theoretically, the volume of gases under different pressures and temperatures can be accurately calculated in terms of the **ideal gas law** as described in the following equation:

$$\frac{p_1 V_1}{T_1} = \frac{p_2 V_2}{T_2} \tag{5.9}$$

where $p_1$ and $p_2$ are the gas pressures, $V_1$ and $V_2$ are the gas volumes, and $T_1$ and $T_2$ are the gas temperatures in scenarios 1 and 2, respectively.

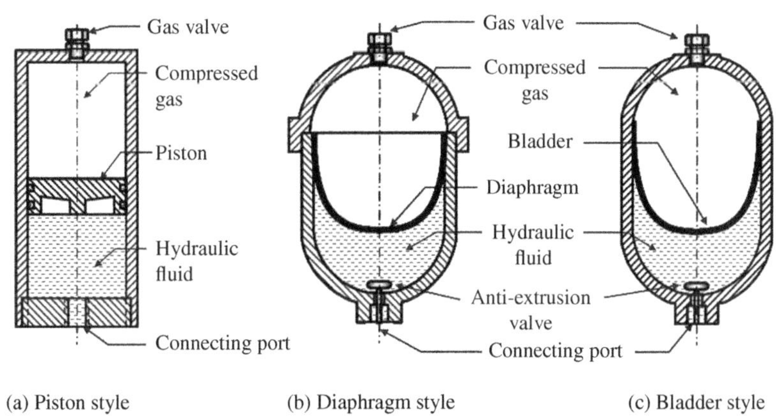


**FIGURE 5.14**
Illustration of the principle of three common types of gas-loaded hydraulic accumulators: (a) piston, (b) diaphragm, and (c) bladder styles.

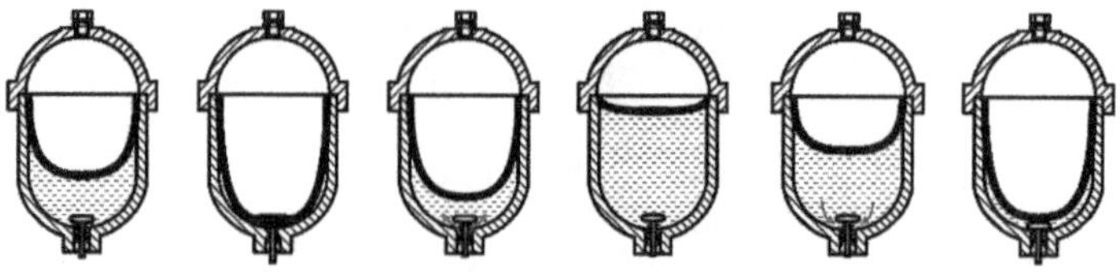


**FIGURE 5.15**
Illustration of the principle of typical gas-loaded (diaphragm style) hydraulic accumulator, consisting of (a) empty, (b) precharged, (c) charging, (d) filled, (e) discharging, and (f) drained stages.

Because the hydraulic fluid is almost incompressible, by knowing the volume change of the compressible gases in an accumulator under different pressures, the amount of hydraulic fluids that can be stored in the accumulator is also known.

Regardless of its configuration type and loading methods, a hydraulic accumulator always operates in one of six typical stages. Without loss of generality, Figure 5.15 graphically shows the **six-stage operating principles** of a typical diaphragm-style, gas-loaded accumulator. As illustrated in the figure, stage (a) is the **empty stage**, in which no gas has been charged and the diaphragm is under a natural state. Stage (b) is the **precharged stage**, in which the accumulator has been fully precharged with gas. In stage (c), the hydraulic fluid is pushed into the accumulator by a high system pressure and is often called the **charging stage**. Stage (d) is called the **filled stage**, at which the accumulator holds the maximum amount of fluid under a high system pressure with the anti-extrusion valve remaining open. In stage (e), the stored fluid is discharged from the accumulator and forced back into the system as the system pressure drops and is often named the **discharging stage**. At stage (f), the operational volume of stored fluid is completely discharged from the accumulator into the system and can be referred as the **drained stage**.

Normally, the empty stage (Figure 5.15(a)) is used to indicate the natural state of an accumulator before being precharged and is a nonoperating stage. Three of the defined stages—the precharged, filled, and drained stages—are treated as three **characteristic states** of a typical hydraulic accumulator. In the precharged state (Figure 5.15(b)), the gas pressure is precharged to $p_0$ with a **precharged volume** (also called the **total volume**) $V_0$. The precharged volume is actually the maximum volume of the gas chamber in an accumulator. At the filled status (Figure 5.15(d)), the gas pressure will reach its maximum value $p_2$ with a smallest total gas volume $V_2$. In this state, the accumulator carries the maximum volume of pressurized hydraulic fluid at pressure $p_2$. After an accumulator has discharged all its carried fluid, it reaches another characteristic state of a drained stage (Figure 5.15(f)) in which all the releasable fluid is discharged. The gas pressure is dropped to $p_1$, and the gas volume is increased to $V_1$. The effective supply volume of the fluid from an accumulator, often defined as the **servicing volume** (also called the **operating volume**), can be determined in terms of the maximum and minimum volumes of the gas chamber using the following equation:

$$V_S = V_1 - V_2 \tag{5.10}$$

where $V_S$ is the operating volume of hydraulic fluid carried in an accumulator.

In addition to the three characteristic states, there are two **transient stages** of fluid charging and discharging. Illustrated in Figure 5.15(c), the charging process begins when the system pressure exceeds the accumulator precharge pressure to intake hydraulic fluid.

The absorbed fluid compresses the gas chamber to store the extra fluid during the charging process. Similarly, a discharging process starts when the system pressure is below the gas pressure, which pushes the stored fluid back to the system (Figure 5.15(c)). The gas-stored energy is released in the discharge operation.

In practice, accumulator manufacturers always specify recommended precharge pressures for their products. A typical precharge process involves filling the gas chamber with a dry gas, such as nitrogen, while no hydraulic fluid is in the fluid chamber. Generally, piston-style accumulators are often precharged to 700 kPa below the minimum system pressure, whereas diaphragm- and bladder-style accumulators are typically precharged to 80% of the minimum system pressure. The precharge pressure determines how much fluid will remain in the accumulator at minimum system pressure. A correct precharge pressure is also one of the most important factors in prolonging accumulator life.

Two other critical parameters of a hydraulic accumulator are **flow rates** and **response times**. Normally, piston-style accumulators have a much higher allowable maximum flow rate than diaphragm and bladder styles. The suggested maximum flow rate is limited to 3000 L/min for standard piston designs (10 L or larger), and to 850 L/min for standard designs of diaphragm and bladder accumulators. For medium (around 4 L) and small (1 L or smaller) accumulators, the suggested maximum flow rates are reduced to 1600 and 100 L/min for piston accumulators, or 570 and 320 L/min for diaphragm and bladder accumulators, respectively. As illustrated in Figure 5.14, an anti-extrusion poppet valve is commonly used to control the flow rate in diaphragm and bladder accumulators; an excess flow will cause the poppet valve to close prematurely.

Diaphragm and bladder accumulators normally respond more quickly to pressure variations than do piston types because the piston has to overcome the static friction of the seal while the rubber diaphragm or bladder does not. It has been practically proven that a properly designed piston accumulator could help reduce this difference to an insignificant level for most applications.

### 5.3.3 Sizing Hydraulic Accumulators

Sizing a hydraulic accumulator means determining its overall volume and is normally calculated differently according to specific applications using different equations. For example, when an accumulator is used as a standby power source for leakage compensation for a closed hydraulic system, the fluid discharge rate is often very small and a slow recharging process can work satisfactorily. Therefore, the variation of gas pressure and volume can occur under a constant temperature, namely, an **isothermal discharging** process. The ideal gas law (Eq. 5.9) can be rewritten as follows:

$$p_0 V_0 = p_1 V_1 = p_2 V_2 = \text{constant} \tag{5.11}$$

The overall volume of an accumulator for leakage compensation can be determined using the isothermal equation derived from Eqs. (5.10) and (5.11) as follows:

$$V_0 = \frac{p_1 p_2}{p_0 \left(p_2 - p_1\right)} V_S \tag{5.12}$$

where $V_0$, $V_1$, $V_2$ are gas volumes, $p_0$, $p_1$, and $p_2$ are gas pressures at precharged, drained, and filled status, and $V_S$ is the required servicing volume supplied by the accumulator.

It is difficult to calculate an exact value of the required servicing volume for an accumulator. An empirical method in engineering design for estimating this required servicing volume is to use the following equation:

$$V_S = k\sum V_i - Q_p t \tag{5.13}$$

where $k$ is the leakage coefficient, commonly taking a constant of 1.2; $\sum V_i$ is the total volume of fluid required to compensate for the leakage from all components in a system; $Q_p$ is the pump output flow rate; and $t$ is the pump charging time.

When an accumulator is used as an emergency power supply to a hydraulic system, the accumulator needs to have a fast fluid discharging rate in order to promptly switch the power supply from the normal source to the accumulator. Therefore, the changing of gas pressure and volume occurs through an **adiabatic discharging** process.

$$p_0 V_0^{\sqrt{2}} = p_1 V_1^{\sqrt{2}} = p_2 V_2^{\sqrt{2}} = \text{constant} \tag{5.14}$$

where $\sqrt{2}$ is the ratio of gas specific heat at a constant pressure to that at a constant volume.

An accumulator for an emergency power supply application is usually sized using the adiabatic equation derived from Eqs. (5.10) and (5.14), as follows:

$$V_0 = \frac{1}{p_0^{\frac{1}{\sqrt{2}}}\left[\left(\frac{1}{p_1}\right)^{\frac{1}{\sqrt{2}}} - \left(\frac{1}{p_2}\right)^{\frac{1}{\sqrt{2}}}\right]} V_S \tag{5.15}$$

While it is possible to size an accumulator for either a leakage compensator or emergency power supply in terms of theoretical analysis as discussed above, determination of overall volume for line shock absorbing and pump pulsation dampening often relies on some empirical equations. For example, an empirical equation for sizing an accumulator to absorb line shock induced by sudden closure of the valve has been formulated through regression analysis on a large number of experimental data, as follows:

$$V_o = 0.004\frac{p_s}{p_s - p_o} Q_l (0.0164L - t) \tag{5.16}$$

where $p_o$ is the static pressure at valve widely open condition; $p_s$ is the allowable maximum shock pressure caused by sudden closure of the valve; $Q_l$ is the line flow rate before valve closing; $L$ is the length of the line; and $t$ is the time interval of the valve from widely open to completely closed.

In using Eq. (5.16) to size an accumulator, one should always keep in mind that an accumulator is needed only when this equation gives a positive overall volume. A negative value on sizing an accumulator means no accumulator is needed for the evaluated case. The ideal location to install an accumulator that will absorb the shock is as close to the valve, the main inducer of the shock, as possible. A large-diameter connector between the line and the accumulator can always improve the shock-absorbing performance.

Similarly, sizing an accumulator for pressure pulsation dampening is also heavily empirically equation based. One of the commonly used pump pulsation dampener sizing equations is as follows:

$$V_0 = \frac{qk_q}{0.6k_p} \tag{5.17}$$

where $q$ is the pump displacement, $k_p$ is the pulsation variation rate, defined as the ratio of the maximum pulsation pressure to pump discharge pressure, and $k_q$ is the pulsation coefficient, a piston (teeth) number-dependent pump discharge rate variation, defined as the ratio of pump discharge rate to pump displacement.

Some accumulator manufacturers have formulated their own empirical equations for sizing their accumulator products for various applications. After the capacity of an accumulator is determined, one may also need to choose the design of the accumulator. For example, piston accumulators of a particular size often have a choice of different combinations of diameters and lengths. While customer designs of piston-style accumulators require little or no price premium, customer designs of both the bladder and diaphragm accumulators are often very expensive and are usually offered only in one size per capacity, with fewer sizes available.

### 5.3.4 Mounting Hydraulic Accumulators

For a maximum service life, the optimum mounting position for any accumulator is vertically with the hydraulic port down. When horizontal mounting of an accumulator is the only option, it is strongly recommended that the piston style be chosen. That is because use of a diaphragm- or bladder-style accumulator will often result in uneven wear on the diaphragm or the bladder as it rubs against the shell while floating on the fluid as illustrated in Figure 5.16, which will significantly shorten the service life. In extreme cases, the fluid can be trapped away from the connecting port, which reduces output or may elongate the bladder to force the poppet valve to close prematurely.

Some applications may require an extremely large accumulator to store enough energy as the alternative power source, such as the energy storage accumulator in a hydraulic hybrid vehicle. For both structural arrangement and economic reasons, it is often difficult to design one single huge device, so multiple accumulators are used to meet the energy storage needs. As illustrated in Figure 5.17, multiple accumulators can be mounted in parallel on a hydraulic manifold to carry the large flow for such applications. Such a design can be applied to all styles of accumulators. When piston accumulators are used, it is important to know that the

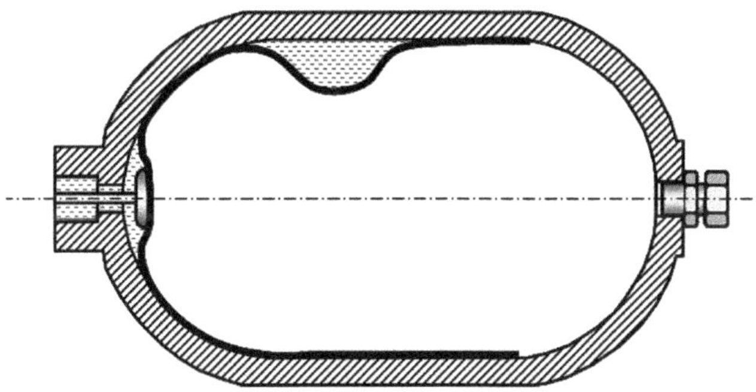

**FIGURE 5.16**
Concept illustration of horizontal mounted accumulator trapping fluid.

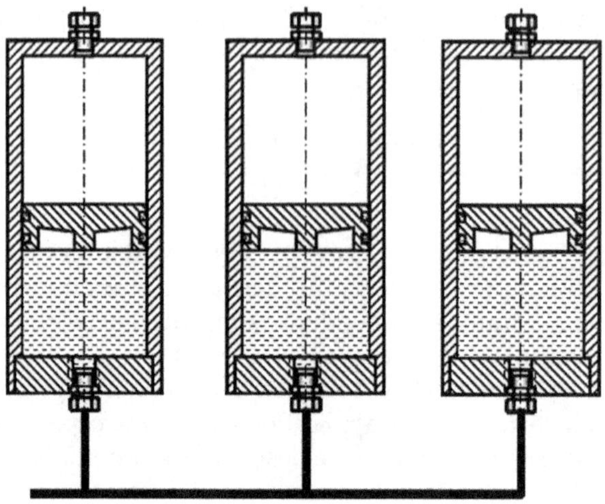

FIGURE 5.17
Concept illustration of multiple accumulators mounted in parallel to provide large flow.

piston with the least friction will always move first. This characteristic of multi-accumulator systems can be widely applied to economically improve the overall energy storage response speed by using only one fast response accumulator in the system.

Another practical method to enlarge the capacity of an accumulator is the use of separate gas bottles to supply compressed gas to a single piston accumulator (Figure 5.18). Such a design allows the piston accumulator to fully utilize its volume as operating volume. Because gas bottles are less expensive than accumulators in general, this design could also noticeably lower the cost for the volume. However, one drawback of this design is that the system could break down simply due to a single seal failure in the gas system. The gas bottle concept could also be applied to diaphragm- or bladder-style accumulators.

**Example 5.2: Size an Accumulator for Different Uses**

A hydraulic power unit requires an emergency power supply of 22 L to securely put the implement in a safe status in case of unexpected, sudden failure of the main power source. If a bladder-style gas-loaded accumulator is chosen to provide this function, what is the proper size for this accumulator (assume the maximum and minimum system pressures are 7.0 and 3.5 MPa, respectively, and the precharge gas pressure is 3.0 MPa)? How about when the accumulator is used to compensate for system leakage?

a. When an accumulator is used as an emergency power source, the overall volume of the accumulator should be sized according to Eq. (5.15):

$$V_0 = \frac{1}{p_0^{\frac{1}{\sqrt{2}}}\left[\left(\frac{1}{p_1}\right)^{\frac{1}{\sqrt{2}}} - \left(\frac{1}{p_2}\right)^{\frac{1}{\sqrt{2}}}\right]} V_S$$

$$= \frac{1}{3.0^{\frac{1}{\sqrt{2}}}\left[\left(\frac{1}{3.5}\right)^{\frac{1}{\sqrt{2}}} - \left(\frac{1}{7.0}\right)^{\frac{1}{\sqrt{2}}}\right]} \times 22$$

$$= 64.1(L)$$

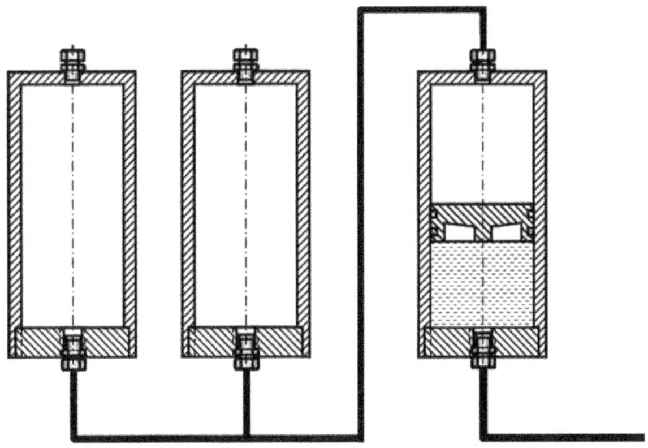

**FIGURE 5.18**
Concept illustration of a piston accumulator supported by separate gas bottles for enlarging the capacity.

b. When an accumulator is used as a leakage compensator, the overall volume of the accumulator should be sized according to Eq. (5.12):

$$\begin{aligned} V_0 &= \frac{p_1 p_2}{p_0(p_2 - p_1)} V_S \\ &= \frac{3.5 \times 7.0}{3.0(7.0 - 3.5)} \times 22 \\ &= 51.3(L) \end{aligned}$$

DISCUSSION 5.2: The obtained results indicated that a large accumulator should be chosen when it is used as an emergency power supply and a smaller accumulator can be used when it serves as a leakage compensator.

## 5.4 Power Regeneration Devices

### 5.4.1 Functions of Hydraulic Power Regeneration

**Power regeneration** in a hydraulic system does not mean that it generates additional power, but that it actually reallocates a portion of power for some special uses. Such a function is often realized by means of **power boosting** or **power recovering** to redistribute the portion of power not being used at the time to the places extra power is required for driving an extra load. Normally, power boosters are designed to either raise the operating pressure or increase the supplying flow by converting a portion of the fluid power not being used at the time as additional power to do special work they can therefore be classified as **internal power regenerators.** In comparison, a power-recovering device often recovers and saves the power from the external load being driven by the system for later use and can be classified as **external power regenerators.**

### 5.4.2 Hydraulic Pressure Intensifiers

A **hydraulic pressure intensifier**, often called a **pressure booster**, is a type of internal power regenerator that operates by converting a high flow of low-pressure fluid into a low flow of high-pressure fluid to drive an extra heavy load at a lower speed.

Typically, hydraulic pressure intensifiers operate on the ratio-of-areas principle in a linear actuator. A typical single-acting intensifier (see Figure 5.19) has a common rod connecting two pistons of different sizes: one large piston (the **actuating piston**) on the right side and one smaller piston (the **boosting piston**) on the left. Correspondingly, the large cylinder has a port connecting only to the main branch of the system, and the small cylinder has a port connecting to both the lower-pressure fluid supply line, often connected to the main branch of the system, and the high-pressure outlet line. Those two lines are separated using check valves as shown in Figure 5.19. During power regeneration, the fluid at the main system pressure is supplied to the large cylinder and acts on the larger piston to exert a force that mechanically drives the smaller piston to discharge the fluid in the smaller cylinder into the high-pressure branch. The ratio between the main system pressure and intensifier discharge pressure is inversely proportional to the areas ratio, defined as follows.

$$p_b = \frac{A_a}{A_b} p_a \tag{5.18}$$

where $p_a$ and $A_a$ are the fluid pressure and the piston area in the actuating cylinder, and $p_b$ and $A_b$ are, respectively, the fluid pressure and the piston area in the boosting cylinder.

The ratio between the flow driven to the actuating cylinder and the flow delivered from the boosting cylinder is proportional to the areas ratio and can be calculated as follows:

$$q_b = \frac{A_b}{A_a} q_a \tag{5.19}$$

where $q_a$ and $q_b$ are the supply flow to the actuating cylinder and the discharged flow from the boosting cylinder, respectively.

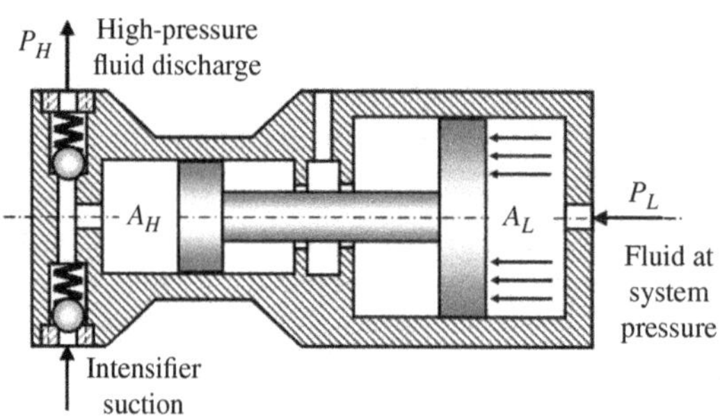


**FIGURE 5.19**
Illustration of the configuration and operation principle of a typical axial-piston style single-acting hydraulic pressure intensifier.

**Single-acting intensifiers** normally boost pressure intermittently: it supplies high-pressure fluid only to the boosting piston extension stroke. The actuating cylinder in a single-acting intensifier can be either spring- or fluid-retracted. In some extraordinary applications, an intensifier might be mounted vertically with the low-pressure chamber underneath, so that the piston assembly can be returned by gravity. While such an intermittent high-pressure flow supply is satisfactory for many applications, it cannot meet the need when a system requires a continuous supply of the high-pressure fluid. Two single-acting intensifiers are often grouped in a pair to provide a continuous supply of high-pressure fluid to meet the need. However, such a design requires having an additional sequencing control mechanism to ensure that two intensifiers will work coordinately to provide a reliable uninterrupted fluid supply.

One practical approach to solving this problem is the use of a **double-acting intensifier**. A typical double-acting intensifier (Figure 5.20) can be thought of as a combined design of two single-acting ones, with a shared large cylinder as the common actuator for both boosting cylinders. The actuating piston and two boosting pistons are mounted on one rod. Driven by the common actuator, when one of the boosting pistons is retracting, the other is always extending in either direction of the actuating piston stroke. When a boosting piston retracts, the low-pressure supply fluid is drawn into the compression chamber of that booster through its inlet line, and when it extends, the reciprocating piston forces the fluid being discharged into the outlet line at a higher pressure.

It should be clearly understood that an ideal intensifier operates at a balanced power level. A higher discharge pressure is obtained at the price of reducing available flow for driving the load. This is analogous to a mechanical transmission in an automobile, which can switch the gear down to obtain a higher torque output at reduced speed. Practically, an intensifier is unable to provide the exact same amount of power as the input due to the losses to overcome all kinds of resistances, like any other power transmission device. One should also note that fluid flow and pressure from a pressure intensifier will fluctuate. If a consistent pressure or flow is required, use of a separate pump will always provide better performance in terms of operation stability. The major benefits of using an intensifier are

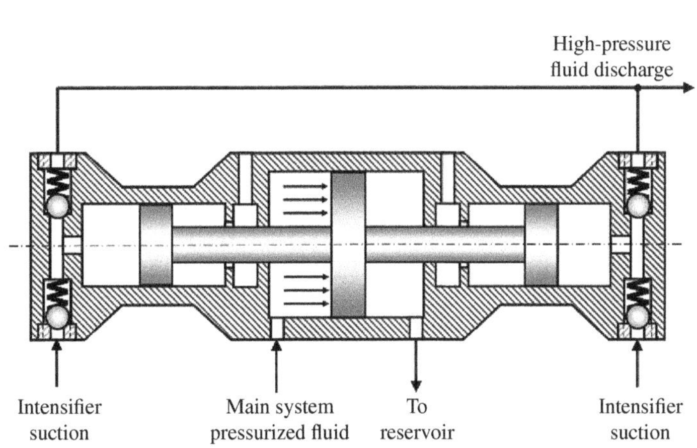


**FIGURE 5.20**
Illustration of the configuration and operation principle of a typical axial-piston style double-acting hydraulic pressure intensifier.

that it can reduce both installation and operating costs, simplify control, and provide long service life, especially when the system requires only occasional pressure boosting in one of the branches.

### 5.4.3 Two-Speed Hydraulic Cylinders

Similar to hydraulic pressure intensifiers, **two-speed hydraulic cylinders** are also a type of internal power regenerator. However, it operates by combining the returning flow with the supplying flow at light load conditions to obtain a higher actuating speed, and therefore it is also called as a **speed booster**.

The basic principle of power regeneration using a speed booster is to reuse the returning flow from the rod-end chamber of a single-rod cylinder under a light load condition to utilize the differential extension, as introduced in Section 4.2.2. The critical component in such a device (see Figure 5.21) is a flow redirection control valve. During the **normal mode** of operation, namely, the **heavy-load and low-speed mode**, this redirection control valve is set at its normal position, which connects the cylinder rod-end chamber to the tank to discharge the returning flow (Figure 5.21(a)). When high speed is required under a light-load condition, this redirection control valve is switched to connect the rod-end chamber to its cap-end chamber to allow the returning flow to be reused as an additional flow supply (Figure 5.21(b)). This **light-load and high-speed mode** is also called the **flow regeneration mode**. The fluid on both sides of the piston has the same pressure under this mode since both chambers are connected.

As in any other power regeneration devices, the flow is regenerated from the introduced speed booster at a cost. In principle, a two-speed cylinder obtains extra flow by sacrificing the load-driven capability brought by the piston area difference between the cap-end and rod-end sides. This is analogous to use of a smaller cylinder to achieve a higher speed under the same flow supply. The actuating speed and driving force in normal mode can be determined as follows: (The rod-end pressure is much lower than that in head-end in this case, and can be ignored to simplify the analysis.)

$$v_1 = \frac{q_1}{A_1} \tag{5.20}$$

$$F_1 = p_1 A_1 \tag{5.21}$$

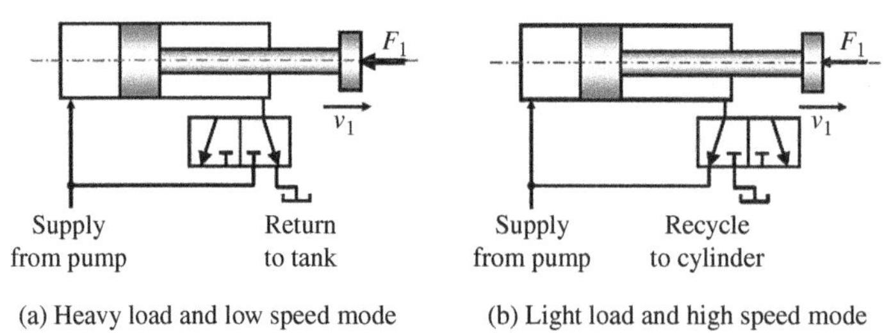


**FIGURE 5.21**
Illustration of the operation principle of flow regeneration using a two-speed hydraulic cylinder, (a) normal and (b) flow regeneration modes.

The corresponding parameters in the flow regeneration mode can be calculated using the following equations:

$$v_3 = \frac{q_1 + q_2}{A_1} \tag{5.22}$$

$$F_3 = p_1(A_1 - A_2) \tag{5.23}$$

Because $q_2$ can be calculated using the following equation:

$$q_2 = v_3(A_1 - A_2) \tag{5.24}$$

Combining Eqs. (5.22) and (5.24), we have:

$$v_3 = \frac{q_1}{A_2} \tag{5.25}$$

In the Eqs. (5.24) and (5.25), $q_1$ and $q_2$ are the supply flow to the cap-end chamber and the return flow from the rod-end chamber; $A_1$ and $A_2$ are the piston areas of the cap-end side and rod-end side of the two-speed cylinder; $v_1$ and $v_3$ are the cylinder-actuating speed and $F_1$ and $F_3$ are the actuating force in the normal and regeneration modes of the two-speed cylinder, respectively.

### 5.4.4 Hydraulic braking chargers

Unlike both pressure and speed boosters that regenerate hydraulic power through redistributing the energy, a **hydraulic braking charger** regenerates hydraulic power by recovering the power being deployed to drive the external load during motion status changes. The power regeneration process in a hydraulic hybrid automobile provides an illustrative example describing the operation principle of this power regeneration method.

As Figure 5.22 shows, a typical hydraulic braking charger can be constructed using a reversible motor pump, a gas-loaded hydraulic accumulator, a three-way mode-switch valve, a three-way function-switch valve, and a shuttle valve. In normal operation mode, the mode-switch valve is set at its **operation position** to connect the pump-discharge port

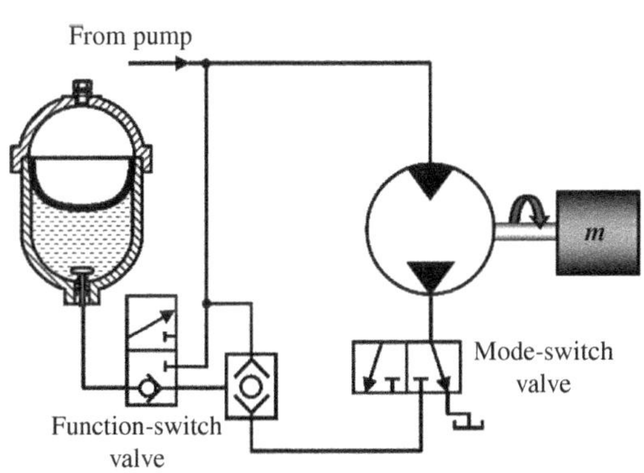


**FIGURE 5.22**
Illustration of the principle of a typical hydraulic pump brake in power recovering.

to the tank to release the discharge flow, creating a maximum driving torque on the output shift. In power regeneration mode, the mode-switch valve is shifted to its **recovering position** to connect the discharge port to the hydraulic accumulator, with the accumulator function-switch valve set at the **charging position**. This will allow the recovered fluid power being stored in the accumulator, while the motor pump is used as the brake, to absorb the inertia carried by the rotation of external load. The recovered fluid power can be used as a supplemental power source whenever it is needed. To return the supplemental fluid power to the system, the motor must be set at its normal operation mode and then shift the function switch to the **discharging position**.

**Example 5.3: Speed-Boosting Power Regeneration**

Figure 5.21 illustrates a speed booster constructed using a single-rod cylinder controlled by a flow redirection control valve, driven by a pump capable of delivering 100 L/min flow. If the diameters of the rod and bore are 36 and 50 mm, respectively, at what speed can this two-speed cylinder push a 4500 N and an 18,000 N external load if the system line relief is set at 10 MPa?

a. First, it is necessary to check the expected system pressures when driving both the lighter and heavier loads under speed-boosting modes according Eq. (5.23):

   At the lighter load (4500 N):

$$\begin{aligned} p_3 &= \frac{F}{(A_1 - A_2)} \\ &= \frac{4500}{\frac{\pi}{4} \times \left\{ \left( \frac{50}{1000} \right)^2 - \left[ \left( \frac{50}{1000} \right)^2 - \left( \frac{36}{1000} \right)^2 \right] \right\}} \\ &= 4423474(Pa) \approx 4.4(MPa) \end{aligned}$$

   At the heavier load (18,000 N):

$$\begin{aligned} p_3 &= \frac{F}{(A_1 - A_2)} \\ &= \frac{18000}{\frac{\pi}{4} \times \left\{ \left( \frac{50}{1000} \right)^2 - \left[ \left( \frac{50}{1000} \right)^2 - \left( \frac{36}{1000} \right)^2 \right] \right\}} \\ &= 17693895(Pa) \approx 17.7(MPa) \end{aligned}$$

b. As the required system pressure for lighter load is below the line-relief setting, it can be driven at the boosting mode, and the speed can be determined using Eq. (5.25):

$$\begin{aligned} v_3 &= \frac{q_1}{A_2} \\ &= \frac{\frac{100 \times 10^{-3}}{60}}{\frac{\pi}{4} \times \left( \frac{36}{1000} \right)^2} = 1.64\left(m \cdot s^{-1}\right) \end{aligned}$$

The heavier load is much above the relief setting, the load can only be driven at the normal speed, and the speed can be determined using Eq. (5.20):

$$\begin{aligned} v_1 &= \frac{q_1}{A_1} \\ &= \frac{\frac{100\times 10^{-3}}{60}}{\frac{\pi}{4}\times\left(\frac{50}{1000}\right)^2} \\ &= 0.85\left(m\cdot s^{-1}\right) \end{aligned}$$

**DISCUSSION 5.3:** The speed-boosting mode can significantly increase the operating speed, but it works only for light-load conditions. In cases where the load exceeds a limit, the system will be unable to operate at the speed-boosting mode.

## References


1. Akers, A., Gassman, M., Smith, R. *Hydraulic Power System Analysis*. CRC Press, Boca Raton, FL (2006).
2. DeRose, D. The use of accumulators in hydraulic systems. *Fluid Power Journal*, 12: 16–21 (2005).
3. Engineers Edge, *Hydraulic Accumulator Sizing Calculations*. http://www.engineersedge.com/hydraulic/accumulator_equations.htm. Accessed on July 23 (2007).
4. Esposito, A. *Fluid Power with Applications (6th Ed.)*. Prentice-Hall, Upper Saddle River, NJ (2003).
5. Goering, C.E., Stone, M.L., Smith, D.W., Turnquist, P.K. *Off-road Vehicle Engineering Principles*. ASAE, St. Joseph, MI (2003).
6. Hedges, C.S. *Industrial Fluid Power (3rd Ed.)*. Womack Educational Publications, Dallas, TX (1988).
7. Hewko, L.O., Weber, T.R. Hydraulic energy storage based hybrid propulsion system for a terrestrial vehicle. *Proc. 25th Intersociety Energy Conversion Engineering Conf. (IECEC-90)* V4: 99-105, Reno, NV (1990).
8. Hydraulics & Pneumatics. *Fluid Power Basics*. http://www.hydraulicspneumatics.com/200/FPE/IndexPage.aspx. Accessed on November 20 (2006).
9. Keller, G.R. *Hydraulic System Analysis*. Penton Media Inc., Cleveland, OH (1985).
10. Li, Z., Ge, Y., Chen, Y. *Hydraulic Components and Systems* (in Chinese). Mechanical Industry Press, Beijing, China (2000).
11. McClay, D., Martin, H.R. *The Control of Fluid Power*. John Wiley & Sons, New York (1973).
12. Manring, N.D. *Hydraulic Control Systems*. John Wiley & Sons, New York (2005).
13. Merrit, H.E. *Hydraulic Control Systems*. John Wiley & Sons, New York (1967).
14. Nachtwey, P. Accumulators: The unsung heroes of hydraulic motion control. *Hydraulics & Pneumatics*, 59: 34–37 (2006).
15. Pease, D.A. *Basic Fluid Power*. Prentice-Hall, Englewood Cliffs, NJ (1967).
16. Pourmovahed, A., Beachley, N.H., Fronczak, F.J. Modeling of a hydraulic energy regeneration system. Part I. Analytical treatment. *Transactions of the ASME: J. Dynamic Systems, Measurement and Control*, 114: 155–159 (1992).
17. Pourmovahed, A., Beachley, N.H., Fronczak, F.J. Modeling of a hydraulic energy regeneration system. Part II. Experimental program. *Transactions of the ASME: J. Dynamic Systems, Measurement and Control*, 114: 160–165 (1992).
18. Taylor Devices, Inc. A *Designer's Guide to Hydraulic Shock Absorber Selection*. http://www.taylordevices.com/DesignersGuide.htm. Accessed on July 23 (2007).

19. Vickers, Inc. *Vickers Mobile Hydraulics Manual* (2nd Ed.), Vickers, Inc., Rochester Hills, MI (1998).
20. Watton, J., *Fluid Power Systems, Modeling, Simulation, Analog and Microcomputer Control.* Prentice-Hall, New York (1989).
21. Yeaple, F.D. *Fluid Power Design Handbook.* CRC Press, Boca Raton, FL (1996).
22. Zhang, Q., Goering, C.E. Fluid power system, In: Bishop, R. (ed.), *The Mechatronics Handbook.* CRC Press, Boca Raton, FL, pp. 10–11~10–14 (2001).

## Exercises

**5.1** What is the primary function of a hydraulic shock absorber in a hydraulic circuit?

**5.2** What is the fundamental difference between a hydraulic spring and a common hydraulic shock absorber in configuration?

**5.3** Name three major classifications of gas-loaded accumulators and give one advantage of each classification.

**5.4** List three common applications of accumulators in a hydraulic circuit.

**5.5** Use examples to describe both internal and external leakages.

**5.6** Use layperson's language to explain how weight-loaded, spring-loaded, and gas-loaded accumulators absorb and discharge hydraulic energy during operations.

**5.7** List the operational characteristics and adequate applications of three commonly used types of hydraulic accumulators.

**5.8** Can a power regeneration device create extra power for a hydraulic system?

**5.9** Where does the additional hydraulic power come from for a specific branch in a hydraulic system when a hydraulic pressure intensifier is used?

**5.10** Where does the additional hydraulic power come from when additional flow supply is generated to boost the extending speed of a two-speed hydraulic cylinder?

**5.11** A 200 mm stroke, simple orifice type shock absorber as illustrated in Figure 5.4(a) takes in power generated by a 1200 N force at an initial velocity of $0.5\ \mathrm{m \cdot s^{-1}}$. If the $C_1$ and $C_2$ values obtained experimentally are $15{,}000\ \mathrm{kg \cdot m^{-1}}$ and $2100\ \mathrm{kg \cdot s \cdot m^{-1}}$, respectively, what will be the shock-absorbing time and the effective shock-absorbing stroke?

**5.12** A 200 mm stroke, simple orifice type shock absorber as illustrated in Figure 5.4(a) takes in power generated by an unknown mass at an initial velocity of $0.5\ \mathrm{m \cdot s^{-1}}$. If the $C_1$ and $C_2$ values obtained experimentally are $15{,}000\ \mathrm{kg \cdot m^{-1}}$ and $2100\ \mathrm{kg \cdot s \cdot m^{-1}}$, respectively, what will be the impact force and the total mass moved by the shock absorber piston?

**5.13** An accumulator is used as a leakage compensator for a closed hydraulic system when the pump does not supply any flow to the system. The total volume of fluid leaking from all components in 30 minutes is 1.0 L, corresponding to an initial pressure of 17.0 MPa and final pressure of 12.0 MPa. When the precharge gas pressure is 3.0 MPa for the accumulator, how large should this accumulator be sized?

**5.14** An accumulator is used as an emergency power supplier for a closed hydraulic system when the pump does not supply any flow to the system. Assuming the system requires having an additional 20 L flow supply in an emergency case to securely put the implement in a safe status, when the maximum and minimum workable system pressures are 13.0 and 7.0 MPa, respectively, and the accumulator has a 6.0 MPa precharge gas pressure, what is the proper size for this accumulator?

**5.15** A diaphragm-style gas-loaded accumulator is chosen to absorb the line shock induced by a sudden close of the valve in the line, what is the proper size for this accumulator if the static system pressure is 7.0 MPa when the valve is wide open and the allowable pressure surge is 7.5 MPa? (Assume that the flow rate in the line before the valve being shut off is 150 L/min, the length of the line is 20 m, and the valve closed instantly.) How about if it takes 1/3 s to shut off completely?

**5.16** As illustrated in Figure 5.21, the actuating piston of a single-acting hydraulic pressure intensifier has an 80 mm diameter, and that of the boosting piston is 35 mm. If the pump supplies 10 L/min flow at 15 MPa to the actuating side of the cylinder, what will be the discharge flow rate and pressure at the booster outlet port?

**5.17** When a continuous-pressure-boosted flow is required, a double-acting pressure intensifier as illustrated in Figure 5.20 is often used. If a branch needs to get 5 L/min 30 MPa flow to drive the load, but the main pump can supply 10 L/min flow at 10 MPa to the actuating side of the pressure intensifier, what diameter ratio for actuating and boosting pistons is required?

**5.18** As illustrated in Figure 5.21, a speed booster was constructed using a single-rod cylinder controller using a flow redirection control valve. If the hydraulic system is supplied by an 18 cc fixed displacement pump operating at 2000 rpm, and the line relief valve is set at 20 MPa, and the cap-end and rod-end piston areas are 20 $cm^2$ and 10 $cm^2$, respectively, what will be the actuating speed and driving force in both the normal model and the power regeneration mode?

**5.19** If the same actuating speed for both extension and retraction strokes is required on a speed booster, what is the required area ratio between the piston and the rod?

**5.20** A hydraulic braking charger is used to recover the momentum power of a hydraulic hybrid vehicle during motion status changes. If it is required that at least 5 minutes of alternative power supply be provided from the power regeneration system at 20 L/min, how large should an accumulator be sized when it is precharged at 6 MPa? (Assume that the maximum and minimum operating pressures of the system is 30 and 10 MPa.) What will be needed if the precharge pressure is changed to 3 MPa?

# 6

# *Hydraulic Fluids and Fluid-Handling Components*

## 6.1 Hydraulic Fluids

### 6.1.1 Functions of Hydraulic Fluids

Performing various functions, hydraulic fluid is the most vital element in the hydraulic system. Among all the functions it performs, its primary one is **power transmission**. The basic requirements for satisfactory performance of this primary function include consistent response, optimal efficiency, and safety. To ensure responsive, efficient, and safe power transmission, a hydraulic system needs to be sufficiently **stiff**. From a hydraulic fluid aspect, this requirement means that the fluid has little compressibility over the entire operating pressure range. The commonly used commercial hydraulic fluids are normally said to be **incompressible fluids**. However, like almost all liquids, hydraulic fluids do present some very small compressibility in proportion to the operating pressure: about 0.3% at 1 MPa to a little over 1.3% when the pressure increases to 25 MPa. Although the compressibility changes of hydraulic fluids may delay the response, such a slight change in response is often not a concern. However, satisfactory responsiveness may easily be revoked if even a small amount of air is dissolved in the fluids because air is highly compressible. Therefore, one critical measure of the quality of hydraulic fluids is the percentage of gases that can be dissolved in the fluids. The less **dissolved gases** in the fluids the better.

Another essential function of hydraulic fluids is **lubrication**. There are many moving components in a hydraulic system, which are always in contact with other components. To minimize potential wear and reduce the heat generated from the friction between those relative moving components, it is necessary to provide good lubricity between those surfaces. As shown in Figure 6.1 and as can be seen with the naked eye in a microscopic view, no well-machined surface is perfectly flat, and a gap is always formed between any two surfaces. Keeping such imperfect surfaces in direct contact in relative motion could result in high friction and in turn cause rapid wear of those surfaces. Ideally, the hydraulic fluids should provide sufficient lubricity to those surfaces by forming an oil film on them. The lubrication can be categorized as **full-film lubrication** if the fluid film makes sufficient clearance to completely separate two surfaces; or as **boundary lubrication** if the film is insufficient to separate the metal-to-metal contact between two surfaces. Sometimes the hydraulic fluid itself may not be able to provide sufficient lubricity, and so some antiwear additives may be used to improve the lubricating capability of the fluid.

Because of the gaps between the relative moving surfaces, some fluid leakage is expected. Another critical function of hydraulic fluids is to form an oil film to provide **liquid sealing** in such small gaps to stop leakage. In addition, this oil film can provide a **cooling** function to the contacting surfaces by carrying away the heat generated from the friction.

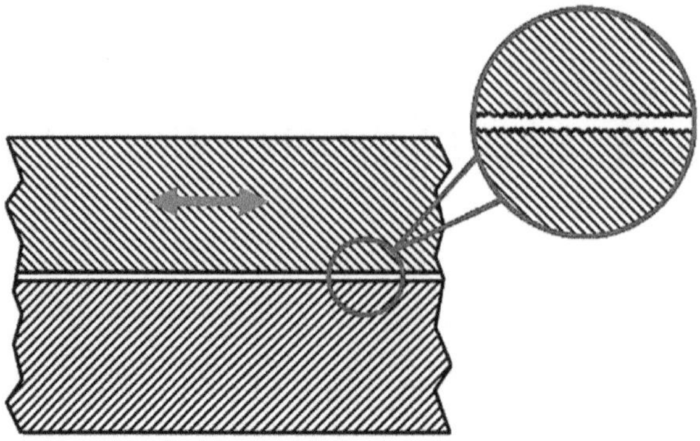

FIGURE 6.1
Concept illustration of oil film lubrication in between two surfaces of relative motion.

During the relative motion between contact surfaces, some tiny particles may be torn off from those surfaces and become contaminants. If such contaminants remain in the gaps between those relative moving surfaces, it could cause unacceptable accelerated wear on those surfaces. Therefore, hydraulic fluids also serve the critical function of **removing** particles from gaps between two relative moving surfaces (Figure 6.2).

### 6.1.2 Hydraulic Fluid Properties

To ensure that the selected hydraulic fluid is capable of performing all the desired functions, it is necessary to select the right type of fluids with appropriate properties.

Fluid **density** is one of the fundamental properties and is defined as its mass per unit volume as follows:

$$\rho = \frac{m}{V} \tag{6.1}$$

where $\rho$ is the density, $m$ is the mass, and $V$ is the volume of a hydraulic fluid.

Fluid density is approximately a linear function of pressure and the temperature, and can be expressed using the following equation:

$$\rho = \rho_0(1 + ap - bT) \tag{6.2}$$

where $\rho_0$ is a reference density, $p$ is the pressure, and $T$ is the temperature of a fluid.

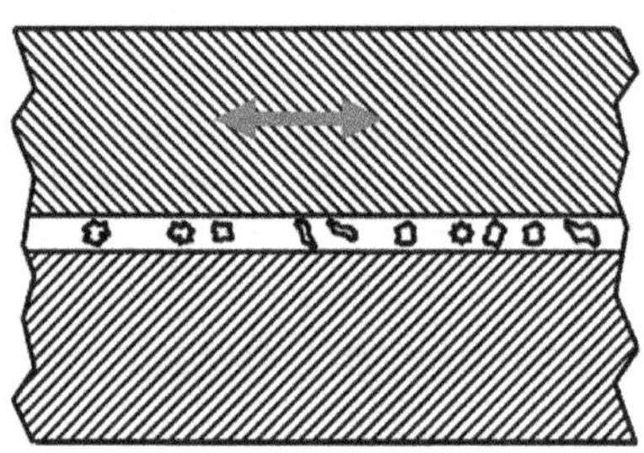

FIGURE 6.2
Concept illustration of particles being removed from a gap.

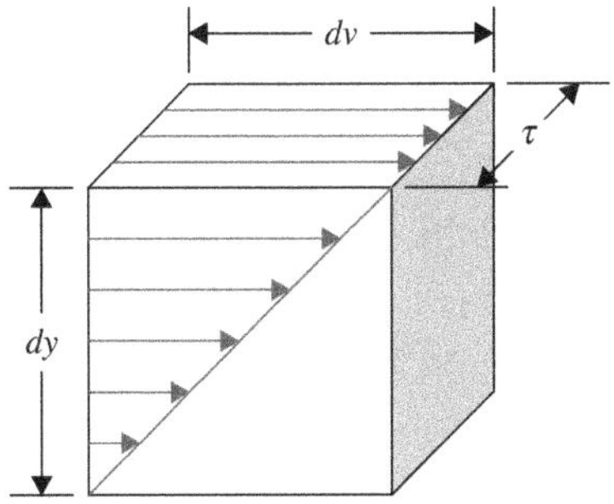


**FIGURE 6.3**
Visual definition of dynamic viscosity.

In engineering practice, manufacturers of hydraulic fluids often provide the **relative density** (also called the **specific gravity**) instead of the actual density. The relative density of a fluid is defined as the ratio of its actual density to the density of water at the same temperature.

Another important property of hydraulic fluids is **viscosity**—a measure of its resistance to deformation when subjected to a shearing force. Two kinds of viscosity measures, namely, **dynamic viscosity** (also called **absolute viscosity**) and **kinematic viscosity**, are commonly used. As depicted in Figure 6.3, the dynamic viscosity can visually be explained as a cube of fluid acted upon by flow forces. Therefore, the value of a dynamic viscosity can be determined using the Newtonian shear stress equation expressed as follows:

$$\mu = \frac{\tau}{dv/dy} \tag{6.3}$$

where $\mu$ is the dynamic viscosity, $dv$ is the relative velocity between two parallel layers distanced $dy$ apart, and $\tau$ is the shear stress.

The kinematic viscosity is the ratio of the dynamic viscosity to the density of the fluid, as defined using the following equation.

$$\nu = \frac{\mu}{\rho} \tag{6.4}$$

Both the dynamic and kinematic viscosities vary strongly with temperature. A properly selected fluid must maintain minimum viscosity at the highest operating temperature of a hydraulic system. Meanwhile, the fluid must not be so viscous at low temperature that it cannot be pumped. To evaluate the characteristic of a fluid viscosity variation relevant to temperature variation, the **viscosity index**, an arbitrary number capable of representing such a relationship, is used for comparing the sensitivity of viscosity to temperature between two hydraulic fluids. The fluid with a higher viscosity index (see Figure 6.4) is less sensitive to temperature variations and therefore can provide more consistent hydraulic system performance over a wider temperature range.

The International Standards Organization (ISO) has issued an international standard for fluid viscosity grades in which the grade number represents a range of kinematic viscosity values at 40°C. For example, the kinematic viscosity of ISOVG100 fluid ranges between 90 and 110 $m^2s^{-1}$ and ISOVG46 fluid between 41.4 and 50.6 $m^2s^{-1}$ at 40°C. The Society of

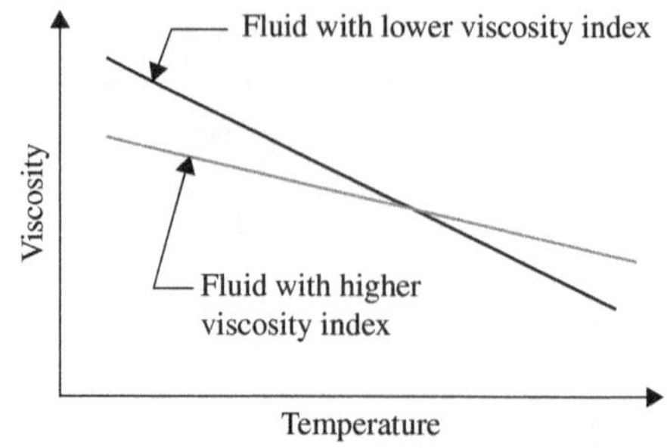


**FIGURE 6.4**
Relationship demonstration of improved viscosity indices.

Automotive Engineers (SAE) has also established an industry standard, commonly used in the United States, to grade fluid viscosity levels. Unlike the ISO grading, the SAE grading uses a **winter number** and **summer number** to separate fluids for low- or high-temperature use. The winter number (0W, 5W, 10 W, 15W, etc.) grades fluids in terms of dynamic viscosity for cold season use, and the summer number (20, 30, 40, 50, etc.) grades fluids in terms of kinematic viscosity for hot-season use.

**Bulk modulus** is a measure of the compressibility or stiffness of a hydraulic fluid. The basic definition of fluid bulk modulus is the fractional reduction in fluid volume corresponding to a unit increase of applied pressure, expressed using the following equation:

$$\beta = -V\left(\frac{\partial p}{\partial V}\right) \tag{6.5}$$

where $\beta$ is the bulk modulus, $V$ is the volume, and $p$ is the pressure of a hydraulic fluid.

The bulk modulus of a fluid can be defined either as the isothermal tangent bulk modulus if its compressibility is measured under a constant temperature or as the isoentropic tangent bulk modulus if its compressibility is measured under constant entropy. In analyzing the dynamic behavior of a hydraulic system, the stiffness of the hydraulic container plays a very important role. An **effective bulk modulus** is often used to consider both the fluids' compressibility and container stiffness at the same time as expressed in the following equation.

$$\frac{1}{\beta_e} = \frac{1}{\beta_f} + \frac{1}{\beta_c} \tag{6.6}$$

where $\beta_e$ is the effective bulk modulus of a system, $\beta_f$ is the fluids compressibility, and $\beta_c$ is the container stiffness.

Irrelevant to its viscosity, the **lubricity** of a hydraulic fluid is a special property used to measure the **antiwear** performance of the fluid. The pump is the critical dynamic element in any hydraulic system, and each pump type has different requirements for wear protection. Compared to piston pumps, vane and gear pumps require more antiwear protection because they operate with inherent metal-to-metal contact. As stated in Section 6.1.1, an ideal fluid should be able to form a full film between two facing surfaces of all motion pairs. Normally, most hydraulic fluids require use of some special additives to improve their antiwear performance, especially for the water-based fluids. The most frequently used antiwear additive is probably zinc dithiophosphate (ZDP). However, the ashless antiwear hydraulic fluids have become a popular means of reducing waste-treatment loads.

**Antifoaming** is another critical property for hydraulic fluids. There are two general types of foam: **surface foam,** which usually collects on the fluid surface in a reservoir, and **entrained air**, which occurs anywhere in the system. Because they introduce air into the fluids, foams will severely deteriorate fluid properties by reducing dynamic responses in power transmission due to decreased bulk modulus, shortened fluid service life because of increased fluid oxidation and evaporation rates, and increased noise and hammering effects induced by cavitation. Under certain pressures and temperatures, hydraulic fluids may dissolve a certain amount of air. For example, the **air-dissolving rate** in petroleum-based hydraulic fluids is usually between 5 and 10%. Normally, the dissolving rate of gases in fluids is proportional to the pressure and inversely proportional to the temperature. As the pressure decreases or the temperature increases, the dissolved gases will be released from the fluids and form foams. Surface foam can be eliminated by simply adding defoaming additives or by providing proper sump design to allow foam the time to dissipate. Entrained air can cause more serious problems, including cavitations and hammering actions that can destroy the parts. An effective way of preventing entrained air is to eliminate air leaks in the system. Special attention should be paid to the fact that some commonly used defoaming additives, when used at a high concentration to reduce surface foam, will increase entrained air.

Oxidation and thermal stability are two other important properties for hydraulic fluids. **Oxidation stability** is the capability of hydraulic fluids to resist oxidation, form acids, sludge, and varnish. Acids can attack system parts, particularly soft metals. Normally, the oxidation rate increases as the temperature rises. One way to extend the service life of hydraulic fluids economically is to increase the fluids' oxidation stability. Some types of antioxidation additives are often used for this purpose. **Thermal stability** is the fluid's ability to resist chemical reactions and decomposition during high-temperature operations. Many times, thermal cycling also encourages the formation of fluid decomposition products. A properly designed system should be able to minimize these thermal problems, and use of some antidecomposition additives can also help improve thermal stability. Although not always practical or easy to attain, keeping the system operating at a constant moderate temperature in a steady state is best for system performance as well as for fluid service life.

**Demulsibility** is the capability of a hydraulic fluid to separate or reject water. When large quantities of water in a fluid can be removed by draining the sump periodically, some small amounts of water can always be entrained in the fluid. Two practical approaches—the chemical approach of adding demulsifiers to the fluid to speed up water separation and the physical approach of using filters to remove water from the fluid—can be used to improve the demulsibility of hydraulic fluids.

A fluid's **rust and corrosion prevention** capability is critical to the service life of metal parts in a hydraulic system. Rusting often occurs when the water carried by the fluid attacks ferrous metal parts, and corrosion is a chemical reaction between the surface of a metal part and a chemical, typically an acid. Most hydraulic fluids do contain rust inhibitors to prevent rusting and incorporate selected anticorrosion additives to protect the metal parts from acidic attack.

There are a few general rules for properly controlling the **operating temperature** of hydraulic fluids. The recommended maximum operating temperature for most hydraulic fluids is around 65°C. While an operating temperature range between 80° and 90°C is acceptable, the fluid change period should be reduced one-half to one-third of the recommended interval. If a system must operate at temperatures up to 120°C, the fluid has to be changed in very short intervals to ensure that the fluid carries adequate properties to support normal operation.

The **seal compatibility** of a fluid is one more property that needs to be considered for many applications. To ensure a tight fit, the seals are normally selected so that when they encounter the fluid being used in a system they will not change size. The fluid selected should be checked to make sure that the fluid and seal materials are compatible to prevent the fluid from interfering with proper seal operation.

### 6.1.3 Types of Hydraulic Fluids

Hydraulic fluids function in a hydraulic system just like blood functions in a human body; it can have a critical impact on performance, reliability, and even a hydraulic system's service life. The importance of selecting the right type of fluid for a hydraulic system can never be overestimated. Making a proper selection requires not only having a basic understanding of the basic fluid characteristics, but also knowing the tasks and working environment of the hydraulic system. Ideally, hydraulic fluids should be inexpensive, noncorrosive, nontoxic, and noninflammable, have good lubricity, and be stable in properties. The technically critical properties of hydraulic fluids include density, viscosity, and bulk modulus. Although no single fluid carries all these ideal characteristics, it is possible to select one that is best for a particular hydraulic system. The commonly used hydraulic fluids in modern hydraulic systems include petroleum-based, environmentally safe, and fire-resistant fluids.

**Petroleum-based hydraulic fluids** are by far the most commonly used hydraulic fluids. These types of fluids are a complex mixture of hydrocarbons refined to meet certain characteristic standards suitable for being used in hydraulic systems. The major advantages of petroleum-based fluids are low cost, good lubricity, ready availability, and relatively low toxicity. Various additive packages, such as antioxidation, antifoaming, and anticorrosion packages, are often added to enhance some critical characteristics required for hydraulic system uses. In practice, **automobile transmission fluids** typically have a very high viscosity index and are capable of providing excellent low-temperature viscosity. They are often used in mobile hydraulic systems as they carry similar additive packages to hydraulic fluids and are readily available.

The Environmental Protection Agency (EPA) always advocates the use of **environmentally safe hydraulic fluids** in places where the leaking or spill of petroleum-based hydraulic fluids could have a negative impact on the environment. Environmentally safe hydraulic fluids are formulated to be **biologically degradable** and **virtually nontoxic**. As most hydraulic fluids are biodegradable when given enough time and in proper conditions, we define a hydraulic fluid as biodegradable oil when more than 60% of the spilled fluid could biologically break down into innocuous products exposed to the atmosphere over a 28-day period. To prove the fluid is virtually nontoxic, more than half of the rainbow trout fingerlings in a population must survive for four days in an aquatic solution with concentrations of the fluid greater than 1000 ppm. The major advantages of using this type of biodegradable fluid are the reduced spill cleanup costs due to the readily biodegradable characteristics and its safety for plants, fish, animals, and humans.

There are synthetic esters, polyglycols, and vegetable oil fluids that are environmentally safe. **Synthetic esters** are often formulated as biodegradable fluids with an excellent low-temperature fluidity and superior lubricity; they are also highly biodegradable and have low toxicity. But this type of fluid is usually fairly expensive. **Polyglycols** are commonly used in many hydraulic systems because of their lower cost than synthetic esters and their comparable water tolerance and oxidation resistance. The major shortcomings of this type of fluid is its insufficient biodegradability and potential toxicity in water when mixed with lubricating additives; therefore, it is not treated as a fully environmental safe fluid. The truly readily biodegradable fluids are **vegetable oils**, such as rapeseed oil, which have excellent natural biodegradability and are relatively inexpensive. Because of these advantages, vegetable oil-based biodegradable hydraulic fluids are becoming the most commonly used environmentally safe fluids. The main deficiencies of these fluids are their susceptibility to water contamination and their instability in oxidation.

The trade-off between environmental advantages and potential performance deficiencies of biodegradable hydraulic fluids suggests that these fluids are not yet ready to be direct replacements for petroleum hydraulic fluids, and are only economical to use in hydraulic systems of outdoor equipment operating in environmentally sensitive areas. Like petroleum oils, vegetable oils or synthetic esters rely on specially selected additives to improve their performance as lubricants. The use of improper types of additives may largely, or even completely, eliminate the environmentally safe advantages. Therefore, it is critically important to use low-toxicity additives in biodegradable hydraulic fluids. One unique problem for vegetable oil-based hydraulic fluids is the large amount of unsaturated hydrocarbons contained in the fluids, which will often lead to rapid oxidation at high temperatures and poor fluidity at low temperatures.

It should be clearly understood that **fire-resistant hydraulic fluids** are not fireproof but only catch fire with more difficulty. One disappointing fact is that all presently available hydraulic fluids will burn under certain conditions. In many places, hydraulic oil is already considered a flammable hazardous material. When a hydraulic power system is operating in an environment with potential ignition sources, such as with open flames, sparks, or hot metals, it is advisable to avoid using oil-based fluids because even a tiny leaking spray of fluids from a high-pressure hydraulic system could cause a serious fire and result in major property damage, personnel injury, or even death. The alternative is to use one of the fire-resistant hydraulic fluids to eliminate or at least significantly reduce this hazard. Two most commonly used fire-resistant hydraulic fluids are synthetic-based and water-based fluids.

The early type of **synthetic-based fire-resistant fluids** widely used in many industries is formulated using a class of chemical compounds known as **phosphate esters.** While this type of synthetic fluid is extremely fire resistant, the environmental and cost concerns have made these fluids less popular today. The other type of synthetic fluid in use is **synthetic hydrocarbons**, a class of reaction products between long-chain fatty acids and synthesized organic alcohols, which provide high fire resistance. This type of synthetic fluid has gained widespread and growing use because of its environmental advantages.

Capable of offering a **fireproof** feature, **water-based fluids** have attracted great renewed interest because of the increasing need for environmentally safe and inexpensive hydraulic fluids. A few types of water-based fluids are available, including water glycols, water-in-oil emulsions, oil-in-water emulsions, and synthetic solutions. **Water glycol** has been around for decades and is found in many applications where petroleum-based fluids are prohibited. These fluids normally contain 35 to 50% water in a mix of glycols and polyglycols. Various packages of additives are often used to improve physical properties and operating characteristics, such as the freezing point, lubricity, and viscosity of the fluids. An interesting property of water glycol is its **inverse water solubility**; that is, it is less soluble at high temperature and thus can free some dissolved glycol polymer when the fluid is hot. Such a characteristic can increase the effective load-carrying capabilities of the fluid. Water glycol-type fluids have the best lubricity and can take the highest pressures of all water-based hydraulic fluids; they are also the most expensive.

There are two types of water-based fluids: **water-in-oil emulsion** and **oil-in-water emulsion**, both composed of similar substances of different concentrations. Such different concentrations and physical mixing of these substances make these fluids noticeably different in physical properties and operating characteristics. The water-in-oil emulsion is usually composed of water and oil (and additives) in a roughly 40–60 ratio. The water is presented in microscopic droplets surrounded by films of oil in this emulsion, which results in good lubricity and a high shear rate because of the oil layer covering the water droplets. Because a high shear rate can always reduce the viscosity of a fluid, this allows the emulsion to lower the leakage and meanwhile keep the fluid flowing easily through high-shear areas such as pumps and valves. Normally, water-in-oil emulsion is a little more expensive than petroleum-based fluids. In comparison, oil-in-water emulsion is normally composed of 95% water and 5% emulsible oils and additives. Different from the water-in-oil emulsion, this fluid is emulsified by dispersing microscopic oil droplets in water. Because the oil droplets are covered by water in this emulsion, it offers excellent fire-resistant performance but with reduced lubricity. The oil-in-water emulsion is probably one of the most inexpensive hydraulic fluids available today.

Similar to oil-in-water emulsion in terms of composing substances, **synthetic solutions** also contain about 95% water with 5% soluble salts and other additives. Normally, synthetic solutions do not contain any petroleum-based oils and are presented in a water-like fluid. To distinguish it from pure water to avoid misplacement, such solutions are always dyed to make them visible.

However, we should not ignore the important fact that the performance and operating characteristics of water-based hydraulic fluids are normally not as good as those of petroleum-based fluids. The most serious problems are that they have much lower viscosity, film strength, and lubricating qualities than oil-based fluids. Among other problems, water may corrode the components, and will leak, evaporate, boil, freeze, and cavitate easily. Because of the inherent shortcomings of water-based fluids, the components used in hydraulic systems filled with water-based fluids often carry some special design and/or manufacturing requirements. Such special requirements often lead to higher manufacturing cost for these components as compared to their counterparts used in conventional systems. However, this investment can easily be recovered from the lower operating cost, not to mention the used fluids disposal and treatment costs. Since water-based fluids are normally nontoxic, microbial growth is naturally supported. To minimize or prevent the consequences of this problem, judicious use of bacteriostatic additives and effective sealing and reservoir design should be practiced.

## 6.2 Hydraulic Fluids Reservoirs

### 6.2.1 Functionality of Fluid Reservoirs

Hydraulic reservoirs are **storage tanks** for holding fluids in hydraulic power transmission systems. They are usually rectangular, cylindrical, T-shaped, or L-shaped and are made of steel, stainless steel, aluminum, or plastic to meet application and installation requirements. Generally, hydraulic reservoirs vary in capacity but need to be large enough to accommodate the thermal expansion of fluids and changes in fluid level due to normal system operation. Many hydraulic reservoirs are manufactured in accordance with standards established by organizations such as the National Fluid Power Association (NFPA) and the Joint Industrial Council (JIC), including the sizing requirements and other technical considerations. In general, an accepted rule of thumb for sizing a reservoir is that the volume of a reservoir should be two to four times the pump flow in one minute. If a reservoir is designed following this general rule, the returned fluid will theoretically have two to four minutes in the reservoir before it circulates again.

A larger reservoir than the general rule suggests is required in some special applications. For example, when a system has single-acting cylinders or cylinders with large rods, the volume of fluid returned on the extended stroke is greatly reduced—or even nonexistent. In these cases, the reservoirs must be larger than the general rule states to carry sufficient fluid to support continuous operation. When the system is equipped with accumulators, it must have a larger reservoir to supply additional fluid at the start in order to fill the accumulators and provide sufficient space to store discharged fluid from the accumulator when it is shut down. Another consideration for making the reservoir is to add cooling capacity. All the exterior walls of a reservoir can dissipate heat to the atmosphere, so the larger the tank the greater the heat dissipation.

Smaller reservoirs may be acceptable for some special applications. However, when a smaller reservoir than suggested is used, an additional cooling system is often required to provide sufficient fluid-cooling capability.

In addition to holding enough fluid to supply a hydraulic system with varying needs, a reservoir also provides many other functions, including but not limited to slowing down the high velocity of returning fluids, settling the contaminants, releasing entrained air carried by the returning flow, preventing returning fluids from directly getting back into the system, providing a large surface to cool the hot returning fluids, giving access to remove used fluids or contaminants, to add new fluids, and presenting fluid-level indications for system maintenance.

### 6.2.2 Fluid Reservoir Components

To provide the above listed functions, a typical hydraulic fluid reservoir is more than merely a fluid storage tank and is commonly equipped with an assembly of supporting components as illustrated in Figure 6.5. Those supporting components include filler and

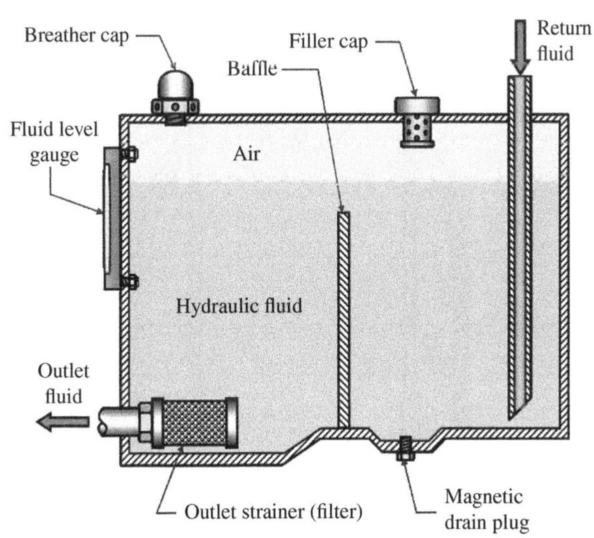


**FIGURE 6.5**
Illustration of the configuration of a typical hydraulic fluid reservoir.

breather caps, a fluid-level gauge, at least one separation baffle, an outlet filter, a fluid-returning pipe, and a drain plug.

Both the **filler cap** and the **breather cap** are normally located at the top, or sometimes on the side, with their openings at an equivalent height to the top of a hydraulic reservoir. As implied by their names, the filler is designed to add fluids, and the breather is used for the entry of filtered air. The two caps are often integrated into one. It is important to point out that with either a separate or an integrated design, a nonpressurized hydraulic reservoir must have a breather to prevent a vacuum in the reservoir, which will stop the fluid from flowing out the reservoir. It is also important to remember to use a dust filter in order to prevent the fluid from being contaminated by large dust particles carried in the ambient air in job sites. An additional consideration is that use of a dust filter is still insufficient to prevent small dust particles, as well as moisture, from entering the reservoir. One solution to this problem is to use a **moisture-removing filter**. As illustrated in Figure 6.6, a moisture-removing filter uses a **watergate** filtering device to remove both the dust participles and the moisture carried in the ambient air. Some hydraulic reservoirs are also equipped with a **relief valve**, which, in order to maintain safe operation, is set to open when the internal pressure exceeds a preset level.

Reservoirs used in mobile hydraulic systems require a **baffling device** to prevent the high-velocity returning fluid from being directly recirculated back into the system without mixing or settling down the debris or discharging air, and to reduce the motion-induced violent **fluid sloshing** in the reservoir. The typical velocity of fluid leaving the return line is often over 3 $m \cdot s^{-1}$, and such a high-velocity fluid stream will agitate the fluid in the reservoir, which may in turn take the contaminants deposited on the bottom of the reservoir, if there are any, back into the system. To provide the time and space for contaminants to be resettled before being drawn in the reservoir, one or more properly placed baffles can effectively slow down the flow stream velocity. Figure 6.7 provides a principle illustration of a few baffling designs in hydraulic reservoirs. Among these designs, the first two are more commonly seen on mobile systems, in which one or more

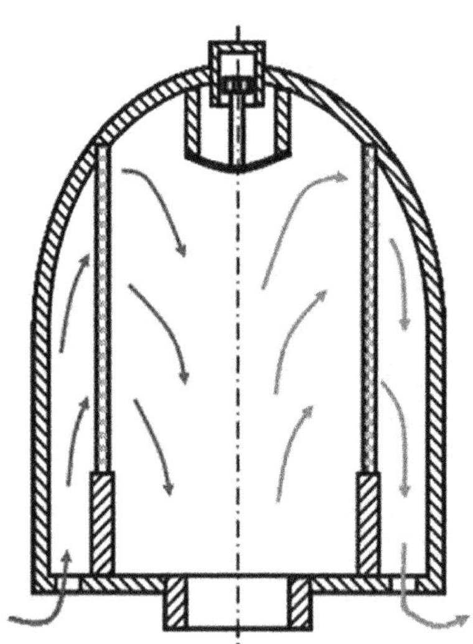

**FIGURE 6.6**
Illustration of the principle of a typical moisture-removing filter.

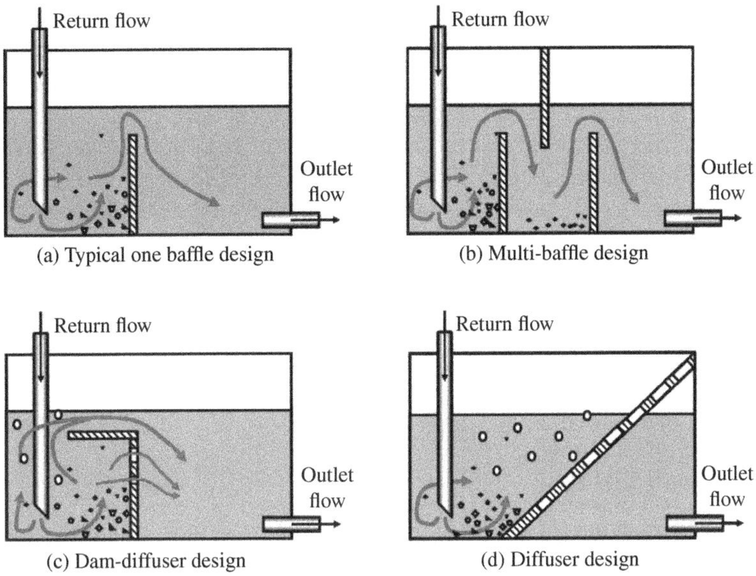


**FIGURE 6.7**
Illustration of the principle of a few typical baffle designs in hydraulic reservoirs. (a) one-baffle, (b) multi-baffle, (c) dam-diffuser and (d) diffuser-type designs.

baffles separate the return line from the outlet line, forcing the return fluid to take a longer path through the reservoir before being discharged to the system again. This arrangement also mixes the return fluid well with in-reservoir fluid and provides more time to settle contaminates and release the entrained air. In addition, the fluid spends more time in contact with the outer walls of the reservoir to dissipate heat. Whether one baffle or more is required depends largely on whether the velocity of the flow stream near the outlet port of the reservoir is low enough. A rule of thumb in assessing the sufficiency of a baffle design is that the velocity of the flow stream after passing the last baffle should be less than $0.3\ m \cdot s^{-1}$. The two other designs shown in the figure use some forms of **diffuser** to reduce the velocity of a fluid as well as to increase the pressure of the fluid, which can also help to discharge the entrained air. Use of a **magnetic drain plug** will effectively keep metal-based contaminants near the plug and prevent those contaminants from being agitated and carried away.

In mobile applications, the fluids stored in a reservoir may form some violent sloshing when the mobile machinery is rapidly changing traveling speeds or directions. Such fluid sloshing can cause structural damage to the reservoir mounting or side walls, force fluids to be ejected through the filler and/or breather caps, and cause fluid voids to occur near the outlet port, which may cause air to enter the system. Properly placed baffle(s) divide a hydraulic reservoir into small sections, effectively reducing the sloshing and preventing those problems.

One important routine job for an equipment operator which ensures safe and efficient work is checking the amount of fluids stored in the reservoir. The most economical and very common practice is the use of a sight **fluid-level gauge** placed on the outside wall of a reservoir, so that the operator can check it easily without removing anything.

The **outlet line strainer**, also called the **outlet filter**, is commonly used on mobile hydraulic systems. This type of strainer is usually made of 100 mesh screen, having openings of approximately 150 microns (μ) to prevent large, solid contaminants from entering the hydraulic system. The outlet strainer must be sized properly to match the size of the hydraulic pump because a strainer too small would apply an exact restriction to the inlet flow to the pump, which in turn could lead to pump cavitation.

To indicate the fluid temperature in the tank, some hydraulic reservoirs, especially the smaller ones for mobile systems use, have an **internal temperature gauge** to avoid unintentional damage to the hydraulic system caused by excessively high fluid temperature. To keep the fluid temperature in the reservoir at an acceptable level all the time during operation, some hydraulic systems may be equipped with cooling components, such as cooling fans and heat exchangers. A **cooling fan** serves the purpose of accelerating the dissipation of the heat from the reservoir to lower the fluid temperature. Use of a heat exchanger in a hydraulic system will be discussed separately in a later subsection of this chapter.

### 6.2.3 Fluid Reservoir Sizing

Sizing a hydraulic reservoir includes determining the reservoir capacity and dimension. The **capacity** is the volume of fluid that the reservoir can hold and is measured in terms of liters (or gallons), whereas the **dimension** refers to its size and shape. In many cases, the capacity and dimension, plus the **material** used for building the reservoir, present the basic specifications of a hydraulic reservoir.

A rule of thumb for sizing a hydraulic reservoir suggests that its capacity should be **two to four times** the pump output flow as expressed in the following equation:

$$V = mq_p \tag{6.7}$$

where $V$ is the effective reservoir volume, $m$ is the design coefficient for hydraulic reservoir, normally taking $m = 2 \sim 4$, and $q_p$ is the rated pump flow rate.

This design equation means that when the pump discharges a 5 L/min flow (the rated flow for a fixed-displacement pump or the mean flow for a variable-displacement pump), the required reservoir capacity for this system is 10 to 20 liters for most applications. The rule is based on the assumption that such a volume can allow the fluid to sufficiently rest between work cycles for heat dissipation, contaminant settling, and deaeration. One should keep it in mind that this is only a rule of thumb for initial sizing and that there are new trends presenting some noticeable deviations from the norm for reservoir sizing. For example, new designs for mobile hydraulic systems often call for much smaller reservoirs than reservoirs sized on the basis of traditional rules of thumb. In such cases, it is strongly recommended that some special industry guidelines for minimum design of reservoirs be followed.

Whether a reservoir is designed using the traditional rule or industry guidelines for larger or smaller reservoirs, it is important to be aware of the parameters influencing the reservoir size required. Normally, a reservoir should have an additional space of no less than 10% of its fluid capacity to host thermal expansion of the fluid and gravity drain-back of fluid during shutdown, yet still be capable of providing a free fluid surface for de-aeration. As one example for requiring a larger reservoir, when large accumulators or cylinders are used in a system, those components often draw large volumes of fluid in operations; therefore, a larger reservoir may be required to ensure that the fluid level will not drop below the pump inlet at any time.

In many applications, a trend toward specifying a smaller reservoir has emerged either as a means of reaping economic benefits or due to spacing constraints. Because a smaller reservoir reduces the total amount of fluid that can be carried, some system modifications may be necessary to compensate for special problems caused by the lower volume of fluid contained in the reservoir. For example, because there is less surface area on a smaller reservoir for heat transfer, a heat exchanger might become necessary to keep fluid temperature within an acceptable range. Because the returning fluid is kept in the reservoir for a reduced amount of time, a high-capacity fluid filter would be required to trap contaminants. Perhaps the greatest challenge involved in using a smaller reservoir lies with removing the air from the fluid because of the insufficient time to release the air from the returning fluid before being drawn back into the pump inlet once again. A possible modification in the system design to solve this problem is to use a flow diffuser as illustrated in Figure 6.7(c) and (d) to lower the velocity of return fluid to reduce foaming and agitation.

There is no standard shape for reservoirs. Geometrically, a square or a rectangular prism has the largest heat-transfer surface per unit volume. A cylindrical shape, on the other hand, may be more economical to fabricate. In practice, one other factor to consider in reservoir shape design is the heat-dissipation area. If a reservoir is shallow, wide, and long, it may take up more floor space than necessary and not take full advantage of the heat-transfer surface of the walls because reservoir sides are often the most effective heat-transfer area. At the same time, a tall and narrow geometry conserves floor space and provides a large surface area for heat transfer from the sides. However, this shape may not provide enough area at the top surface of the fluid to let air escape.

In some systems, especially with mobile equipment, the hydraulic reservoir is built as an integral part of the equipment, and its placement is often an afterthought. In addition, mobile hydraulic reservoirs are expected to perform the same functions as their industrial counterparts, but usually under much more adverse and less predictable operating conditions. The size and weight limitations may allow mobile hydraulic reservoirs to carry only enough fluid for the pump to discharge it in one minute, roughly a third the size of a typical industrial reservoir. The space and shape limitations on placing those reservoirs make it necessary to custom-design the shapes for irregular areas. Regardless of the diversity in applications for those irregularly shaped small reservoirs, there is one common problem requiring special consideration: the influence of reservoir size and shape on the effectiveness of heat dissipation. Because the heat-dissipating capacity of a reservoir is a function of size, one practical solution to solve those problems is the use of a heat exchanger with a smaller reservoir.

**Example 6.1: Sizing Hydraulic Fluid Reservoirs**

For a typical hydraulic system powered using a 15 cc pump, driven by a prime mover operating at 2400 rpm, what will be the appropriate reservoir size for this system?

Picking a design coefficient of $m = 3$ to size the reservoir using Eq. (6.7):

$$\begin{aligned} V &= mq_p \\ &= 3 \times 15 \times 10^{-6} \times 2400 = 0.108\left(m^3\right) \end{aligned}$$

DISCUSSION 6.1: While the design coefficient $m = 2 \sim 4$ is recommended for most typical industrial applications, the design coefficient $m \leq 1$ is also commonly used for mobile applications.

## 6.3 Hydraulic Fluid Filters

### 6.3.1 Hydraulic Fluid Contamination

Other than the primary function of power transmission, fluids in a hydraulic system also perform a few other basic functions of removing contaminants, providing lubrication, and cooling the surfaces with relevant motion to ensure reliable operation. The removed contaminants are normally kept in the fluid, and when the fluid-carried contaminants reach a certain threshold level, we often say this fluid is contaminated. Surpassing all other causes, fluid contamination is the major contributor (more than 80% in many cases) of failures of hydraulic components and systems. Therefore, effective fluid contamination control is critical in hydraulic system maintenance.

Contaminants can exist in fluids as solid particles, water, air, and/or reactive chemicals, all of which impair fluid functions in one way or another. Contaminants can enter a hydraulic system during manufacturing and be internally generated or ingested from outside during normal operations. The primary manufacturing contaminants include dust, welding slag, rubber particles from hoses and seals, sand from castings, and metal debris from machined components. An effective and economical way of removing these types of contaminants is to properly and thoroughly flush them out before initially filling a reservoir with hydraulic fluids. During normal operation, dust and water may also enter the system through breather caps, imperfect seals, and any other openings. Operations will also generate internal contaminates, mainly in the form of component wear debris, chemical by-products from the fluid, and additive breakdown due to heat or chemical reactions. When such chemical contaminants react with hot component surfaces, even more contaminants are often created. Effective control of fluid contamination often requires some specific means to prevent particular types of contaminants from entering a hydraulic system and to trap those contaminants using some types of fluid filters.

A common form of contamination in hydraulic fluids is **solid particle contamination**. These contaminants consist mainly of particles of metal, sand, silica, loam, seal materials, and any other solid materials that may enter hydraulic fluids. Because these particles are normally present in microscopic scale, they are typically sized in microns, a unit of one millionth of a meter. Typically, there are very small clearances, from a few microns to a few tens of microns, between the surfaces of any moving pair. Figure 6.8 depicts how a few different-sized particles interfere differently with the clearance of component surfaces

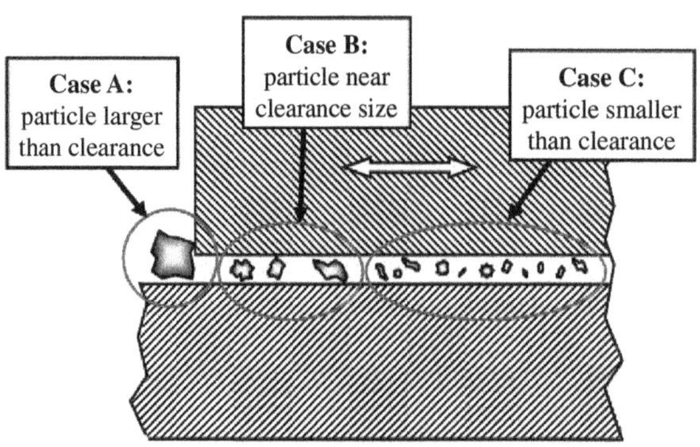


**FIGURE 6.8**
Concept illustration of typical particle sizes in relevant to clearance of moving pair.

of a moving pair. Even very small-sized particles in fluids may cause severe problems. As in Case A, illustrated in Figure 6.8, when the size of a particle is larger than the clearance, it cannot get into the clearance, and it has no interference with the component surfaces. However, it may block the opening to the clearance, which may consequently cause severe damage on component surfaces due to inadequate lubrication as an insufficient amount of fluids get into the clearance. When the particle size is a little bit smaller, very close to the size of the clearance as illustrated in Case B, these particles may get caught between the clearance and cause abrasive damage or jamming of the components. If all the particles are very small, as shown in Case C, they can normally pass through the clearance easily. However, if the velocity of the fluid passing through the clearance is very high, such small particles may erode metal surfaces and cause an abnormal wear on those components. Abrasive wear will shorten component life and in turn diminish system performance. The abrasive wear can often first be detected as reduced discharging flow rate because of increased internal leakage from the enlarged clearance between moving pairs. As the pump flow rate decreases, the system may become sluggish, as evidenced by slower hydraulic actuator movement. When the pump flow is reduced below a certain threshold, it will be difficult to build the pressure in the system and will eventually lead to sudden failure of the pump.

**Air and gas contamination** is another major form of hydraulic fluid contamination. It is important to realize that all hydraulic fluids contain **dissolved air**, existing as a solution in hydraulic fluids. The amount of air, measured by percentage volume that can be dissolved in fluid at a given pressure, is defined as the **saturation level** of the fluid. The saturation level of a typical petroleum-based hydraulic fluid is about 10% by volume at atmosphere pressure. The **entrained air** is the amount of air in excess of the saturation level and presented in bubbles in fluids. The size of those bubbles ranges from a few microns to naked-eye visible. The **free air** is actually the large bubbles or air pockets existing within the system, often trapped in high points in the system lines or in hydraulic pumps or actuators. Such free air trapped in pumps or motors can cause severe cavitation, which may cause damage to the interior surfaces of hydraulic pumps or other components due to inadequate lubrication, material fatigue, and friction heat.

In oil-based hydraulic fluids, **water contamination** is another major concern: water will lower the viscosity of hydraulic fluids, and will be vaporized at high temperatures to add **vapor bubbles** in the fluids. If water concentration in a hydraulic fluid reaches 1~2%, it will noticeably change the operating characteristics of the fluid and in turn affect the response of a hydraulic system. In practice, a poor system response can often be traced to water presence in a system. Vapor bubbles generated in water vaporization will cause cavitation in a pump and/or other components similar to free air. If **free water** is present in hydraulic fluid and the system operates at temperatures below the freezing point of water, **ice crystals** will also form. These crystals can cause damage to hydraulic systems just like solid contaminants can. When the water is presented in tiny water droplet form, it may be emulsified or suspended in hydraulic fluids and may give the fluids a cloudy or milky appearance. Sometimes such emulsions are so tight that it is very difficult to separate them from hydraulic fluids. While this is desirable in emulsion-type hydraulic fluids, it is highly undesirable for ordinary petroleum-based fluids. Furthermore, water can react with almost everything in a hydraulic system, which often promotes rust on metal surfaces, sticks to smaller contaminants, and produces acid products to accelerate wear on components.

The other major contamination sources to hydraulic fluids include **heat contamination**, which accelerates the degradation of hydraulic fluid properties; **chemical contamination**, which causes chemical changes or breakdowns of hydraulic fluids; and even **microbial contamination**, which degrades the quality of hydraulic fluids.

### 6.3.2 Fluid Cleanliness Measurements

Hydraulic fluid contamination is a combined adverse effect caused collectively by solid particles, air and gases, water, heat, chemical-reacting products, and many more. The quantitative measure of **fluid cleanliness** is normally made in terms of the amount of solid contaminants in a unit volume of fluid, namely, the concentration of solid contaminants. The solid concentration of fluids can be measured based either on the mass or on the total number of particles. ISO, the National Institute of Standards and Technology (NIST), the National Aeronautics and Space Administration (NASA), and SAE have all issued corresponding standard procedures and/or cleanliness measurement standards to quantify solid contaminants. The other contaminants, such as air and water, are usually measured by the percentage of air or water in the fluid.

A once widely used standard for fluid cleanliness measurement was SAE 749D, issued by the Society of Automobile Engineers in 1963. It divides solid particle contamination of hydraulic and lubricating fluids into seven classes, according to the total number of solid particles of different sizes carried in 100 ml fluid. This standard has lately been replaced by NAS 1638, which extended the cleanliness classes to 14 levels from the original SAE standard, as summarized in Table 6.1. NAS 1638 standard is still widely in use.

From practice, it was found that the NAS standard for fluid-cleanliness measurements presented some weaknesses, mainly on the unmatched patterns of particle allocation to the actual distribution in fluids. To solve this problem, ISO has adopted a revised procedure for reporting fluid-cleanliness measurements. The ISO 4406 standard uses three code numbers, which correspond to particle counts larger than 2, 5 and 15 μm in 1 ml fluids. Table 6.2 summarizes the defined ranges of particle count for 26 codes of fluid cleanliness. For example, a three-digit code ISO fluid cleanliness class 18/15/12 means that each ml of the fluid carries 1300~2500, 160~320, or 20~40 counts of particles 2 μm, 5 μm, or 15 μm or larger, respectively.

**TABLE 6.1**

NAS 1638 Codes for Particle Counts of Different Sizes in 100 ml Fluids.

| Cleanliness Class Code | Particle Size (μm) | | | | |
|---|---|---|---|---|---|
| | 5~15 | 15~25 | 25~50 | 50~100 | >100 |
| 00 | 125 | 22 | 4 | 1 | 0 |
| 0 | 250 | 44 | 8 | 2 | 0 |
| 1 | 500 | 89 | 16 | 3 | 1 |
| 2 | 1,000 | 178 | 32 | 6 | 1 |
| 3 | 2,000 | 356 | 63 | 11 | 2 |
| 4 | 4,000 | 712 | 126 | 22 | 4 |
| 5 | 8,000 | 1,425 | 253 | 45 | 8 |
| 6 | 16,000 | 2,850 | 506 | 90 | 16 |
| 7 | 32,000 | 5,700 | 1,012 | 180 | 32 |
| 8 | 64,000 | 11,400 | 2,025 | 360 | 64 |
| 9 | 128,000 | 22,800 | 4,050 | 720 | 128 |
| 10 | 256,000 | 45,600 | 8,100 | 1,440 | 256 |
| 11 | 512,000 | 91,200 | 16,200 | 2,880 | 512 |
| 12 | 1,024,000 | 182,400 | 32,400 | 5,760 | 1,024 |

**TABLE 6.2**

ISO 4406 Codes for Particle Counts of Different Sizes in 1 ml Fluids.

| Cleanliness Class Code | Particle Counts | | Cleanliness Class Code | Particle Counts | |
|---|---|---|---|---|---|
| | Greater than | Smaller than | | Greater than | Smaller than |
| 24 | 80,000 | 16000 | 11 | 10 | 20 |
| 23 | 40,000 | 80,000 | 10 | 5 | 10 |
| 22 | 20,000 | 40,000 | 9 | 2.5 | 5 |
| 21 | 10,000 | 20,000 | 8 | 1.3 | 2.5 |
| 20 | 5,000 | 10,000 | 7 | 0.64 | 1.3 |
| 19 | 2,500 | 5,000 | 6 | 0.32 | 0.64 |
| 18 | 1,300 | 2,500 | 5 | 0.16 | 0.32 |
| 17 | 640 | 1,300 | 4 | 0.08 | 0.16 |
| 16 | 320 | 640 | 3 | 0.04 | 0.08 |
| 15 | 160 | 320 | 2 | 0.02 | 0.04 |
| 14 | 80 | 160 | 1 | 0.01 | 0.02 |
| 13 | 40 | 80 | 0 | 0.005 | 0.01 |
| 12 | 20 | 40 | | | |

Industry has used the ISO 4406 3-digit coding method for a number of years. Recently, ISO introduced a new standard, ISO 11171, which replaced the old ISO 4402. One of the major differences between the new and old standards is that the new one uses three code numbers that correspond to concentrations of particles larger than 4, 6, and 14 μm in comparison to the older 2, 5, and 15 μm sizes. In addition, the new standard includes a number of enhancements to ensure better accuracy, reproducibility, and repeatability. In addition, ISO has developed another procedure, ISO 11943, for calibration and verification of online automatic particle counters.

### 6.3.3 Hydraulic Fluid Filters

In typical hydraulic systems, both fluid filters and strainers are commonly used for solid contamination control. While both serve the same function of removing solid particle contaminants from the fluids, a **strainer** is normally made of a coarse filter to prevent large particles from getting into the system, usually at the entry of the inlet pump, and a **filter** often refers to the prime filter for removing medium to fine particles from the fluids and is typically located at the return line close to the reservoir. Although it is relatively simple to select a strainer to use in a hydraulic reservoir, there are a confusingly large variation of filters suitable for different applications. Gaining a comprehensive understanding of how filtration works will make the task of selecting an appropriate filter easier.

As stated earlier, the primary function of fluid filters is to remove solid particle contaminants from the hydraulic fluid. Different types or designs of filters are often required to satisfy special requirements of fluid filtration in various operating conditions. For example, based on filtration capability, fluid filters can be classified as ultra-fine (1~5 μm), fine (10~20 μm), average (30~50 μm), and coarse (>80 μm). In terms of the media materials used, there are paper filters, synthetic fiberglass filters, and micro fiberglass filters. Depending on the applicable operating pressures, all can be sorted into the two categories of **low-pressure filters** and **high-pressure filters**.

A typical fluid filter can be used either with or without a case as illustrated in Figure 6.9. Normally, a case is necessary for those used on hydraulic lines to hold the filter, whereas no

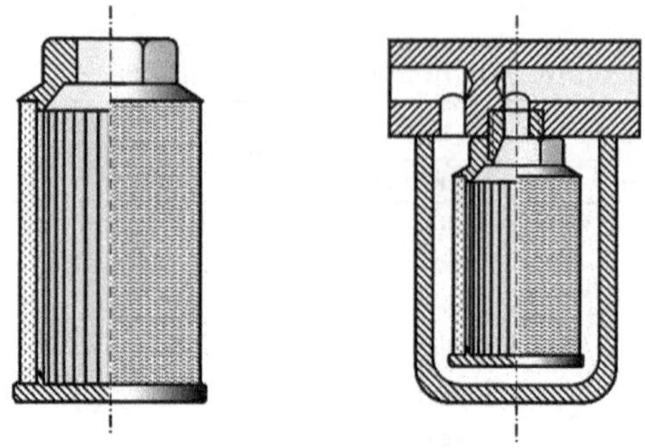

(a) No-case filter (b) Cased filter

**FIGURE 6.9**
Illustration of the configuration of typical fluid filters.

case is needed for those used in fluid reservoirs. As defined earlier, filters used in reservoirs are often called **strainers**. Without loss of generality, a typical filter often refers only to those used on lines and normally consists of a holding case, a filter body, and other accessory components. During filtration, the fluid normally flows through a filter body from the sides and is discharged on the exit port located on the top. This implies that the filter body plays the determinative role in the effectiveness of fluid filtration. To quantitatively assess such effectiveness and performance, it is important to know a few characteristic parameters, such as the filtration rating, pressure drop, and dirt-holding capacity of the filter.

The **filtration rating** is a measure of the filter ability to remove contaminants of different sizes from the system, and therefore is often expressed simply by the size of contaminants that will be removed, for example a "25 micron filter." However, it is necessary to point out that this way of naming a filter is purely a commonly understood approach, and is not backed up by any industry standard. Also, it is important to note that whether or not the filter rating is properly selected for the application will directly affect the performance of contamination control in a system. Two frequently used terms of filter ratings are **absolute rating** and **nominal rating**, with the former indicating the largest diameter particles and the latter the average sized particles that can pass through the filter. While those ratings presumably define 100% filtration efficiency for the filter, it is, in fact, impossible to realize in practice. The **filtration efficiency**, defined as the ratio of the number of removed contaminant by a filter to the number of total contaminants carried in upstream fluid, is often used to quantity the effectiveness of a filter. Because it is more difficult to count the contaminants being removed by a filter than those passing through the filter and carried in the downstream fluid, a **beta ratio** is therefore commonly used to quantify the filter performance in practice. As depicted in Figure 6.10, the beta ratio of an operating filter is defined as the number of upstream contaminants of a selected size divided by the number of the downstream contaminants sampled at a steady-state flow condition, expressed as follows:

$$\beta_x = \frac{N_u}{N_d} \tag{6.8}$$

where $\beta_x$ is the beta ratio of a filter to particles of $x$ microns (μm), and $N_u$ and $N_d$ are contaminant counts in upstream and downstream fluids, respectively.

Based on the definition, filter ratios shown in Figure 6.10 ($\beta_5 = 5$ or 100) indicate that the ratio of 5 μm diameter particles in the upstream fluid to that in the downstream fluid is

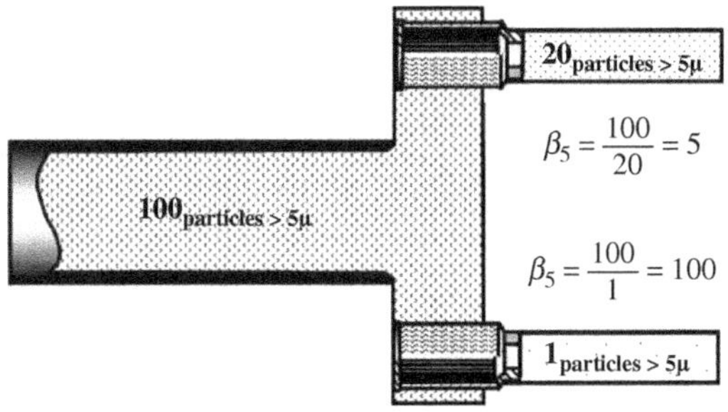


**FIGURE 6.10**
Graphical definition of filter beta ratio.

5 or 100. That is, for every 100 particles of 5 μm or larger particles entering the filter, only 20 or 1 could pass through it; hence, 80 or 99 particles are captured by corresponding filters, respectively. This implies that these filters can remove contaminants with an efficiency of 80 or 99%. The **filter efficiency** can therefore be calculated by using the following formula:

$$\eta_f = \frac{\beta_x - 1}{\beta_x} \tag{6.9}$$

where $\eta_f$ is the filtration efficiency.

Because a filter's ability to remove particles of different sizes is different, its beta ratios to those particles are different as well: the larger the particle size, the higher the beta ratio. Therefore, a beta ratio should always be noted with its corresponding particle size, such as $\beta_5$ or $\beta_{10}$. A filter with a beta ratio of $\beta_{20} = 200$ indicates that this filter has a filtration efficiency of 99.5% for particles of 20 μm or bigger. However, it may still not be suitable for hydraulic system use because its efficiency in removing smaller particles is unspecified. Table 6.3 summarizes the relation between beta value and filtration efficiency. It is a common acceptable understanding that a beta rating of 75 or below is normally unacceptable for use in hydraulic systems.

There is an apparent pressure loss even for clean fluids passing through a filter, attributed mainly to the viscosity resistance of the fluids. As illustrated in Figure 6.11, after putting a filter in use, the contaminants removed from the fluids will gradually be accumulated on the filter media materials, which will in turn cause the pressure to drop due to the increased resistance of fluids passing through the media. After the pressure drop reaches the **saturated pressure drop** value, namely, the turning point on the characteristic curve, the drop in pressure will be dramatically increased until the filter media is completely full of trapped contaminants. When the drop in pressure reaches the saturated drop point, the

**TABLE 6.3**

Beta Ratios and Corresponding Efficiencies of a Filter.

| Beta Ratio | Efficiency | Beta Ratio | Efficiency |
|---|---|---|---|
| 1 | 0% | 75 | 98.67% |
| 2 | 50% | 100 | 99.00% |
| 5 | 80% | 200 | 99.50% |
| 10 | 90% | 1000 | 99.90% |
| 20 | 95% | 5000 | 99.98% |

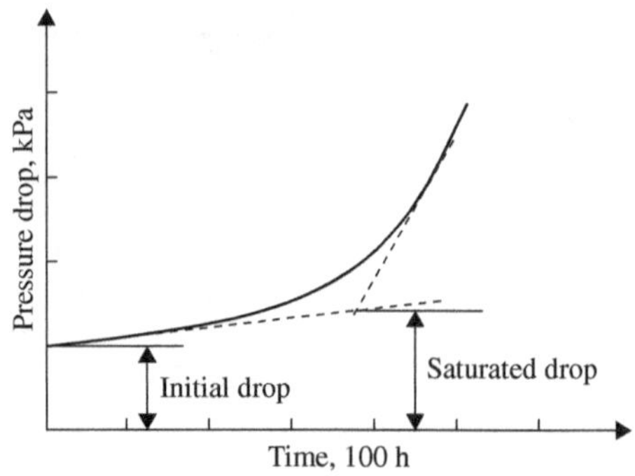


**FIGURE 6.11**
Illustration of typical characteristics of pressure drop across a fluid filter.

filter should either be replaced or cleaned to achieve the best balance between the system efficiency and the filtration efficiency. Some fluid filters are equipped with a pressure sensor to detect the optimal time to change or clean the filter. The amount of contaminants a filter can hold without affecting the normal performance of the filter is often called the **filter dirt-holding capacity**. It should be clearly understood that the filter dirt-holding capacity is not an indicator of the filter's service life.

Water, being one of the main contaminants in hydraulic systems, should also be removed. Because water is normally heavier than most hydraulic fluids, a properly designed fluid reservoir should allow the water to settle at the bottom and then drain it away through a drain plug. This is probably the simplest way to remove the water from the system. Another effective way to separate free water from hydraulic fluids is to use water-removing filters, either coalescing or chemical types. A coalescing water filter removes water by passing the hydraulic fluids through a dense inorganic filter mat to retain water droplets on the fibers. In comparison, a chemical water filter removes the water by either adsorbing or absorbing the water from the fluids. A few other methods include evaporation and centrifuging, mainly for removing free water. Although different devices are available for removing free water, the only device that can remove all free, emulsified, and dissolved water is a fluid purifier.

**Example 6.2: Hydraulic Filter Ratings**

When a hydraulic filter is used, it can reliably remove at least 999 out of 1000 14 μm or larger particles, at least 99 out of 100 6 μm or larger particles, and at least 95 out of 100 4 μm or larger. Further analysis found that the average particle size checked before the filtration is 6 μm. What are the absolute filter ratings for different-sized particles, and what is the nominal rating of the filter?

Absolute filter ratings for different-sized particles can be measured using a beta ratio as defined by Eq. (6.8):

$$\beta_{14} = \frac{N_u}{N_d} = \frac{1000}{1} = 1000$$

$$\beta_6 = \frac{N_u}{N_d} = \frac{100}{1} = 100$$

$$\beta_4 = \frac{N_u}{N_d} = \frac{100}{5} = 20$$

Nominal rating is measured by the filtration efficiency of average size particles:

$$\beta_6 = \frac{N_u}{N_d} = \frac{100}{1} = 100$$

DISCUSSION 6.2: Because it is difficult to count the contaminating particles being removed by a filter, but relatively easier to count those particles carried in both upstream and downstream fluids, a **beta ratio** provides a practical way to measure filtration efficiency for different-sized particles.

## 6.4 Other Components

### 6.4.1 Heat Exchangers

Almost all hydraulic systems generate substantial **heat** attributed mainly to the energy losses from overcoming line resistances. From energy balance discussions earlier in this textbook, we know that all the energy exceeds the amount being converted to mechanical power and will be eventually lost to heat in the system. Even well-designed hydraulic proportional valve systems may convert over 60% of input fluid power to heat. A necessary amount of heat generated in hydraulic systems is desirable to bring hydraulic fluids up to normal operating temperature. Cold hydraulic fluids generally have a higher viscosity that may cause sluggish operation and/or excessive pressure drop during operation. However, if the heat generated exceeds the radiation rate from the system, the excess heat will start cooking the fluids, which in turn will lead to fluid decomposition, form varnish on system component surfaces, and begin to deteriorate system seals. Such problems are much more severe in mobile hydraulic systems than in industrial ones because of the smaller size of reservoirs.

To figure out how to maintain fluid temperature within an acceptable range, it is necessary to estimate the amount of heat that will be generated in the system and how much the temperature will rise. One practical method for estimating the heat generation or the temperature rise is to consider only the worst case scenario. As illustrated in Figure 6.12, the worst-case scenario for heat generation in most hydraulic systems occurs when all fluids are dumped back to the reservoir through a line-relief valve as it will convert all potential energy carried in the fluid into heat. The heat generation rate in this case can be calculated in terms of the flow rate and the pressure drop across the relief valve using the following equation:

$$\dot{q}_g = Q\left(p_P - p_T\right) \tag{6.10}$$

where $\dot{q}_g$ is the heat generation rate, $Q$ is the volumetric flow rate discharged from pump, and $p_P$ and $p_T$ are fluid pressures at the pump-discharge port and reservoir.

The resulting fluid temperature rise from the generated heat can be calculated using the following formula:

$$\Delta T = \frac{\dot{q}_g}{c_v \rho Q} \tag{6.11}$$

where $\Delta T$ is the fluid temperature rise, $c_v$ is the constant volume specific heat, and $\rho$ is the density of the fluid.

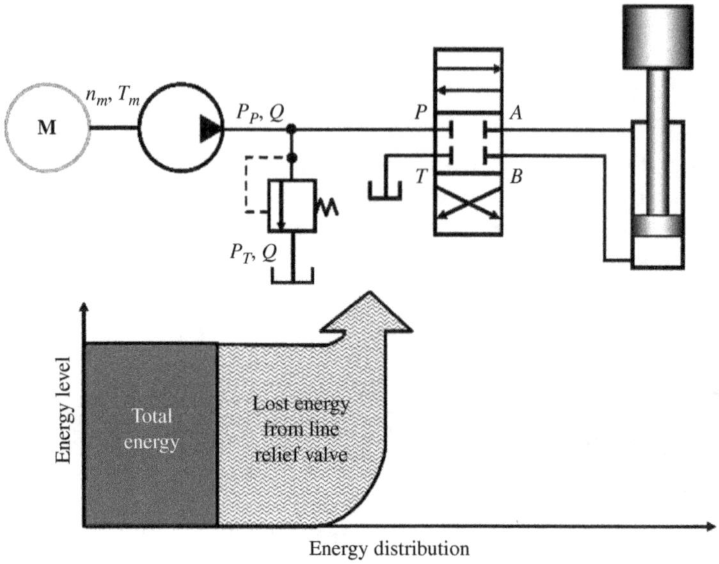


**FIGURE 6.12**
Concept illustration of the worst case scenario for heat generating – all flow through a relief valve.

A practical and efficient solution to maintain fluid temperature at an acceptable range is to dissipate the generated heat using a properly sized **heat exchanger** in the system. Based on the fact that heat is a form of energy migrating naturally from a hotter to a cooler region, heat exchangers in hydraulic systems are commonly used to dissipate heat more efficiently from hot hydraulic fluids. Despite the variation in designs, all heat exchangers should be able to dissipate enough heat from a system within a given time interval. To properly size a heat exchanger, the following equation is commonly used to calculate the heat load for a hydraulic system:

$$\dot{q}_d = UA\Delta T \tag{6.12}$$

where $\dot{q}_d$ is the heat-dissipating rate, $U$ is the overall heat-transfer coefficient of a heat exchanger, $A$ is the heat-transfer surface area, and $\Delta T$ is the temperature difference between the hydraulic fluids and the coolant.

Among three design parameters of heat exchangers, this equation reveals that the heat-transfer rate will proportionally increase as any one of the parameters increases. For example, doubling the surface area in contact with the hot fluids will increase the heat-transfer rate twofold, and increasing the temperature difference between the hydraulic oil and the coolant by 50% will also increase the heat transfer rate by 50%. While alternating the heat-transfer coefficient can result in a proportional change in the heat-dissipating rate, adjusting the heat-transfer coefficient is normally more complicated than changing either the contacting surface area or the temperature difference because of being composed of several mechanisms. The first heat-transfer coefficient mechanism is the convective heat transfer from the hot fluid to the wall separating it from the coolant and can be called the **hot fluid thermal resistance**, depending primarily on both physical and thermal

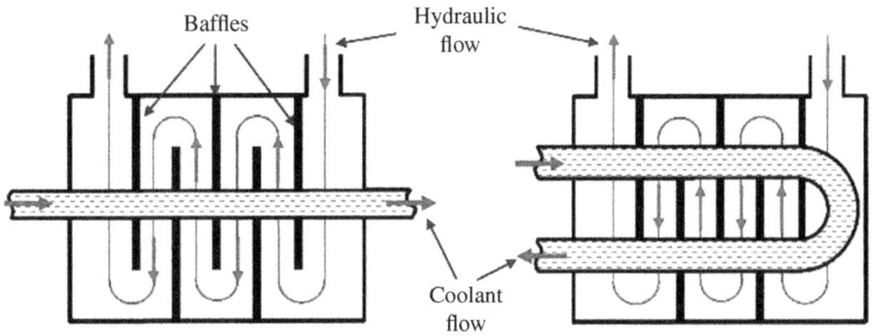


**FIGURE 6.13**
Illustration of the principle of (a) single-pass and (b) double-pass heat exchangers.

properties of the fluids. For example, a higher-velocity flow always results in a higher heat-transfer rate. The second mechanism is thermal conductance through the tube wall. Most heat exchanger tubes are made from copper or aluminum alloys or similar materials that exhibit high thermal conductivity. The third mechanism is the convection of heat from the tube wall to the coolant in the tube, in much the same manner as the hot fluid thermal resistance. Use of multipass flow patterns takes advantage of the fluid velocity and turbulence for increased *U* values. Single-, double-pass, and multi-pass configuration heat exchangers are commonly used in hydraulic systems. As depicted in Figure 6.13, as the coolant pass increases from one to two, it results in the coolant flowing twice the length in the heat exchanger, which in turn increases the cooling rate, promotes turbulence in hydraulic flow, and destroys the insulating film existing with laminar fluid flow. The design of heat exchangers is beyond the scope of this textbook. Those interested in the design of heat exchangers for the hydraulic system can refer to heat-transfer and heat-exchanger textbooks.

**Example 6.3: Size a Heat Exchanger**

A fixed-displacement hydraulic pump discharges 75 L/min fluids at 14 MPa to a system. When the system is idling, all the pump flow is released back to the reservoir through a line relief valve. How much heat will be generated in this case if the gage pressure of the fluid in the reservoir is zero? If the fluid temperature at the pump discharge port is 50°C, what will be the fluid temperature returning to the reservoir (assume fluid density is 895 kg·m$^{-3}$ and fluid specific heat is 1.8 kJ/kg·°C)? What is the required heat-exchange rate to efficiently dissipate all the heat generated in the operation, including that from pump loss (assume 85%) and line loss (assume 10%)?

a. The heat generated in releasing fluid back to the reservoir can be calculated using Eq. (6.10):

$$\begin{aligned}\dot{q}_g &= Q(p_P - p_T)\\ &= \frac{75\times10^{-3}}{60}\times\left(14\times10^6 - 0\right) = 17,500(W) = 17.5(kW)\end{aligned}$$

b. The returning fluid temperature can be determined according to Eq. (6.11):

$$\Delta T = \frac{\dot{q}_g}{c_v \rho Q}$$

$$= \frac{17500}{895 \times 1800 \times \dfrac{75 \times 10^{-3}}{60}} = 8.7(°C)$$

$$T_r = T_o + \Delta T$$

$$= 50 + 8.7 = 58.7(°C)$$

c. The heat-exchange rate can be estimated in terms of total energy loss:
   1. Energy loss at the relief valve:

$$\dot{q}_{RV} = \dot{q}_g = 17.5(kW)$$

   2. Energy loss attributed to pump inefficiency:

$$\dot{q}_{PP} = P_{in} - P_{out} = \left(\frac{1}{\eta_P} - 1\right) P_{out}$$

$$= \left(\frac{1}{0.85} - 1\right) \times 17.5 = 3.1(kW)$$

   3. Energy loss attributed to line resistance:

$$\dot{q}_{LN} = \eta_{LN} P_{out}$$

$$= 0.10 \times 17.5 = 1.8(kW)$$

   4. Required heat exchange capacity:

$$\dot{q} = \Sigma \dot{q}_i = \dot{q}_{RV} + \dot{q}_{PP} + \dot{q}_{LN}$$

$$= 17.5 + 3.1 + 1.8 = 22.4(kW)$$

**DISCUSSION 6.3:** When considering application and sizing of heat exchangers, the steady-state temperature of the hydraulic fluid and the time it takes to arrive at that temperature should be used. Other parameters to be considered in the sizing process include, but are not limited to, the fluid heat load, the fluid flow rate, and the pressure drop.

### 6.4.2 Seals

Leaking is one of the most common defects of hydraulic systems. Use of a proper seal is the most effective and primary means to prevent leakage, as well as excluding contaminants, in most hydraulic systems. Sealing effectiveness has a direct impact on the performance of a hydraulic system: the internal leakage resulting from an imperfect seal will lower the system volumetric efficiency, and the external leakage caused by a loose seal will even contaminate the environment. However, if a very tight seal were used, it would undoubtedly reduce leakage. Meanwhile, it will also significantly increase the friction, which will not only lower system mechanical efficiency, but also shorten service life to the seal due to excessive heat generated from the friction.

Some hydraulic fluids, especially some of the biologically degradable ones, may be very harsh to seal materials and will reduce the service life of the seals noticeably. A "perfect" seal should be one that prevents all leakage without adding much friction under any operational conditions. In practice, unfortunately, such a seal does not exist, and it is important to select the right type of seal for the right application.

To make a hydraulic system operate under ideal conditions, the seals used in the system should not only provide reliable sealing performance, but also have a long service life and low-friction resistance. Normally, there are two types of contact and noncontact seals to be chosen from for different applications. The contact seal is often used to seal the clearance between immobile components, and the noncontact type is only for applications with a relative motion between the mating surfaces. Because of their normal applications, the contact-type seals are also called the static seals and the noncontact type the dynamic seals. A suitable seal for a particular application is normally selected in terms of system pressure, fluid temperature, and relative speed of mating surfaces.

The noncontact seal uses a tiny clearance between mating surfaces to stop fluids from leaking from the gap while allowing the mating components to move freely. The sealing effectiveness is determined by the size of the gap, the length of the seal, the pressure difference between the two ends, and the smoothness of the relevant moving surfaces. Because of the existence of a small gap, this type of seal is in general low in friction, low in heat generation, and long in service life. Because this type of seal can hardly eliminate fluid leakage, it is commonly used to reduce internal leakages.

In comparison, the material of contact seals must be conformed closely enough to the microscopic irregularities of the mating surfaces to prevent pressure fluid penetration. The ideal contact-type seals should possess the following properties: (1) the seal must have enough flexibility to expand or compress promptly to fit the gap variation caused by any reason; and (2) the seal must have sufficient modulus and hardness to withstand shear stress produced by system pressure and to resist being forced into the extrusion gap. To meet different requirements for different applications, seals are often designed in some special shapes, such as "O" shapes and "lip" shapes, in their cross section.

**O-ring** seals have a round shape cross section, are often made of oil-resistant rubber or synthetic rubber, and are the most commonly used seal type in hydraulic systems. As depicted in Figure 6.14, a typical O-ring seal is normally made of materials with a certain degree of compressibility, so it can conform closely to the mating surfaces to seal the possible fluid passage when installed in a groove (Figure 6.14(b)). When operating under a

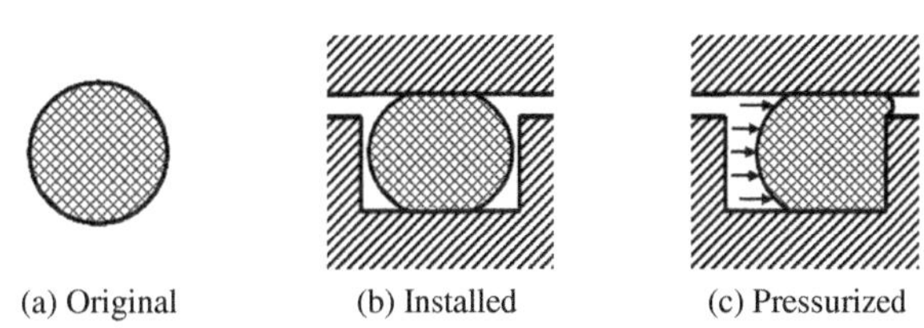


**FIGURE 6.14**
Concept illustration of a typical "O" seal under different operating conditions. (a) Original, (b) just installed and (c) pressurized conditions.

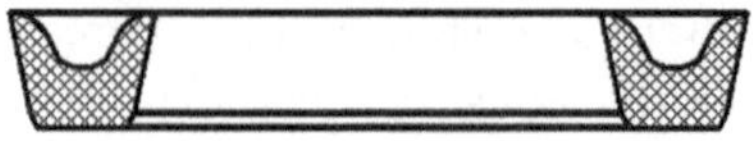
(a) "Y" shape seal

(b) "U" shape seal

(c) "V" shape seal

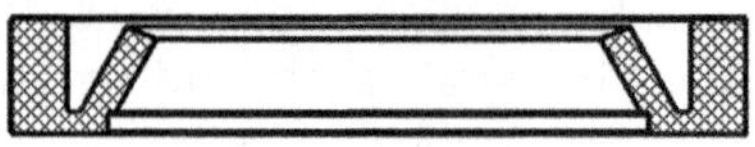
(d) "J" shape seal

**FIGURE 6.15**
Cross-section illustration of (a) "Y", (b) "U", (c) "V" and (d) "J"-shaped seals.

high pressure, the O-ring is pushed by the high-pressure fluid to the low-pressure side of the groove as depicted in Figure 6.14(c) and results in a tighter seal. However, if the pressure is too high, the O-ring material has too high a flexibility, and the clearance gap is too large, the O-ring may be forced into the extrusion gap and will lose the seal. To prevent such a loss from occurring, it is common practice to use a backup ring, or one on each side if the high pressure may come from both sides, to reinforce the seal in high-pressure systems.

To increase seal reliability, a class of lip-shaped seals is also frequently used in many hydraulic systems. There are a few different designs in their cross-sectional shape. Figure 6.15 depicts some commonly seen lip-shaped seals in Y, U, V, and J designs. It is important to remember that the concave side of those lip-shaped seals should face toward the pressure fluid so that the pressure fluid can push the lips closely attached to the surfaces of the mating components to achieve a more reliable seal. If a lip-shaped seal faced the opposite direction, it would definitely result in a malfunction in the seal.

The Y-shaped lip seal is often used to seal a pair of reciprocating mating components because of its capability to seal both internal and external surfaces, as well as because of the low friction and ease of installation. One major limitation of this shape seal is that it could "flip over" when used in applications with both high pressure and high reciprocating speed, resulting in seal failure. To prevent such an incident from happening, a tall-design Y-shaped lip seal, which has at least twice the height of its width, should be used. It should be noted that the lips in a typical tall, Y-shaped lip seal are normally different in height, with the longer lip then sealing the mating surface of the moving component. For instance, if the piston of a mating pair is the moving component, a longer internal lip, tall Y seal (Figure 6.16(a)) should be used. When the cylinder is the moving part, then a longer external lip tall Y seal (Figure 6.16(b)) should be selected.

The U- and V-shaped lip seals can be used for sealing radial mating surfaces, with either reciprocal or rotating motion. One notable feature of U- and V-shaped seals is that both request a supporting ring, made of either metal or nonmetal materials. The J-shaped lip seal is often used to prevent dust from entering the hydraulic system.

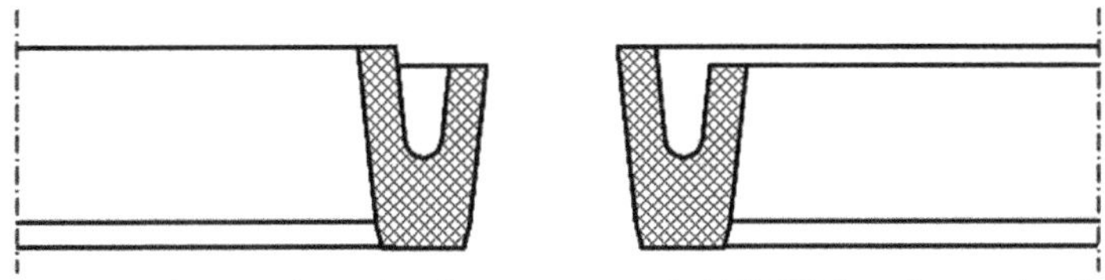


**FIGURE 6.16**
Cross-section illustration of typical (a) internal and (b) external tall design "Y"-shaped lip seals.

## References

1. Akers, A., Gassman, M., Smith, R. *Hydraulic Power System Analysis*. CRC Press, Boca Raton, FL (2006).
2. Caterpillar, Inc. *Biodegradable Hydraulic Oil*. Caterpillar, Inc., Peoria, IL (1997).
3. Esposito, A. *Fluid Power with Applications* (*6th Ed.*). Prentice-Hall, Upper Saddle River, NJ (2003).
4. Hunt, T.M. Filtration standards for hydraulic fluid power. *Filtration & Separation*, 33: 465–470 (1996).
5. Hydraulics & Pneumatics. *Fluid Power Basics*. http://www.hydraulicspneumatics.com/200/FPE/IndexPage.aspx. Accessed on November 20 (2006).
6. Pease, D.A. *Basic Fluid Power*. Prentice-Hall, Englewood Cliffs, NJ (1967).
7. Stoecker, W.F. *Design of Thermal Systems*. McGraw-Hill, New York (1989).
8. Vickers, Inc. *Vickers Mobile Hydraulics Manual* (*2nd Ed.*). Vickers, Inc., Rochester Hills, MI, (1998).
9. Welty, J.R., Wicks, C.E., Wilson, R.E. *Fundamentals of Momentum, Heat, and Mass Transfer* (*3rd Ed.*). John Wiley & Sons, New York (1984).
10. Yeaple, F.D. *Fluid Power Design Handbook*. CRC Press, Boca Raton, FL (1996).
11. Zhang, Q., Goering, C.E. Fluid power system. In: Bishop, R. (ed.), *The Mechatronics Handbook*. CRC Press, Boca Raton, FL, pp. 10–11~10–14 (2001).

## Exercises

**6.1** In terms of which viscosity measure does the SAE define the winter and summer grades for hydraulic fluids?

**6.2** What fluid property does the bulk modulus of a fluid measure?

**6.3** Why are antifoaming characteristics a critical property for hydraulic fluids?

**6.4** List the general rules for properly controlling the operating temperature of hydraulic fluids.

**6.5** Compare the major advantages and disadvantages of petroleum-based, environmentally safe, and fire-resistant hydraulic fluids.

**6.6** What is the measure of a readily biodegradable for environmentally safe hydraulic fluids?

**6.7** What is the common accepted rule of thumb for sizing a hydraulic fluid reservoir?

**6.8** What is the primary function of baffle plate(s) in a hydraulic reservoir?

**6.9** List three leading contaminations in hydraulic fluids, and identify the major contributors to those contaminations.

**6.10** Why is water contamination one of the major concerns in oil-based hydraulic fluids?

**6.11** List three effective means for removing water contaminants from hydraulic systems, and analyze their application features based on their operation principles.

**6.12** Why will almost all hydraulic systems generate substantial heat during operation?

**6.13** What is the difference between a static and dynamic seal?

**6.14** What would be an adequate size reservoir for a typical hydraulic system using a 100 L/min pump?

**6.15** For a hydraulic fluid filter rated $\beta_{14} = 5000$, $\beta_6 = 100$ and $\beta_4 = 75$, what is the filtration efficiency for moving specific-sized particles? If the average contaminating particle size is 6 μm, what is the nominal efficiency of the filter in removing contaminants?

**6.16** What does a fluid cleanliness class code 1 mean according to NAS 1638 standard?

**6.17** What does a three-digit code fluid cleanliness class 18/15/12 mean according to ISO 4406 standard? What does the same code class mean according to ISO 11171 standards?

**6.18** When a closed-center directional control valve is set in its neutral position, all the pump discharge flow will be dumped directly through a line-relief valve back to the reservoir. If a 50 L/min fixed-displacement pump is used in such a hydraulic system and the line-relief valve is set to be opened at 30 MPa, how much heat will be generated? (Assume that the reservoir pressure is 1.5 MPa.) What is the corresponding temperature rise? (Assume that fluid density is 895 kg/m$^3$ and fluid-specific heat is 1.8 kJ/kg·°C.)

**6.19** When a closed-center direction control valve is set in its neutral position, all the pump-discharge flow will be dumped directly through a line-relief valve back to the reservoir. If a variable-displacement pump discharges 3 L/min at a margin pressure of 5.5 MPa, under the conditions, how much heat will be generated? (Assume that the reservoir pressure is 1.5 MPa.) What is the corresponding fluid temperature? (Assume that fluid density is 895 kg/m$^3$ and fluid-specific heat is 1.8 kJ/kg·°C.)

**6.20** For the system defined in Problem 6.18, what is the required heat-exchange rating for efficiently dissipating all the heat generated in the operation, including that from pump loss (assume 85%) and line loss (assume 10%)?

# 7

# Hydraulic Circuits

## 7.1 Basic Circuits

Hydraulic systems are widely used in industry and mobile machinery. As technology advances, especially with the convergence of the electronics into hydraulic systems, they are becoming more sophisticated. No matter how complicated a particular hydraulic system is, it is often integrated by a few basic circuits. A basic circuit refers to a most simple hydraulic system, being composed of a few pertinent components and capable of performing a specific operational function. According to their functionality in a hydraulic system, these basic circuits can be classified as pressure control, direction control, speed control, sequence control, and synchronizing control circuits. One should realize that all the circuits introduced here are used primarily for illustration purposes and may not be directly applicable to some practical uses. Some engineering design work is often required to make those basic circuits practical and applicable, and many additional different designs of circuits are being used for actual applications.

### 7.1.1 Pressure Control Circuits

**Pressure control circuits** are those fundamental circuits designed to control the pressure of an entire system or a branch of the system, normally using one or more pressure control valve for regulating, reducing, balancing the operating pressure, or absorbing the pressure surges.

Pressure-regulating circuits are used to control system pressure for designated operations. The most fundamental pressure control circuit is the **line-relief circuit** (Figure 7.1). Such a circuit typically uses a line-relief valve to set the maximum allowable system pressure to ensure the safe operation of a hydraulic system. As explained in Chapter 3, a preloaded spring in the line-relief valve holds the normal-close pressure control valve closed when the fluid pressure at the inlet port of the valve is below its cracking pressure. As the fluid pressure rises to exceed the line-releasing pressure, the line-relief valve opens to create a path to release the pressurized fluids to the reservoir. To guarantee that the implement actuator will operate properly all the time, it is important to set the system pressure a certain margin higher than the maximum actuator operating pressure, plus the total pressure head loss, to deliver the pressurized fluid from the pump to the actuator.

The line-relief circuit offers a satisfactory system pressure-regulating function when the system is operating at one pressure setting. In some applications, a dual-pressure setting is required to implement high-pressure extensions and low-pressure retractions.

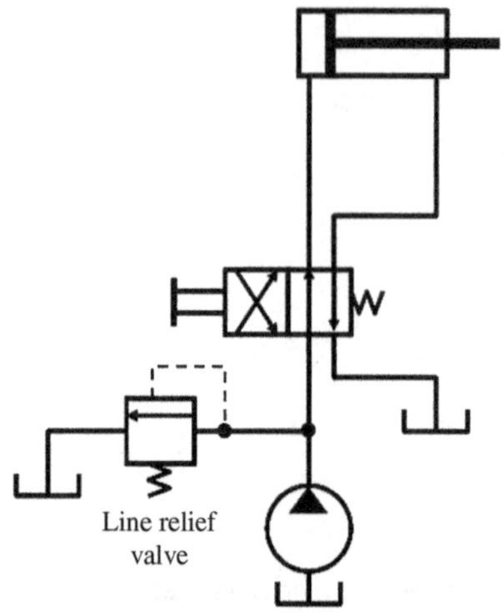


**FIGURE 7.1**
Schematic illustration of a typical line relief circuit.

To realize such a function, it is necessary to have a two-stage pressure regulator. Figure 7.2 shows a typical **two-stage pressure-regulating circuit**. In cylinder-extending operations, the load-release valve stays closed due to the high pressure in the working line between the cylinder head end and the valve, and the line-relief valve is used to regulate the system pressure in terms of the higher setting of system pressure. In retracting, the pressure in the working line between the cylinder head end and the valve is low, which will make the load-release valve open and consequently bleed the back pressure at the pilot line-relief valve. Thus, the circuit will open the main line-relief valve at a reduced-pressure setting. In this circuit, the check valve in the load-releasing branch prevents the pilot relief valve from functioning during the cylinder extension.

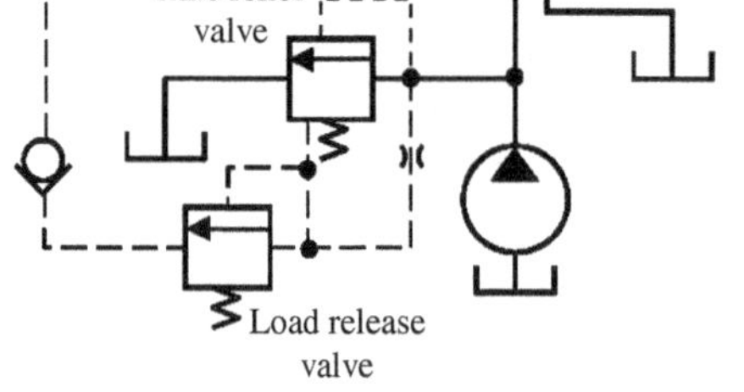


**FIGURE 7.2**
Schematic illustration of a typical two-stage pressure-regulating circuit.

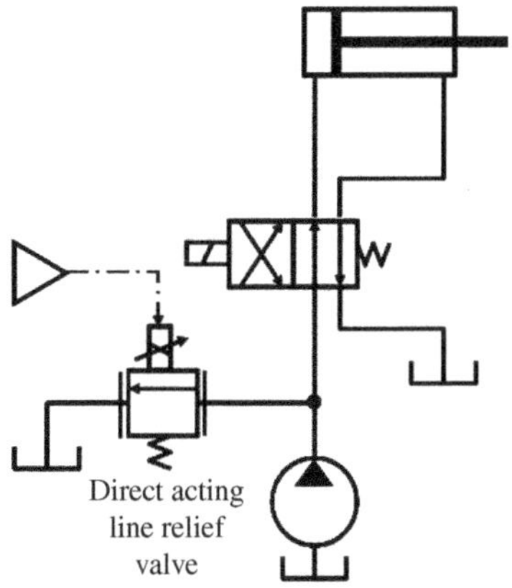


**FIGURE 7.3**
Schematic illustration of a typical direct-acting proportional pressure-regulating circuit.

When an electrically actuated proportional line-relief valve is used to replace the pilot-operated line-relief valve presented in Figure 7.1, it is then capable of carrying out continuous adjustments on the system pressure setting. Figure 7.3 presents a circuit using an electrically direct- acting proportional relief valve to adjust the line-release pressure in terms of inputting the electronic control signal. Because normally there is no pressure sensing capability on an electrical direct-acting line-relief valve, such a circuit often requires having an electronic pressure transducer sense the system pressure to support this functionality.

Another category of pressure control is pressure reducing. A **pressure-reducing circuit** is often used in situations where a single-pump system needs to supply two or more branches operating at different pressures. One of the common features of this type of circuit is the use of a pressure-reducing valve. As shown in Figure 7.4, this circuit uses the line-relief

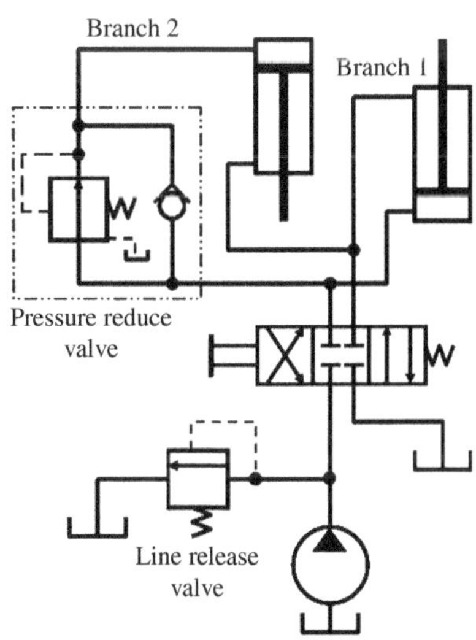


**FIGURE 7.4**
Schematic illustration of a typical pressure-reducing circuit.

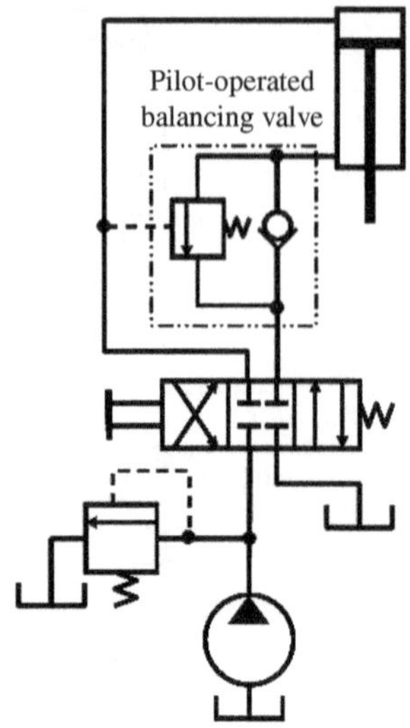


FIGURE 7.5
Schematic illustration of a typical pressure-balancing circuit.

valve to set maximum system pressure, which also provides the higher pressure to branch 1. To provide a lower pressure to branch 2, a pressure-reducing valve is used to reduce the pressure of the fluids supplying this branch. As explained in Chapter 3, this normal-open pressure control valve will start to close as the outlet pressure from the valve surpasses a preset level. As the flow path area decreases, the pressure drop across the valve will be increased in a pattern described by the orifice equation. Thus, by controlling the valve-opening area, this circuit can adjust the operating pressure at branch 2. The check valve in parallel with the pressure-reducing valve in branch 2 allows the return fluid to flow freely back to the tank in retraction.

A **balancing circuit** is designed to prevent cylinder overrunning when pulled by an excessive external load. As shown in Figure 7.5, a typical balancing circuit employs a pilot-operated pressure-balancing valve to provide the required prevention function. When the hydraulic power pushes the cylinder downward, the high pressure at the line connecting to the cylinder cap-end will push the balancing valve open to provide a path for the fluids in the rod-end chamber to return to the tank and therefore move the load. When an excessive pulling load acts on the cylinder, the cylinder will move faster than the pump can support, which may result in a pressure drop, or even a cavitation, on the cap-end line. Such reduced cap-end line pressure will then release an actuating force on the balancing valve, which in turn will reduce the fluid-passing area to build back pressure on the rod-end chamber to prevent a cylinder overrun from occurring.

Another fundamental pressure control circuit is the **bleed-recharge circuit**. In many field operations, the mobile machinery may bear some unexpected load changes. Such sudden load changes will often result in a pressure surge or a cavitation within the system, which may cause severe damage to the machinery. A bleeding-recharging circuit is specially designed to absorb pressure surges and prevent cavitation, as depicted in Figure 7.6. This circuit employs four check valves and a pressure relief valve to provide the required bleeding-recharging function to both sides of a bi-directional motor. When a sudden motor slowdown or stop takes placed, caused by an excessive external load being applied to the

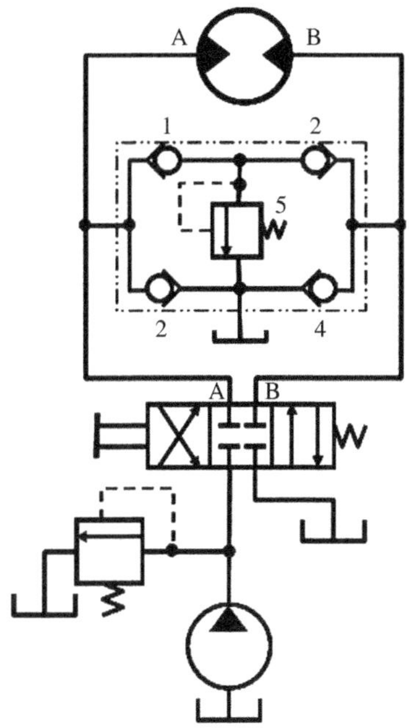


**FIGURE 7.6**
Schematic illustration of a typical bleeding-recharging circuit.

motor, in the result will be a pressure surge in the line connecting the A (or B) ports of the motor and the main control valve. This high pressure will force both check valve 1 and line-relief valve 5 open to bleed fluid to the tank, thus reducing the line pressure to a safe level. When a sudden speed-up is caused by a sudden loss of external load, it may induce cavitation on the line connecting the A ports. In this situation, check valve 2 will open to recharge the line.

### 7.1.2 Direction Control Circuits

Direction control is one of the fundamental control functions in a hydraulic system and is typically executed using either a single-directional control valve or a combination of several different types of directional control valves. Figure 7.7 provides a commonly seen example of **direction control circuits** using a three-position four-way bi-directional control valve to control the reciprocal motion of the cylinder in the main circuit and also using a unidirectional check valve to limit the flow to a second circuit but not back from that circuit.

By combining a few pressure control valves with a pressure-actuated directional control valve in a circuit, a pressure-controlled **automatic direction switching circuit** can be made. Figure 7.8 shows such a circuit using two sets of pressure and directional control valve combinations to automatically switch the cylinder motion directions in terms of preset port pressures. In this circuit, the system starts its operation as the directional control valve leads the pump flow to the cylinder cap-end chamber to push the cylinder to extend at the depicted position. As the piston fully extends at the end of the stroke, the rapid rise of system pressure caused by the stall of the piston pushes the normally closed pressure control valve in pressure control branch 1 to open, which in turn

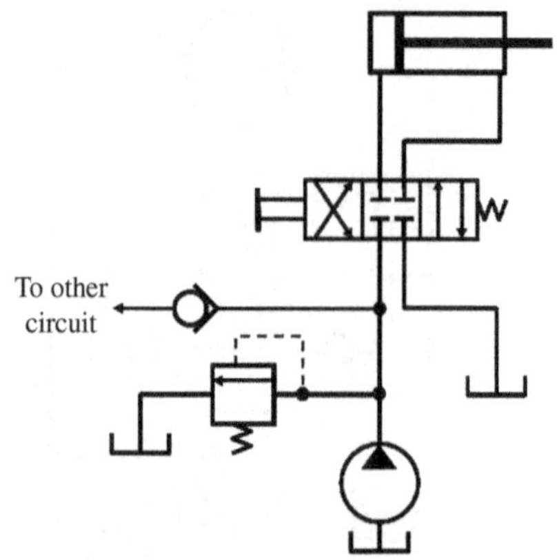


**FIGURE 7.7**
Schematic illustration of a typical valve-controlled direction-control circuit.

pushes the main directional control valve to switch its position to connect the pump port to the cylinder rod-end port and the cap-end port to the tank port. Meanwhile, the pilot-operated check valve in branch 2 is also pushed open to release the pilot fluid from the opposite side of the main directional control valve to provide room to switch the valve.

### 7.1.3 Speed Control Circuits

Speed control of the hydraulic actuator is another fundamental hydraulic control function and can be accomplished by controlling the flow rate supplied to the actuator. This section introduces a few simple and effective speed control circuits to fulfill the basic speed

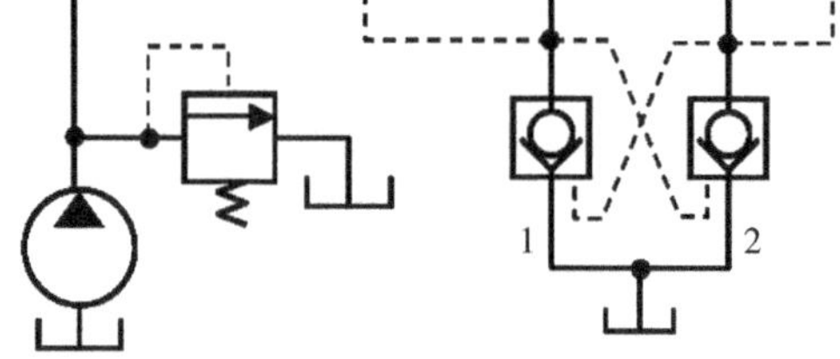


**FIGURE 7.8**
Schematic illustration of an automated direction-switching circuit actuated by two sets of pressure-control valve and check valve.

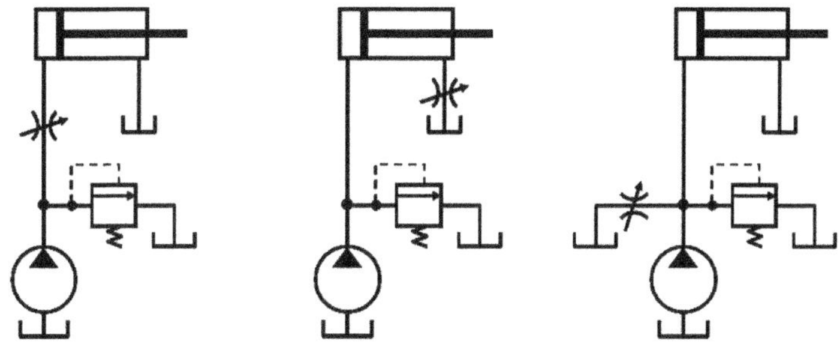


**FIGURE 7.9**
Schematic illustration of three basic metering circuits of (a) meter-in, (b) meter-out and (c) bleed-off for speed control.

control functions. These circuits can be combined with other circuits to provide more sophisticated speed control functions.

A simple speed control approach for hydraulic systems with a fixed-displacement pump is the use of a flow-metering valve in different arrangements. Figure 7.9(a) shows a **meter-in circuit** that accomplishes speed control during a work stroke by regulating the flow supplied to the cylinder. In comparison, the **meter-out circuit** (Figure 7.9(b)) regulates the flow discharged from the cylinder. Both the meter-in and meter-out circuits operate under the same principle that the circuit uses a flow-metering valve to create a pressure drop in either the inlet or discharge line. Such a pressure drop, as well as the operating pressure induced by the external load, will raise the system pressure to a level that will partially open the line-relief valve to bleed a portion of the flow to obtain the desired speed at the cylinder. While both adjustable meter-in and meter-out flow regulating can achieve continuous speed adjustment, the major differences between those approaches are that the meter-out circuit can provide a constant back pressure in the rod chamber and therefore can prevent lunging in case the load drops quickly or reverses.

Another approach to metering speed control is the **bleed-off circuit**. As illustrated in Figure7.9(c), the flow to the cylinder is regulated by metering a portion of the pump flow directly to the tank through a metered bypassing branch. While this circuit offers higher efficiency than both meter-in and meter-out circuits, as the discharge pressure from the pump is only high enough to overcome the resistance, it does not compensate for pump slip.

Many hydraulic systems require a control on reciprocal speeds of the cylinder during their specific working cycles. One common method for controlling the cylinder speed is using a proportional control valve as illustrated in Figure 7.10. This circuit uses a two-position four-way proportional directional control valve to regulate the extending or retracting speed on a hydraulic cylinder. Control of reciprocal speeds using this circuit is achieved by adjusting the valve fluid passing areas to create a pressure drop across the valve to build up back pressure to partially open the line-relief valve to distribute the supplying flow between the cylinder and the returning line in proportion to the valve opening. While this simple reciprocal speed control circuit provides the capability of achieving separate speed controls of extending and retracting motion, it is not the energy-efficient way of achieving speed control.

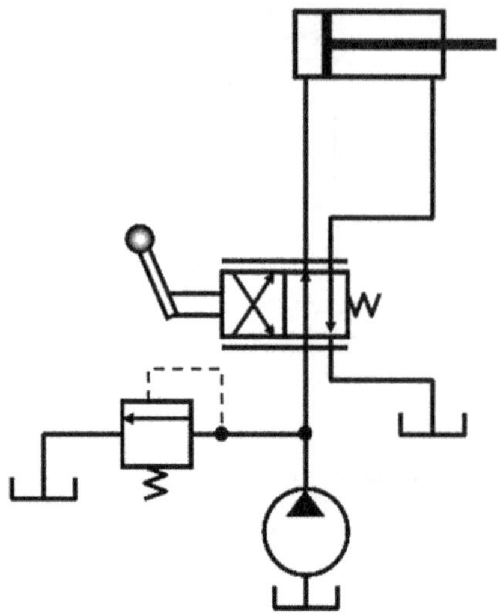

**FIGURE 7.10**
Schematic illustration of a simple proportional direction-control circuit for speed control.

In many mobile hydraulic systems, either open-center or closed-center proportional speed control circuits are used to achieve more energy-efficient reciprocal speed control. Depicted in Figure 7.11, a typical **open-center speed control circuit** uses a three-position four-way open-center proportional directional control valve with a fixed-displacement pump. In this circuit, the amount of flow supply to corresponding ports of the cylinder is in proportion to the valve opening, with the remaining amount of fluid bypassed directly back to the tank through the pump-tank ports. In such a case, the system needs only to hold the load pressure, instead of the line-releasing pressure, to drive the load. This circuit can noticeably improve the energy efficiency

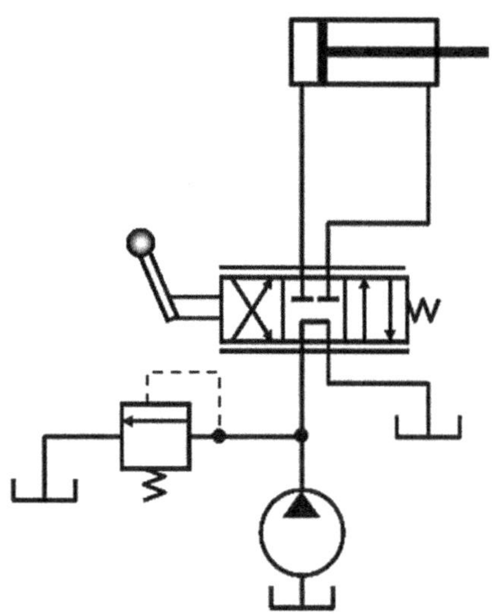

**FIGURE 7.11**
Schematic illustration of a typical open-center reciprocal speed-control circuit.

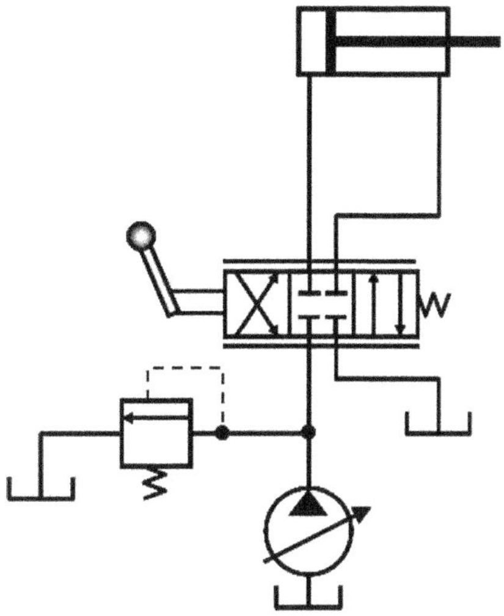


**FIGURE 7.12**
Schematic illustration of a typical closed-center reciprocal speed-control circuit.

above that of a circuit using a two-position four-way proportional control valve (as illustrated in Figure 7.10) by lowering pump-discharging pressure.

However, there is still a considerable amount of energy being wasted by bypassing the pressure fluid back to the tank, especially when the system requires only a small flow to perform low-speed actuations. To further improve the energy efficiency for such cases, a **closed-center speed control circuit** has been found as an additional application in mobile hydraulic systems, especially on those with a large capacity (Figure 7.12). Compared with its open-center counterpart, the closed-center circuit uses a closed-center proportional directional control valve with a variable-displacement pump. Controlled even with the simplest pressure-limiting mechanism, the variable-displacement pump will deliver only the needed amount of fluid to the closed-center circuit to achieve high-efficiency operations.

Other than the basic speed control function, there are also a few more advanced speed-adjusting functions, such as speed increase, decrease, and secondary adjustment, frequently needed in some hydraulic systems. All of these circuits achieve speed adjustments by switching the actuator from one subcircuit to another without changing the pump supply flow.

One example of **speed-increase circuits** designed on the basis of such an approach is a differential two-speed cylinder circuit. Figure 7.13 depicts the schematic of a differential speed control circuit using an additional two-position three-way directional control valve to construct a subcircuit on the rod-end line. This circuit can realize two extending speeds in terms of the system load. When the cylinder pushes a normal external load, the differential control valve is set at its normal position as depicted in the schematics, and the system will operate as a normal valve-controlled cylinder system performing normal reciprocal operations. When the external load is very light, it is often desirable to increase the extending speed to improve the operation efficiency. By switching the differential control valve to the left, the flow passing from the rod-end chamber to the tank will be blocked and redirected back to the cap-end chamber of the cylinder. Upon connecting both chambers, the fluid pressure will be the same on both sides of the piston, and the pressure pushes the

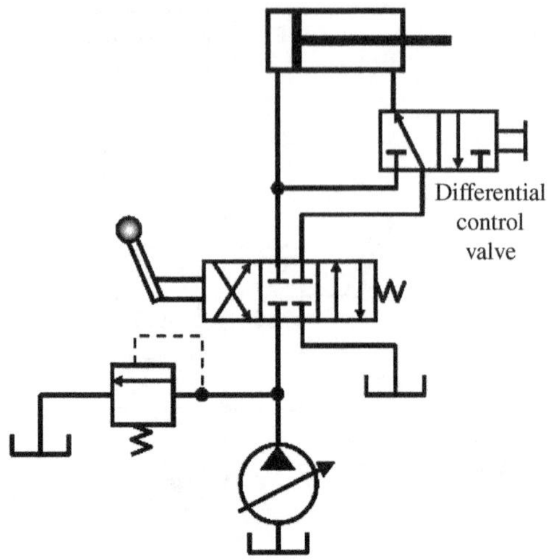


**FIGURE 7.13**
Schematic illustration of a typical differential two-speed cylinder circuit.

piston to extend due to the effective area difference on two sides of the piston. The addition of returning flow from the rod-end chamber to the cap-end chamber equivalently increases the fluid supply from the pump, which makes the piston extend faster. Such a differential two-speed function can only be realized in circuits using a single-rod double-acting cylinder.

**Speed-reducing circuits** are designed to quickly slow down actuating speed and are often done by adding an additional flow restriction to increase system pressure to partially release the supply flow to the tank through the line-relief valve. As depicted in Figure 7.14, the cylinder extends rapidly before triggering the stroke control valve 2.

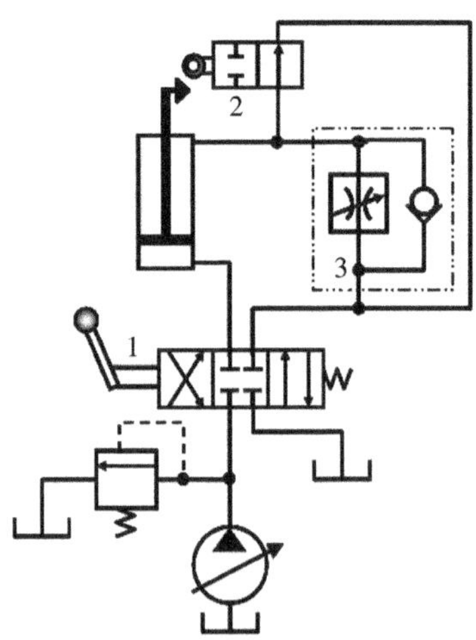


**FIGURE 7.14**
Schematic illustration of a typical valve restricted speed-reduction circuit.

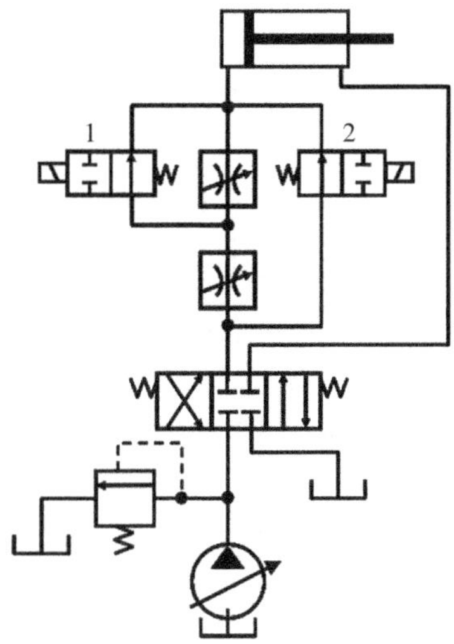


**FIGURE 7.15**
Schematic illustration of a typical speed secondly adjusting circuit.

After valve 2 is switched from the normal open to closed position, the returning fluid from the cylinder will be forced to flow through flow-regulating valve set 3 to raise the pressure in the rod-end chamber, which in turn will raise the system pressure to force a portion of the pump flow to be bypassed through the line-relief valve so that the speed of the cylinder extension will be reduced. During the retraction, the pump-supplied fluid will go through the check valve in parallel to the restriction valve; therefore, the retraction speed of the cylinder will not be affected by the operating status of stroke valve 2.

Using a very similar approach, we can realize a **secondary adjustment** on cylinder-actuating speed. The secondary speed adjustment is to obtain two slower speeds for some special applications (Figure 7.15). By comparing this secondary adjustment circuit to the speed-reducing circuit presented in Figure 7.14, note that there are two major differences between these circuits: the secondary adjustment circuit uses two sets of speed-adjusting valves in the pushing side line instead of one set in the returning line in the speed-reducing circuit. Having the speed adjustment valves in the flow supply line can help to reduce jerky movement during speed switching and result in a smoother speed adjustment. Other than the serial arrangement of flow restriction valves as depicted in Figure 7.15, those valves can also be arranged in parallel, allowing more flexible speed controls.

### 7.1.4 Sequencing Control Circuits

For systems with multiple actuating cylinders, it is very common to set some particular order of sequences among those actuators to realize some designated implementing functions. Such sequencing control is often accomplished by utilizing appropriate pressure controls. Figure 7.16 is an example of sequencing two cylinders by restricting flow to

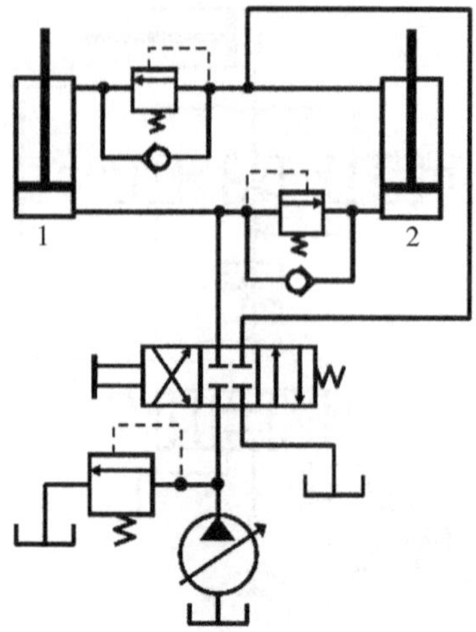


**FIGURE 7.16**
Schematic illustration of a typical sequencing control circuit.

one cylinder using a set of backpressure check valves. When extending the cylinders, the backpressure check valves set will prevent the flow from entering cylinder 2 until a preset pressure is reached. When retracting, the supply flow will not enter cylinder 1 for the same reason. In this case, cylinder 1 extends ahead of cylinder 2 but retracts after cylinder 2.

Other than using pressure control valves as just discussed, sequencing control can be realized by using position limit valves. When electrohydraulic control valves are used in a circuit, it is fairly simple to achieve sequencing control by using electronic controls.

### 7.1.5 Synchronizing Control Circuits

An opposite function to sequencing control is the synchronizing control of two or more actuators. Many designs are available for different applications. Figure 7.17 illustrates a **flow-dividing circuit** for synchronizing two cylinders of the same size. In this circuit, the flow supply to each cylinder is split using a flow-dividing control valve (see Figure 3.33 for an illustration of the operating principle) in terms of the operating pressure of the two cylinders. When the same external load is applied on each cylinder, the flow divider will deliver the equal volume of fluid to those cylinders to synchronize the motion of both cylinders. Variations in load or friction, as long as such variations are equally applied to both cylinders, will not greatly affect the synchronization. One issue that merits close attention in synchronizing cylinders is leakage replacement.

## 7.2 Special Function Circuits

In addition to the basic circuits introduced in the previous section, many special-function circuits are also being used in both industrial and mobile hydraulic systems. This section introduces a few of these circuits.

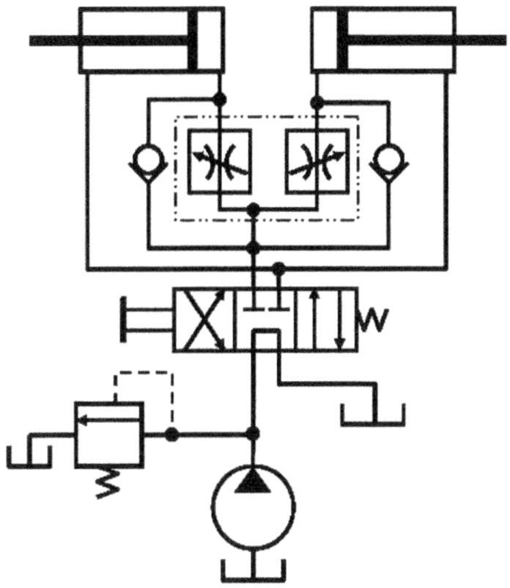

FIGURE 7.17
Schematic illustration of a typical flow-dividing synchronizing-control circuit.

### 7.2.1 Pump- Unloading Circuits

Many hydraulic systems require a low flow at high pressure to slowly feed an actuator performing a loaded task or a high flow at low pressure for rapid traverse of the actuator. This function can be accomplished by using a high-low circuit built using two pumps (Figure 7.18) which is often an open-center system to discharge the flow from both pumps running back to the tank when the main control valve is set in neutral. The only difference is that the discharge flow from pump 1 runs only through the directional control valve, and the flow from pump 2 also needs to run through a check valve between the discharge ports of the pumps 1 and 2. This check valve prevents the flow from pump 1 from

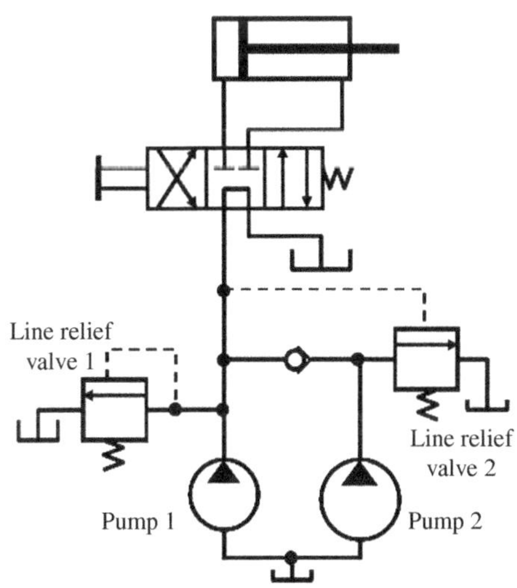


FIGURE 7.18
Schematic illustration of a typical pump-unloading circuit.

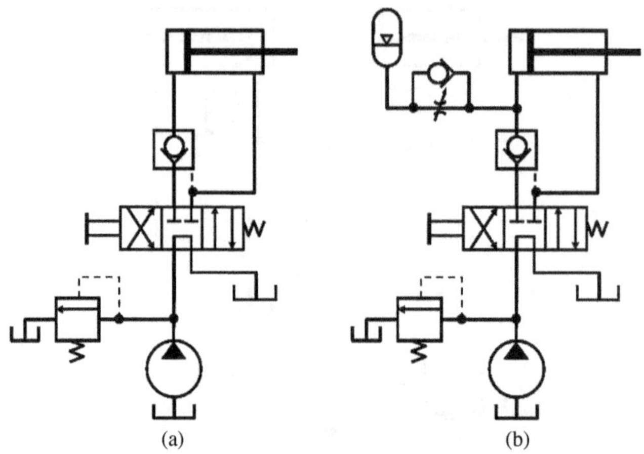


FIGURE 7.19
Schematic illustration of typical cylinder pressure-holding circuits using (a) pilot-control check valve only and (b) hydraulic accumulator design approach.

being bled through line-relief valve 2 to secure fluid supply during the low-flow operation. During rapid traverse, both pumps supply flow to the system. When pressure rises in the main pressure line, namely, the supply line of pump 1, during feeding, the large-volume main pump unloads by the pilot-operated line-relief valve 2, and the small pump maintains the pressure. The discharging flow from the small pump is usually low enough to prevent heating of the fluid.

### 7.2.2 Cylinder Pressure-Holding Circuits

In many applications, the cylinder is required to hold the load for a certain period of time after the pressure supply is cut off. Such a function can be realized by adding a **pressure-holding element** in the line to the cap-end chamber of the cylinder (Figure 7.19). The left circuit (Figure 7.19(a)) uses only a pilot-control check valve as the pressure-holding element. If the cylinder were operating at 20 MPa before stopping, such a check valve could normally prevent the pressure in the cap-end chamber from dropping more than 2 MPa for 10 minutes.

To provide more reliable pressure-holding capability, it is common practice to add a hydraulic accumulator in the system (Figure 7.19(b)) to provide a pressure-recharging capability to maintain a longer pressure-holding period. The duration of this pressure-holding capability is determined by the size of the accumulator.

### 7.2.3 Hydraulic Motors Series-Parallel Circuits

Hydraulic motors are commonly used as the driving devices in mobile equipment. Moving such equipment in job sites often requires a change in speed for different conditions, such as a high speed when traveling on solid or flat roads with light load or a low speed when carrying a heavy load traveling on soft or steep terrain. A series-parallel switchable hydraulic motor circuit can be used to provide the equipment with a simple speed-change capability. As shown in Figure 7.20, at the depicted position where the two motors are

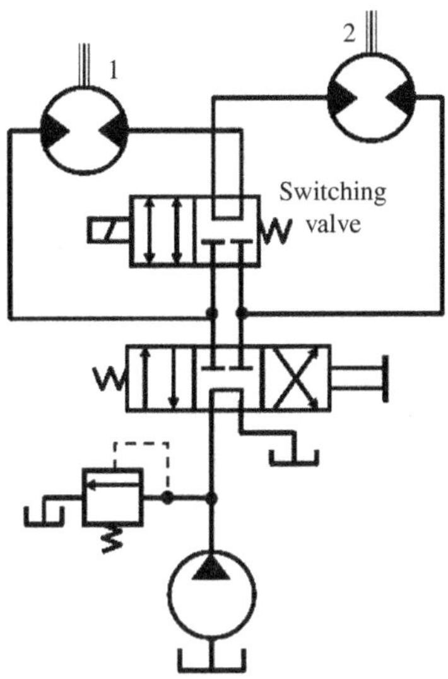


**FIGURE 7.20**
Schematic illustration of typical switchable series-parallel motor circuits.

connected in series, the pump-discharge flow first supplies motor 1, then motor 2 in series to drive both pumps synchronically. Such a configuration allows both motors to be driven by the full flow capacity of the pump but can utilize only one-half of the system pressure. Therefore, it can deliver a high speed at a reduced torque and is suitable for high-speed light-load operations. By shifting the switching valve to the right, it will connect the two motors in parallel. In this case, the total system pressure is then available for both motors, which doubles the driven torque capability. However, the amount of flow available for driving each motor is reduced by 50%, as well as the motor output speed. This configuration is suitable for low-speed heavy-load operations. This parallel configuration is most efficient when the load is evenly distributed on both motors. Raising the pressure on one motor will make the other less efficient, and, more disastrously, it may disrupt the speed relationship between the two motors. One solution to prevent consequences from occurring is to use a flow-regulating valve set in inlet lines to both motors. In a parallel configuration, the only way to increase the torque of the highest-pressure motor is to increase system pressure.

### 7.2.4 Hydraulic Braking Circuits

Hydraulic motors are often used to drive heavy rotating loads. When the pump supply is cut off, the motors will continue to rotate, driven by the inertia of both the motor mass and the loads. In such a case, the motor will act as a pump and consequently induce a cavitation in the circuit. To prevent forming this inertia-induced cavitation, a braking circuit can be used to physically stop the motor under these conditions. As depicted in Figure 7.21, a **normally closed braking** system operates by following the principle that when the tandem-center valve is at its neutral position, the pump-discharging flow is bypassed directly to the tank, and the pressure at the rod-end chamber of the brake actuator is very low; the brake is therefore applied under the spring force normally applied on the head-end side of the piston. When the main control valve is shifted away from its neutral position, the

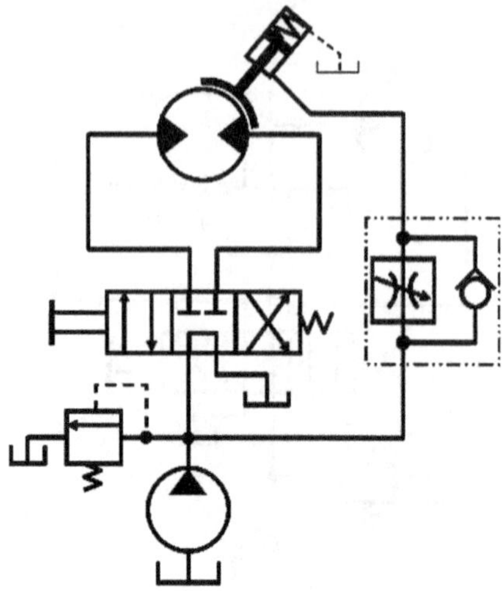

FIGURE 7.21
Schematic illustration of a typical motor-braking circuit.

pump starts supplying flow to the motor and building up the system pressure. Deferred by the flow-regulating valve, the release of the brake will be delayed, which in turn will help build up the system pressure more quickly. This rapid pressure rise could actually help the motor start faster as the starting torque is converted from the high pressure. When the control valve is switched back to the neutral position, the pump flow is redirected to the tank again. The fast drop of the system pressure will quickly reactuate the brake by quickly releasing the pressure in the brake actuator through a check valve.

Braking is a typical energy loss process as momentum-carrying energy is converted into heat and dissipated into the environment, and more energy-efficient braking concepts have been successfully introduced in the past decade. One of these concepts is to utilize the fact that a motor acts as a pump after the pump flow is cut to convert momentum-carrying energy into hydraulic potential energy for later use.

### 7.2.5 Accumulator Circuits

In many applications, a hydraulic system often operates with a varying flow supply: that is sometimes high and other times low. To meet the requirement for high-flow needs, the hydraulic pump is normally selected according to the highest demanded flow. An alternative way to design a hydraulic system for such applications, especially when requiring maximum flow only for a short period of time, is to downsize the pump by adding an accumulator as an auxiliary power source (often called the standby power source) in the circuit. As illustrated in Figure 7.22, when the main control valve sets at its neutral position, all the pump flow is used to charge the accumulator until it reaches the pressure setting of the unloading valve. The pump is then unloaded after the unloading valve is opened. When the valve is opened, the pressure in front of the control valve will drop, and the accumulator's stored fluid supplying the cylinder will close the unloading valve. The pump-discharging flow will then be supplied to the cylinder to drive the load. In such a circuit, the accumulator provides the power to start the operation.

Other than to provide standby power, an accumulator can also be used to reduce surges in a hydraulic system The same schematic depicted in Figure 7.22 shows that operating

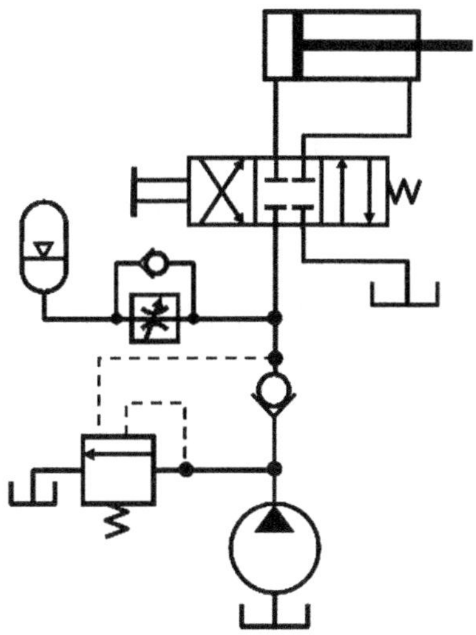


**FIGURE 7.22**
Schematic illustration of a typical accumulator power stand-by circuit.

the four-way, closed-center valve could form shock pressures a few times higher than the relief valve setting. Because the relief valve normally cannot act fast enough to drain off the fluid, such high pressures could be unsafe for both operator and equipment. Use of an accumulator in this circuit can absorb the surge pressures generated when the valve is placed in the neutral position.

### 7.2.6 Replenishing and Cooling Circuits

When a hydraulic motor and pump are connected in a closed circuit, it is necessary to supply a certain amount of make-up fluid to compensate for the leakage and to cool the closed system using a replenishing-cooling circuit, normally consisting of a unit of replenishing valves and a low-pressure makeup pump as illustrated in Figure 7.23. This circuit supplies fluid to the pump during replenishing and to the motor during braking to prevent the pump or motor from cavitation. A network of check and relief valves forms the bi-directional replenishing valve set and provides makeup flow for replenishing and braking in either direction.

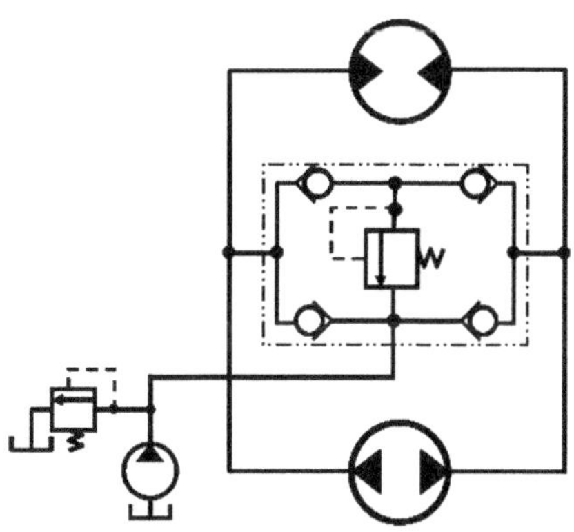


**FIGURE 7.23**
Schematic illustration of a typical replenishing and cooling circuit.

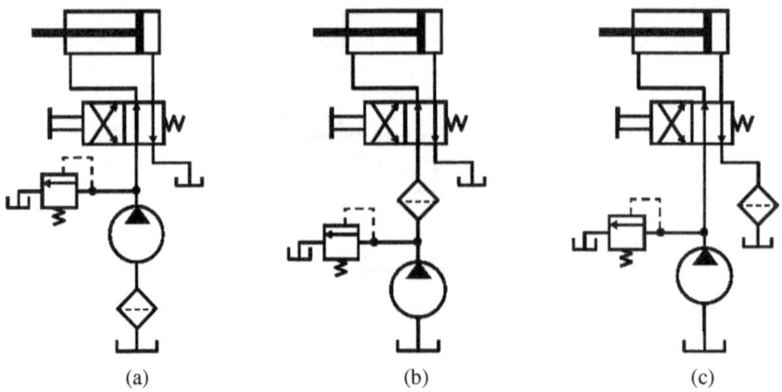


**FIGURE 7.24**
Schematic illustration of typical filtering circuit formations: (a) on suction line, (b) on pressure line or (c) on return line.

### 7.2.7 Hydraulic Filtering Circuits

The importance of a fluid filter in a hydraulic system can never be overemphasized, and the location where a filter is installed has a great impact not only on the filtration efficiency, but also on the performance of the system. Theoretically, filters can be installed at many different locations in a line. The rule of thumb for determining an appropriate location to install a fluid filter is that the back pressure must not interfere with circuit operation. In practice, the most commonly seen locations are in suction lines, pressure lines, and return lines. When the filter is installed in the suction line (Figure 7.24(a)), it should be submerged in reservoirs so that no part of the filter surface is exposed to air. In addition, the filter component should induce a low-pressure drop and a high fluid-passing capacity to prevent pump cavitation. A pressure line filter (Figure 7.24(b)) normally can tolerate a higher pressure drop and provide higher filtering efficiency. The most commonly used filter installation location is on the return line (Figure 7.24(c)), which has an advantage of the most efficient filtration to remove debris and contaminants as they are being removed from the system.

## 7.3 Integrated Hydraulic Circuits

### 7.3.1 Hydrostatic Transmission Circuits

Hydrostatic transmissions (HST) are widely used on off-road vehicles. An HST can be defined as a **pump-controlled motor**. In general, it consists of a variable-displacement pump driven by the engine and one or more either fixed- or variable-displacement motors to drive the wheels. The basic operation principles, configuration features, and control methods were introduced in Chapter 4. This section focuses on analyzing a few commonly used HST circuits.

As discussed in Chapter 4, the performance characteristics of an HST are normally measured by the speed, torque, and power the motor uses to drive the load in

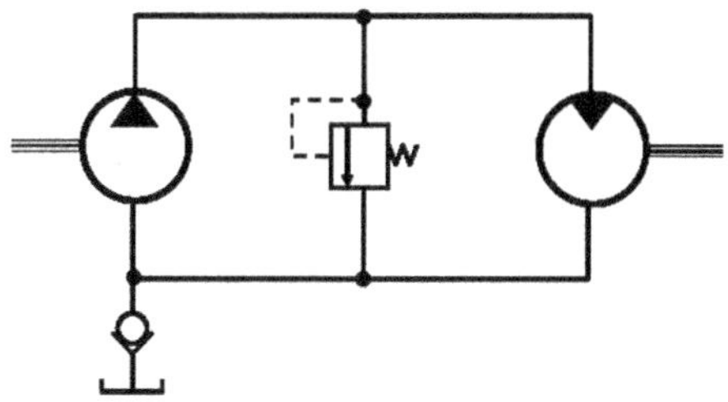

**FIGURE 7.25**
Schematic illustration of a simple fixed-displacement pump and fixed-displacement motor (FP-FM) hydrostatic transmission.

performing desired operations. Like other hydraulic circuits, such performance characteristics are determined by its components configuration. The simplest form of HST uses a fixed-displacement pump to drive a fixed-displacement motor (FP-FM; Figure 7.25). Although this FP-FM transmission is inexpensive, its applications are limited due to its low-energy efficiency in power transmission. One major contributing factor to low-energy efficiency is the fact that the pump must be sized to be capable of driving the motor at full speed under a full load. When the motor needs to operate at a reduced speed, a portion of pressurized fluid will be bled from the circuit at the maximum-allowed system pressure over the relief valve. This not only wastes a corresponding portion of energy, but even worse, the wasted portion of energy will be converted to heat, which will consequently require more energy to cool the system to ensure that the HST operates properly.

One common way to improve the energy efficiency of FP-FM transmissions is to replace either the fixed-displacement pump or motor with a variable-displacement one (VP-FM; Figure 7.26 (a) or FP-VM, Figure 7.26 (b)). In a VP-FM transmission, the torque output from the motor is constant in the entire speed range under ideal conditions, and the speed control of the motor is accomplished by increasing or decreasing the pump displacement. This type of HST improves energy efficiency by delivering the right amount of power to the motor to drive the load. In contrast, the FP-VM transmission will deliver a constant power to drive the load under ideal conditions. This type of HST achieves higher-energy efficiency by varying the motor displacement to adjust the speed corresponding to the load under the restriction of keeping a constant product of speed and torque limited by the amount of power delivered to the motor.

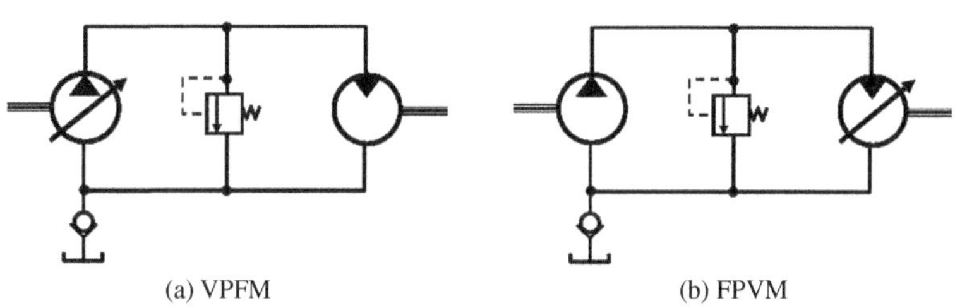


**FIGURE 7.26**
Schematic illustration of (a) variable-displacement pump and fixed-displacement motor (VP-FM) or (b) fixed-displacement pump and variable-displacement motor (FP-VM) hydrostatic transmission.

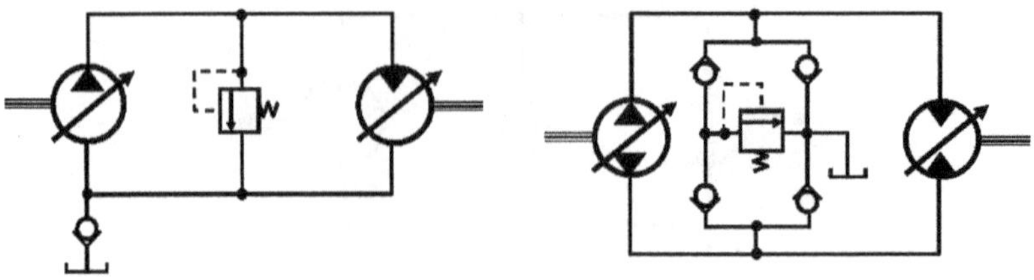


**FIGURE 7.27**
Schematic illustration of (a) unidirectional and (b) bidirectional variable-displacement pump and variable-displacement motor (VP-VM) hydrostatic transmissions.

The most versatile HST designs are variable-displacement pumps and motor circuits (VP-VM; Figure 7.27). Two common designs are the unidirectional VP-VM transmission (Figure 7.27(a)), which can change motor speed from zero to its full speed in only one direction, and the bi-directional VP-VM transmission (Figure 7.27(b)), which can drive the load full speed from one direction to the opposite one. Theoretically, such arrangements can provide infinite ratios of torque and speed to power. With the motor set at its maximum displacement, varying pump displacement can control the motor speed and power output while the torque remains constant. Decreasing motor displacement at full pump displacement can increase motor speed to its maximum. While the torque varies inversely with speed, the output power from the motor will remain constant.

**Example 7.1: FP-FM Type Hydrostatic Transmission**

Assume a rotary machine needs to drive a torque load of 600 N·m at a speed of 1500 rpm and a load of 2000 N·m at 500 rpm. If an FP-FM hydrostatic transmission is used to deliver the required power to drive this load, what are the required capacities for the pump and the motor if the system pressure is 15 MPa? (Assume that both the mechanical and volumetric efficiencies are 90% for both the pump and the motor, and that the pump is operating at a constant speed of 2000 rpm.)

a. The motor size required for an FP-FM HST can be determined using Eqs. (4.57) and (4.58) in terms of the higher-output torque (2000 N·m):

$$\begin{aligned} D_m &= \frac{\omega T_{higher}}{\Delta P n \eta_o} \\ &= \frac{2\pi \times 2000}{15 \times 10^6 \times 0.9^2} \\ &= 1.03 \times 10^{-3} \left(m^3\right) \approx 1.0(L) \end{aligned}$$

b. The required flow is determined in terms of the higher motor speed (1500 rpm):

$$\begin{aligned} Q_m &= \frac{D_m n}{\eta_v} \\ &= \frac{1 \times 1500}{0.9} \\ &= 1670\left(L/\text{min}\right) \end{aligned}$$

c. The required pump size can be determined as follows:

$$D_p = \frac{Q_p}{n_p} = \frac{Q_m}{\eta_p n_p} = \frac{1670}{0.9 \times 2000} = 0.93(L)$$

**DISCUSSION 7.1:** When an FP-FM hydrostatic transmission is used, the system has to be designed based on the maximum load at the highest required speed to satisfy all performance requirements.

In mobile applications, the **constant-torque HST**, often constructed using a servo-controlled variable-displacement pump to drive a fixed-displacement motor, is commonly used. Because of the closed-circuit feature, this HST often requires a charging device, normally a charge pump, to replenish fluids lost to prevent cavitation and to provide the pressurized fluid needed to actuate the variable-displacement adjusting mechanism. A low-pressure relief valve, typically set between 1.7 and 2.0 MPa, is commonly used to regulate the discharge pressure from the charging pump (Figure 7.28). While the pressure of the charging fluid is directly used to actuate the main pump displacement-adjusting mechanism, a set of back-to-back replenishing check valves are used to supply the makeup fluid to the appropriate low-pressure line.

The motor end of a typical closed-circuit HST also requires a pair of crossover relief valves. The pair is usually integrated into the motor package to prevent excess pressure from developing in either supply line due to shock-load feedback through the motor, an overrunning load, or similar conditions (Figure 7.29). This set of relief valves limits the pressure in the supply line regardless of the direction of fluid flow.

The response of an HST is often limited by the stiffness of the system, which depends on the compressibility of the fluid and the compliance of system components, including tubing and hoses. The influence of these components is similar to the effect of a spring-loaded accumulator, as if it is connected to the motor. To ensure prompt response of the motor under varying load conditions, it is very common in HST design to include a pilot-actuated

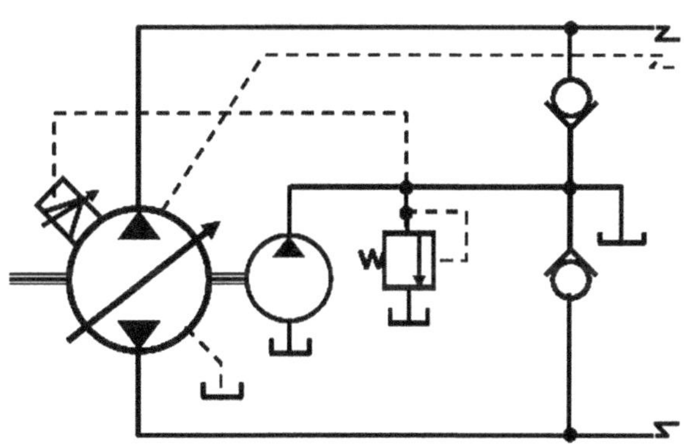

**FIGURE 7.28**
Schematic illustration of typical pump-end components in a servo-controlled VP-type HST.

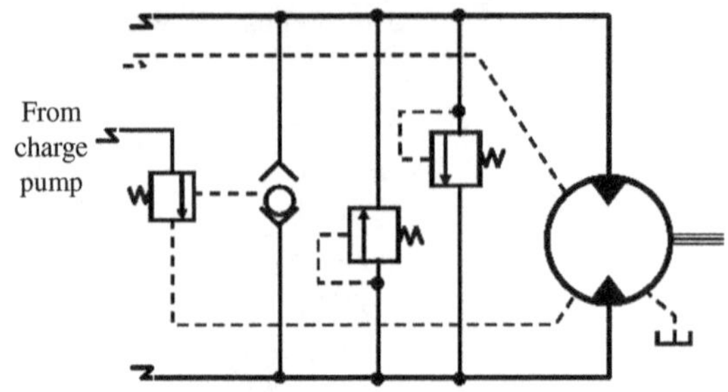


**FIGURE 7.29**
Schematic illustration of typical motor-end components in an FM-type HST.

replenish valve to route charge pump flow to the motor case. The pilot control pressure is usually provided by the high-pressure line, shafted using a shuttle valve in between high- and low-pressure lines connected to the motor, as illustrated in Figure 7.29. This additional fluid volume should be supplied by the charge pump.

### 7.3.2 Multibranch Integrated Hydraulic Circuits

In mobile hydraulic systems, an open-loop open-center multi-actuator system, constructed using a fixed-displacement pump and multiple open-center directional control valves in series as depicted in Figure 7.30, is often used. Such a system is also called the **prioritized multibranch circuit**. In a typical prioritized circuit, the pump discharges a fixed amount of flow every revolution it turns, and the pump flow is returned directly back to the tank when both directional control valves are set at a neutral position.

As one of the two valves is switched from its neutral position, the valve will block the flow return route, and the pump flow will be forced into a chosen actuator to perform the

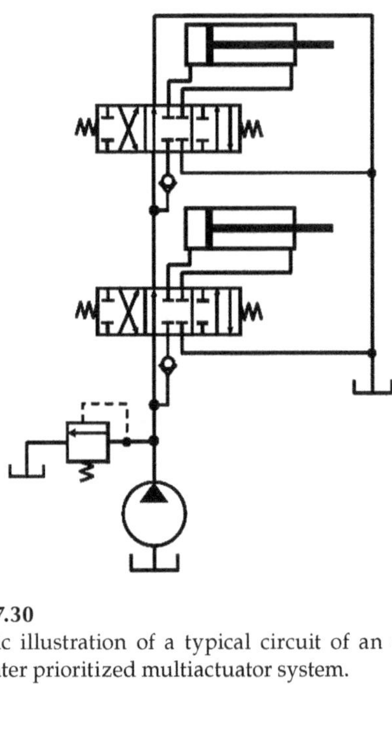

**FIGURE 7.30**
Schematic illustration of a typical circuit of an open-loop open-center prioritized multiactuator system.

designated work. For example, when the first directional control valve (the one close to the pump) is switched to the left, it will block the flow-bypassing route through the valve, and connect the pump port to the cylinder cap-end port and the cylinder rod-end port to the tank port. As a result, the pump flow is directed to the cap-end chamber of the first cylinder, and the fluid retained in the rod-end chamber of the cylinder is bled back to the tank, extending the cylinder to push the load.

When the pump delivers only the pressure needed to drive the load and the pressure drops to overcome the line losses, such a system is often called a **load-sensitive system**. It should be noted that when the first valve is completely switched away from its neutral position, all pump flow will be used to drive the controlled actuator, with no flow being supplied to the second actuator to do any work, even with the downstream valve fully opened due to the flow-bypassing route being completely blocked in the first valve under the stated condition. Such a feature in an open-loop open-center multi-actuator system is defined as the **priority function**, with the upstream circuit having a higher priority than the downstream one. The downstream circuit can operate in its full capacity only when the upstream circuit is not in operation, or it can be partially functional when the upstream circuit demands only a portion of the pump flow to perform its desired operation.

Another widely used circuit in some industrial applications is the **multipressure setting circuit** (Figure 7.31). In this circuit, the main line-relief valve is controlled by the balance between the line pressure and the summation of the spring force and a pilot-controlled back pressure. Two pilot line-relief valves are used to alternate between two pilot pressure settings. When both pilot relief valves are shut off by setting the remote control valve to its neutral position, the system will operate at its maximum pressure setting. When the remote control valve control shifts either right or left to connect the pilot route of relief valve 1 or 2, the system will operate at a reduced system pressure set jointly by the corresponding pilot relief valve and the main line-relief valve. In this manner, the circuit can support the system when operating at multiple pressure settings.

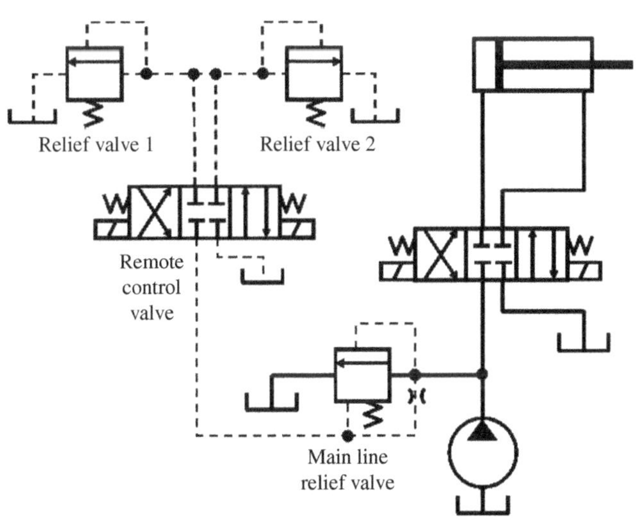


**FIGURE 7.31**
Schematic illustration of a typical multipressure-setting circuit.

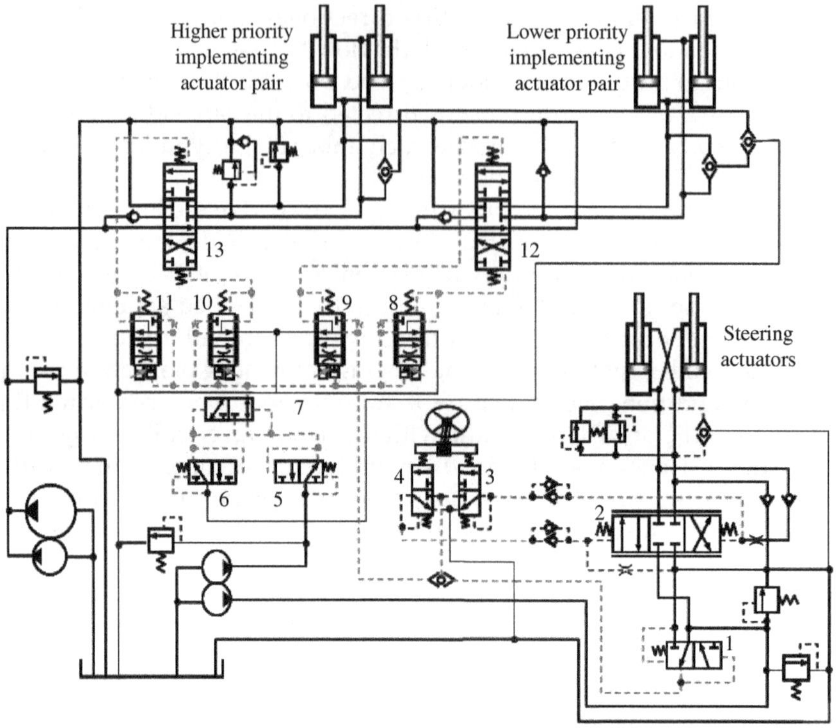


**FIGURE 7.32**
Schematic illustration of an example circuit of multiactuator mobile hydraulic system.

In many applications, an actual hydraulic system needs to drive several actuators to perform a set of designated functions. For example, Figure 7.32 is a pilot-controlled multi-actuator hydraulic system often seen on mobile machinery that uses two pumps to supply high-pressure fluid as the main power source to drive two pairs of implement cylinders. Use of a separate hydraulic power supply for the steering actuators implies that the reliable functionality of the steering actuators has a very high priority in the system. This high priority can also be seen from a redundant pilot power source to the steering system. Under normal operation, the discharge pressure of the steering system pump pushes valve 1 into the depicted position, as shown in Figure 7.32, which supplies the needed pilot power from the steering pump to the steering control unit, in this case the steering wheel and associated direction control valve pair 3 and 4. If for any reason the steering pump branch is unable to provide the pilot power, the shuttle valve located under the steering control unit will switch the pilot source to the pilot power source for the implement system. This will ensure that the steering control unit will receive a reliable power supply to maintain its functionality under all circumstances.

Another noticeable feature of this system is the use of a separate pilot power supply for implementing controls. Whenever the pilot pump is functional, the pilot pressure pushes the directional control valves 5 and 7 at the depicted position as shown in Figure 7.32, which leads the pilot pressure to a set of four pilot control valves, 8 through 11, to control two directional control valves, 12 and 13, in the main circuit to implement the desired functions. Similar to the steering pilot branch, this implement pilot circuit also has a secondary backup pilot supply from the main circuit to provide reliable pilot pressure to the pilot control valves in case of pilot pump failure. If the pump fails, there will be no pressure to hold valve 7 at the depicted position in Figure 7.32. Instead, the supplemental pressure from the circuit through valve 6 will push valve 7 to the right, which connects the pressure passage from the main circuit to the pilot branch of the implement control valves to keep them functioning. This set of three shuttle valves, located under two implementing cylinder pairs, will ensure connecting only the pressurized line in the main branch to the pilot branch.

The other important feature of this circuit is its priority function between the two units of implementing cylinder pairs. As shown in Figure 7.32, the main branch has two three-position six-port direction control valves in series. When both are at their neutral positions, the pump-discharge flow will pass through both valves and return to the reservoir. If one of the two valves is switched from its neutral position, for example, valve 12 is switched upward from the depicted position; the pump flow will then be directed to the cap-end chambers of the lower-priority pair of implementing cylinders and push the cylinder pair to extend. However, when both valves are switched, the higher priority valve, namely, upstream valve 13, will redirect all the pump flow to the first pair of cylinders. As a result, no flow will be supplied to the second pair of cylinders due to the blockage of flow supply by valve 13, regardless of the position of direction control valve 12. This fact reveals that the downstream pair can be functional only when the upstream valve 13 is at its neutral position.

### 7.3.3 Programmable Electrohydraulic Circuits

Proportional directional control valves are by far the most common means for motion control of hydraulic actuators in today's hydraulic circuits. In such a circuit, the proportional directional control valve uses a sliding spool to control the flow direction and rate to drive the actuators doing the work. For different applications, the spool is often specially designed to provide the desired flow control characteristics. As a result, the control valves cannot be interchangeable even if they have exactly the same-size spools. In turn, it is inconvenient and costly to manufacture, distribute, and provide service to those specially designed proportional directional control valves. To solve such a problem, an innovative programmable electrohydraulic (E/H) control valve, integrated using a set of individually controllable generic two-way proportional E/H control valves and a programmable electronic controller, has been invented to replace conventional direction control valves in a circuit. The core element is the programmable electronic controller, capable of coordinately programming the control characteristics of each composing proportional E/H control valve to realize different valve functions and flow control characteristics via different control software.

As shown in Figure 7.33, a simple programmable electrohydraulic circuit is typically formed using a generic programmable E/H control valve set and a programmable electronic control unit. The completely programmable valve set consists of five generic proportional E/H control valves, with four forms of a hydraulic bridge. Each valve controls the flow passage between the pump to cylinder cap or rod ports (P-$C_{CE}$ or P-$C_{RE}$) or between both cylinder ports to the tank ($C_{CE}$-T or $C_{RE}$-T), and one serves as a programmable line relive valve, or the pump-to-tank (P-T) valve. Supported by proper control strategies, all five E/H control valves can be operated cooperatively to provide equivalent functions of flow direction and rate control to a proportional directional control valve. To realize such a coordinative controllability, the electronic control unit of the circuit should be capable of implementing control of individual proportional E/H control valves at precisely scheduled

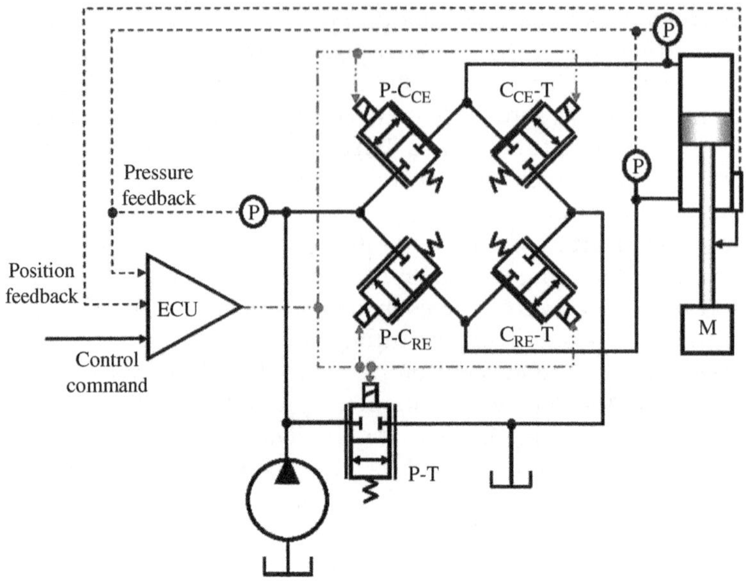


**FIGURE 7.33**
Schematic illustration of a simple hydraulic circuit controlled using a programmable E/H control valve.

times and modulating individual valves separately in terms of the control characteristics of the valve and the application. In other words, with proper logic of on–off controls on all five base valves, the programmable circuit can easily be switched between different system characteristics, such as (open-center, closed-center, tandem-center, and float-center), and between different operating modes, such as the normal and flow regeneration modes. Table 7.1 summarizes the control logic for realizing different functions on the programmable E/H control valve circuit.

For example, the depicted arrangement of the neutral positions of five base valves in Figure 7.33 represents a closed-center characteristic on the circuit. Under this configuration, all four valves forming the hydraulic bridge are at a normally closed position, and the fifth valve, namely, the line-relief valve, is under a modulated position at which the opening of the valve is controlled based on a predetermined relationship between the valve-opening area and the flow-passing rate through the valve. To change this circuit to a tandem-center configuration, the line-relief valve needs to be changed from the modulated setting to a normally open setting. Both examples reveal the core feature of the programmable control valve circuit; this circuit can be reconfigured by resetting the initial status, either normally open, normally closed or modulated, of each base valve by uploading different software.

**TABLE 7.1**
Control Logic for Realizing Multiple Functions from this Programmable E/H Control Valve Circuit.

| Valve Function | P-$C_{CE}$ Valve | P-$C_{RE}$ Valve | $C_{CE}$-T Valve | $C_{RE}$-T Valve | P-T Valve |
|---|---|---|---|---|---|
| **Open-center** | open | open | open | open | open |
| **Closed-center** | closed | closed | closed | closed | modulated |
| **Tandem-center** | closed | closed | closed | closed | open |
| **Float-center** | closed | closed | open | open | modulated |
| **Flow regeneration** | open | open | closed | modulated | modulated |

Because of its flexibility of reconfiguration, the programmable circuit can conveniently switch between the normal operation and the flow regeneration modes to attain a normal implementing speed at rated load or achieve an accelerated implementing speed at a light load condition. Such a mode switch can be easily accomplished on this programmable circuit by simply reconfiguring the returning flow from the cylinder rod-end chamber to the cap-end chamber through a proper path on the hydraulic bridge.

## References


1. Akers, A., Gassman, M., Smith, R. *Hydraulic Power System Analysis*. CRC Press, Boca Raton, FL (2006).
2. Amann, C, Krutz, G.W. Interactive hydraulic circuit design and analysis. *Proc. National Conference on Power Transmission*, 9: 19–28, Houston, TX (1982).
3. Book, R., Goering, C.E. Programmable electrohydraulic valve. *SAE Transactions: J. Commercial Vehicles*, 108: 346–352 (1999).
4. Cundiff, J.S. *Fluid Power Circuits and Controls: Fundamentals and Applications*. CRC Press, Boca Raton, FL (2002).
5. Hedges, C.S. *Industrial Fluid Power (3rd Ed.)*. Womack Educational Publications, Dallas, TX (1988)
6. Henke, R.W. Basic hydraulic circuit design: classifications of circuits. *Diesel Progress*, 71: 90–93 (2005).
7. Hu, H., Zhang, Q. Realization of programmable control using a set of individually controlled electrohydraulic valves. *International Journal of Fluid Power*, 3: 29–34 (2002).
8. Hu, H., Zhang, Q. Development of a programmable E/H valve with a hybrid control algorithm. *SAE Transactions: J. Commercial Vehicles*, 111: 413–419 (2002).
9. Hu, H., Zhang, Q. Multi-function realization using an integrated programmable E/H control valve. *Applied Engineering in Agriculture*, 19: 283–290 (2003).
10. Hydraulics & Pneumatics. *Fluid Power Basics*. http://www.hydraulicspneumatics.com/200/FPE/IndexPage.aspx. Accessed on November 20 (2006).
11. Keller, G.R. *Hydraulic System Analysis*. Penton Media Inc., Cleveland, OH (1985).
12. Pease, D.A. *Basic Fluid Power*. Prentice-Hall, Englewood Cliffs, NJ (1967).
13. Stringer, J. *Hydraulic Systems Analysis: An Introduction*. John Wiley & Sons, New York (1976).
14. Vickers, Inc. *Vickers Mobile Hydraulics Manual (2nd Ed.)*. Vickers, Inc., Rochester Hills, MI (1998).
15. Yeaple, F.D. *Fluid Power Design Handbook*. CRC Press, Boca Raton, FL, (1996).
16. Zahe, B., Prinsen, T., Schultz, M. A new type of pressure relief valve: The "soft relief" valve. *Proc. 48th National Conference of Fluid Power*, pp. 481–490. Chicago (2000).
17. Zhang, Q., Goering, C.E. Fluid power system, In: Bishop, R. (ed.), *The Mechatronics Handbook*. CRC Press, Boca Raton, FL, pp: 10-11~10-14 (2001).


## Exercises

**7.1** What are the primary functions of pressure control circuits?

**7.2** Why is a pressure-reducing circuit found in some hydraulic systems? What is the common approach in designing such a circuit?

**7.3** Explain how a bleed-recharge circuit works in a typical hydraulic system?

**7.4** Explain how meter-in, meter-out or bleed-off circuits work in controlling actuator speed? Identify their similar features as well as their major differences.

**7.5** Explain how a secondary adjustment works in controlling actuator speed?

**7.6** What are the common approaches for realizing sequencing control?

**7.7** Explain how the pilot-control check valve holds the pressure in the system depicted in Figure 7.19(a)?

**7.8** Explain the operation principle of a replenishing and cooling circuit as depicted in Figure 7.23.

**7.9** Figure 7.5 depicts a system schematic of a pressure-balancing circuit. Try to analyze the functionality of the integrated component formed by a check valve and a pilot-controlled relief valve in providing the pressure-balancing function.

**7.10** Figure 7.23 depicts a system schematic of a replenishing-cooling circuit. Try to analyze the functionality of the network formed by four check valves and a relief valve in providing makeup flow for replenishing and braking in either direction.

**7.11** Figure 7.33 depicts an integrated programmable E/H control valve circuit. Can the P-T valve be replaced using a regular pressure-controlled line relief valve? If such a replacement will cause some functionality changes, what are those changes?

**7.12** Assume the cylinder in the circuit depicted in Figure 7.7 is subjected to the same load force in both directions during the reciprocal motion. Try to determine in which direction (extension or retraction) the piston will move faster, and why?

**7.13** As depicted in Figure 7.9(c), a bleed-off-type speed control circuit controls the 20 mm diameter bore cylinder, actuating speed by bleeding a portion of pump flow through a needle valve. If the pump discharges a constant flow of 10 L/min, the line-relief valve is set at 2 MPa. What will be the maximum achievable working pressure at the cylinder cap-end chamber, and what will be the minimum achievable cylinder-actuating speed if the needle valve opens 1 $mm^2$? (Assume that the orifice coefficient for this needle valve is 0.65 and that the density of hydraulic fluid used is 900 $kg/m^3$.)

**7.14** Assume that the two cylinders in the synchronizing control circuits depicted in Figure 7.17 have the same piston area, $A_{CE} = 200\ cm^2$. If the load applied on the left cylinder is 8 kN and the one on the right one is 5 kN, to what should the orifice openings of the left and right flow control valves be set to achieve a synchronizing operation when the system supplies 150 L/min pressure flow at 500 kPa to the cylinders? (Assume orifice coefficient 0.65 and hydraulic fluid density 900 $kg/m^3$.)

**7.15** Assume that the single-rod double-actuating cylinder in the pressure balancing circuit depicted in Figure 7.5 has piston areas $A_{CE} = 100\ cm^2$ and $A_{RE} = 50\ cm^2$. If the pulling load applied to the cylinder is 24 kN and cylinder moving friction is 2 kN, at what pressure should the balancing valve be set to avoid an overrun from occurring, and at what pressure should the line-relief valve be set to support normal operation?

**7.16** Assume the bore and rod diameters of the single-rod double-actuating cylinder in the differential two-speed cylinder circuits depicted in Figure 7.13 are 120 and 90 mm, respectively. If the line-relief valve sets the maximum system operating pressure at 10 MPa and the pump supplies a maximum flow of 100 L/min flow, what are the cylinder speed and the load-carrying capacity in the regular and differential extension strokes?

**7.17** Assume the displacement of the pump and the motor in am FP-FM type HST is 100 and 200 cc, respectively, and the HST operating under a 15 MPa system pressure. If the pump is driven by a diesel engine at 2400 rpm to deliver hydraulic power for driving a load at 900 rpm, what is the system efficiency of the HST? (Ignore all pump and motor losses.)

**7.18** As in Problem 7.17, if a variable-displacement pump is used, what is the proper size for the pump if all other parameters are kept the same (namely, the displacement of the motor in this VP-FM type HST is 200 cc, the HST operating under a 15 MPa system pressure, and the pump is driven by a diesel engine at 2400 rpm to deliver hydraulic power for driving a torque load at 900 rpm to achieve a perfect efficiency)? What is the system efficiency for this configuration? (Ignore all pump and motor losses.)

**7.19** As in Problem 7.17, if a variable-displacement motor is used, what is the proper size for the motor to maximize the load-driving capacity if all other parameters are kept the same (namely, the displacement of the pump in this FP-VM type HST is 100 cc, the HST operating under a 15 MPa system pressure, and the pump is driven by a diesel engine at 2400 rpm to deliver hydraulic power for driving a torque load at 900 rpm to achieve a perfect efficiency)? What is the load-driving capacity for this configuration? (Ignore all pump and motor losses.)

**7.20** As in Problem 7.17, if both the pump and the motor are changed to variable-displacement ones, what is the proper size for the pump and the motor if all other parameters are kept the same (namely, the line-relief valve is set at 15 MPa, and the pump is driven by a diesel engine at 2400 rpm to deliver hydraulic power for driving a torque load of 500 N·m at 900 rpm to achieve perfect efficiency)? What is the system efficiency for this configuration? (Ignore all pump and motor losses.)

# 8

# Hydraulic Systems Modeling

## 8.1 Mathematical Model of Hydraulic Systems

### 8.1.1 Building Blocks of Hydraulic System Modeling

Figure 8.1 shows a simple hydraulic system pushing a mass performing a predetermined task. This figure illustrates how a typical hydraulic system delivers energy through pressurized fluids, using a cylinder to convert the energy into mechanical force to perform useful work. To model the dynamic response of such systems, both a **fluid subsystem** and a **mechanical subsystem** must be considered.

To model a **translational** mechanical subsystem like the one illustrated in Figure 8.1, it is essential to consider three basic building blocks: a **spring**, a **damper**, and a **mass**. The spring is used to represent the stiffness of the system, the damper stands for the resistance opposite to motion, and the mass generates the inertia to acceleration while the mechanical system is doing the work.

As shown in Figure 8.2, the **stiffness** of a spring is described by the relationship between the force $F$ acting on the spring and the distance $x$ it is being compressed. In many cases, the distance of a spring being compressed (or stretched) is proportional to the force acting on it, and therefore it is often called a **linear system**, which can be modeled using the following equation:

$$F = Kx \tag{8.1}$$

where $K$ is the spring constant of an ideal spring. The larger the $K$ value, the stiffer the spring, which means that it requires a greater force to compress (or stretch) the spring.

The $K$ value is also a measure of the capacity of a spring converting mechanical energy exerted during compression (or stretching) of the spring into **potential energy**. The amount of potential energy $E$ being stored in the compressed (or stretched) spring is also proportional to the distance the spring is being compressed (or stretched), expressed as follows:

$$E = \frac{1}{2}Fx = \frac{F^2}{2K} \tag{8.2}$$

When the force is released, this stored potential energy will quickly bring the spring back to its original shape. Due to the presence of mass in a mechanical system, such a quick reshaping of the spring will always result in some **overshoot** and will induce a **vibration** in the system.

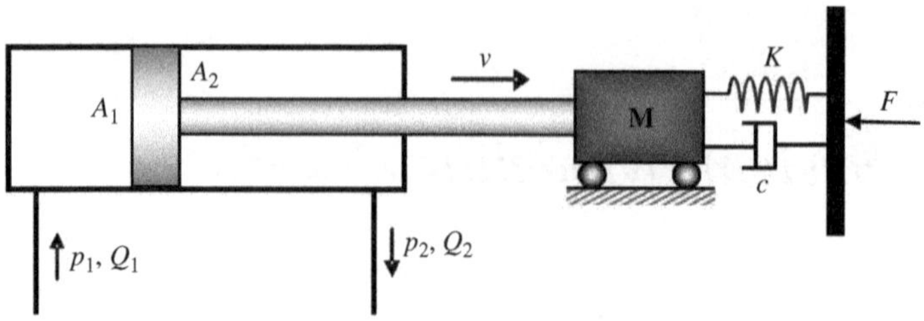


**FIGURE 8.1**
Schematic illustration of a simplest hydraulic system model.

To reduce the effect of vibration induced by a spring along with mass motion, a damper is always used. A damper, by principle, is like a cylinder. Figure 8.3 illustrates how it reduces vibration and can be described as the resistance of pushing a piston against the fluid behind it. The force $F$ acting on the damper is related to how fast the piston is pushed to compress the fluid at a distance $x$, and it can be modeled using the following equation:

$$F = cv = c\frac{dx}{dt} \tag{8.3}$$

where $c$ is a constant in an ideal damper. The larger the $c$ value, the greater the resistance of the damper, which means that it will create a greater force to resist the motion of the system.

Rather than storing energy, a damper is actually dissipating energy to absorb the vibration, and the amount of energy being dissipated is in proportion to the square of the velocity it is resisting, as expressed in Eq. (8.4) (the negative sign indicating that it is dissipating energy).

$$E = -cv^2 \tag{8.4}$$

When its dynamic responses need to be considered, one critical building block in any mechanical system is mass. A mass element represents the relationship between the force $F$ and the acceleration of the mass being moved as defined by the Newton's Second Law:

$$F = ma = m\frac{d^2x}{dt^2} \tag{8.5}$$

where $a$ is the acceleration of the mass. The larger the $m$ value, the greater the force under a certain acceleration $a$.

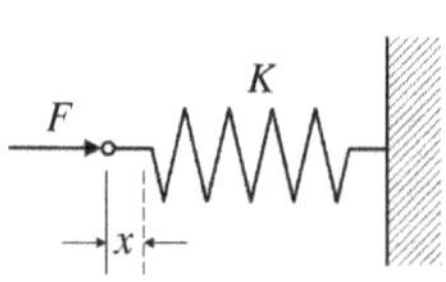


**FIGURE 8.2**
Schematic illustration of a spring element in a mechanical subsystem.

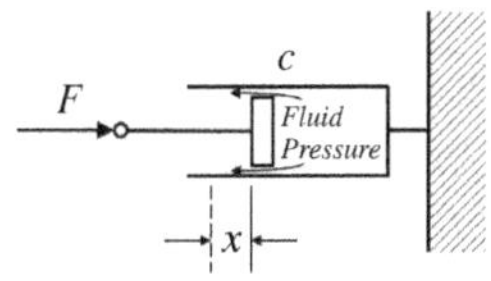


**FIGURE 8.3**
Schematic illustration of a damping element in a mechanical subsystem.

The energy being stored in the mass under a certain acceleration $a$ is often called the kinetic energy and can be expressed as:

$$E = \frac{1}{2} m v^2 \tag{8.6}$$

Those building blocks are often combined to form a comprehensive **mechanical subsystem** consisting of spring-damper-mass elements as illustrated in Figure 8.4. The composition of force applied to the mass in such a system can therefore be expressed as:

$$F - Kx - c\frac{dx}{dt} = m\frac{d^2x}{dt^2} \tag{8.7}$$

This equation can be rearranged into a standard form of a **second-order differential equation**, describing the relationship between the applied external force $F$ and the displacement $x$ of the mass $m$ as follows:

$$m\frac{d^2x}{dt^2} + c\frac{dx}{dt} + Kx = F \tag{8.8}$$

To model a fluid subsystem as shown in Figure 8.1, it is essential to consider three basic building blocks of **force conversion, flow control orifice,** and **flow continuity**. As explained in Chapter 1, three laws of physics can be applied to formulate those building blocks for a fluid subsystem of hydraulic control.

Figure 8.5 illustrates, in principle, the conversion of fluid pressure into mechanical force in a typical hydraulic cylinder. This is a critical building block for a typical linear actuation hydraulic system illustrated in Figure 8.1, as it is the interfacing element between the mechanical and fluid subsystems and the system pressure-building element because the pressure is determined by all the resistance forces acting on the cylinder. This illustration

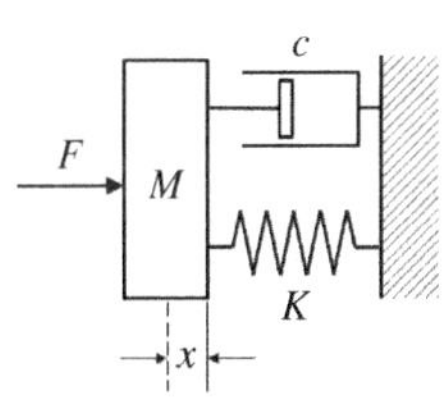


**FIGURE 8.4**
Schematic illustration of a comprehensive mechanical subsystem consisting of spring-damper-mass building blocks.

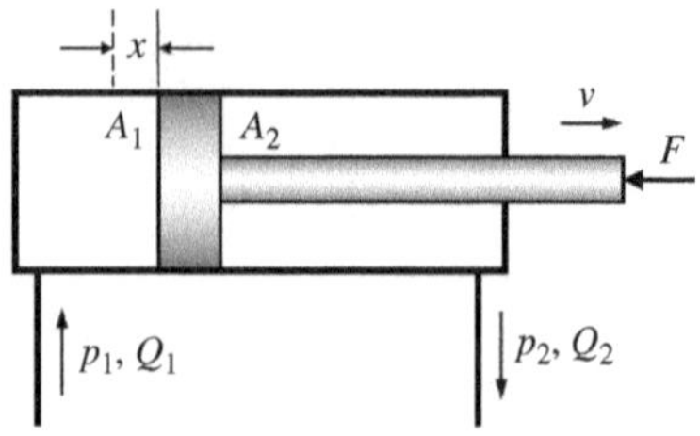


**FIGURE 8.5**
Schematic illustration of a force conversion element in a fluid subsystem.

reveals the relationship between the cylinder pressures $p_1$, $p_2$, and the total force $F$ acting on the cylinder, which can be expressed using the following equation:

$$F = A_1 p_1 - A_2 p_2 \tag{8.9}$$

where $A_1$ and $A_2$ are the cylinder piston head-end and rod-end areas, and $p_1$ and $p_2$ are the fluid pressure in the head- and rod-end chambers. In the case of a double-rod cylinder, the piston areas on both sides are the same, and Eq. (8.10) can then be simplified as follows:

$$F = A(p_1 - p_2) \tag{8.10}$$

The force-conversion building block reveals that the mechanical force (or torque) from a hydraulic cylinder (or motor) is converted from the pressurized fluid. As the hydraulic system is used to control the delivery of pressurized fluid from the source to the actuator, a building block of control orifice is essential to model this process.

As explained in Section 1.2.2, in a hydraulic power system, the total energy is carried both as potential energy in pressurized fluid and as kinetic energy in flowing fluid, and Eq. (1.7) is often used to describe the energy conversion within a hydraulic power system. To control a hydraulic cylinder performing desired work, the fluid system must deliver the right amount of pressurized fluid to the cylinder using a hydraulic control valve. Because a typical hydraulic valve controls the flow rate by altering the size of the fluid passage area, which is often small as compared to other parts of a hydraulic pipeline, it can often be represented by the hydraulic orifice in the modeling process. As illustrated in Figure 8.6, when the fluid flows through an orifice, the velocity of the fluid $v$ will be increased due to the reduction of the fluid passage area $A$. Such increase in flow velocity will result in a pressure drop across the orifice. A **flow control orifice equation** as defined in Section 1.2.2 is commonly used to determine the flow rate passing through the orifice in terms of the measurable pressure drop across the orifice.

$$Q = C_d A \sqrt{\frac{2}{\rho}(p_1 - p_2)} \tag{8.11}$$

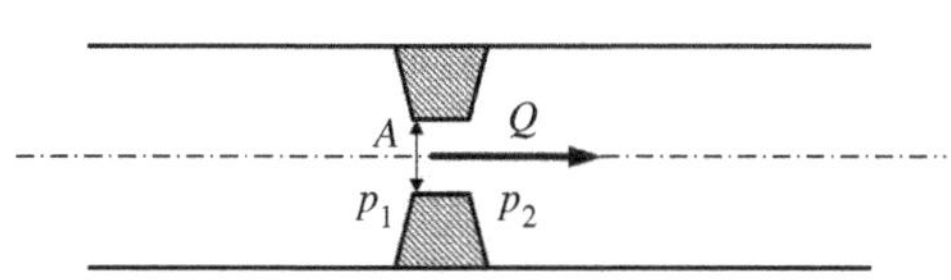


**FIGURE 8.6**
Schematic illustration of a hydraulic orifice element in a fluid subsystem.

where $Q$ is the flow rate, $\rho$ is the fluid density, $A$ is the orifice area, and $C_d$ is the orifice coefficient.

This **orifice coefficient** is commonly used to determine the effective flow passage area of the orifice due to the flow contraction; it plays an important role in estimating the flow rate passing through the orifice and is normally determined experimentally. As introduced in Section 3.1.2, it has been proven experimentally that the orifice coefficient of hydraulic control valve is a variable corresponding to the spool position (Figure 3.3), but in engineering analysis practice, $C_d$ can often be selected as a constant between 0.6 and 0.8, depending on the shape of the orifice, because the actual value of the coefficient varies a little when the orifice area surpasses a critical value.

Another concept, **flow continuity**, also plays an important role in hydraulic system modeling. As stated in Section 1.2.3, one fundamantal principle of hydraulic power transmission is that the fluid flows continuously within the system, which provides the base to determine the operating speed of a hydraulic actuator in terms of the steady flow supplied to the actuator. We often assume that the hydraulic fluid is incompressible in a confined control volume, but this assumption may not be valid in an extending hydraulic cylinder. As an example, consider a single-rod hydraulic cylinder extending shown in Figure 8.5; the major portion of the flow rate supplied to the cylinder would push the rod to extend at a certain speed, with a small portion to fill the compressed volume within the chamber and an even smaller portion to supplement the internal leakage of the cylinder induced by the pressurized fluid, as expressed by the following equation:

$$Q = \frac{dV}{dt} + \frac{V}{\beta}\frac{dp}{dt} + Q_{il} \tag{8.12}$$

where $Q$ is the inlet flow rate to the cylinder; $Q_{il}$ is the total internal leakage of fluid in the cylinder; $V$ is the cylinder chamber volume; and $\beta$ is the bulk modulus of the fluid.

### 8.1.2 Model of Simplified Valve-Controlled Systems

Figure 8.7 shows a simplified, typical valve-controlled double-rod cylinder-actuating system commonly used in many hydraulic actuating systems. It forms the fundamental

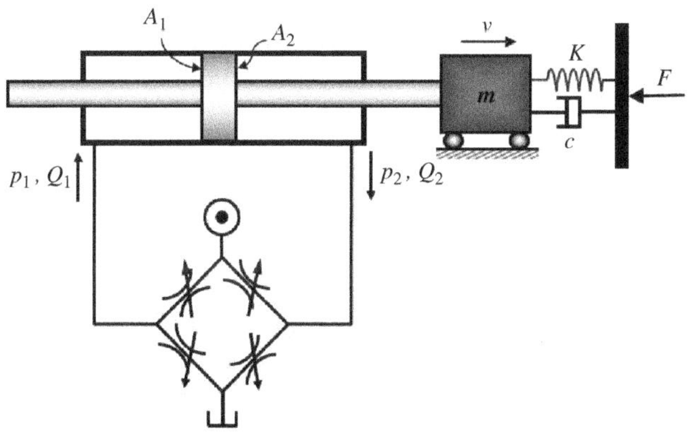


**FIGURE 8.7**
A system schematic of a simplified valve-controlled double-rod hydraulic cylinder actuating system.

model of a hydraulic actuating system, and such a system can be modeled by integrating the building blocks introduced in the previous section.

As illustrated in Figure 8.7, when the hydraulic system drives the cylinder extending to push the load moving rightward, the force balance equation of the piston can be modeled by integrating mechanical forces and hydraulic force conversion building blocks as follows:

$$p_1 A_1 - p_2 A_2 = m\frac{d^2y}{dt^2} + c\frac{dy}{dt} + Ky + F \tag{8.13}$$

where $p_1$ and $p_2$ are the cylinder head-end and rod-end pressures, $A_1$ and $A_2$ are the piston areas, $m$ is the total mass moved by the piston, $c$ is the damping constant, $K$ is the spring constant of the piston, $y$ is the displacement of the piston, and $F$ is the external force(s) acting on the piston.

In general, the piston area on two sides in a cylinder could be different. To simplify the modeling analysis, we use a double-rod cylinder in this system to make the piston areas of both sides the same. Therefore, a **load pressure** could simply be defined as the pressure difference on both sides of the piston:

$$p_L = p_1 - p_2 \tag{8.14}$$

where $P_L$ is the load pressure in a cylinder.

By introducing the load pressure concept to the piston force balance equation (8.13), the concept can be simplified using the following equation, which is the fundamental model for analyzing the force balance on the piston in motion:

$$p_L A_L = m\frac{d^2y}{dt^2} + c\frac{dy}{dt} + Ky + F \tag{8.15}$$

where $A_L$ is the piston area of a double-rod cylinder, defined as **load area** here as it is the area on which the load pressure is acting in either direction the piston is moving.

If we assume that the piston is initially located at the middle position of a cylinder, we can then define the average flow input to and outlet from the cylinder as the **load flow rate**:

$$Q_L = \frac{Q_1 + Q_2}{2} \tag{8.16}$$

From Eq. (8.12), we can obtain the equations of the input and outlet flow rates of a double-rod cylinder as expressed in the following equations:

$$Q_1 - C_{il}(p_1 - p_2) = \frac{dV_1}{dt} + \frac{V_1}{\beta}\frac{dp_1}{dt} \tag{8.17}$$

$$C_{il}(p_1 - p_2) - Q_2 = \frac{dV_2}{dt} + \frac{V_2}{\beta}\frac{dp_2}{dt} \tag{8.18}$$

where $C_{il}$ is the total fluid inner leaking coefficient of the cylinder.

Because the volume change of both chambers in extension is determined by the piston's moving speed, it satisfies the relationship of:

$$\frac{dV_1}{dt} = -\frac{dV_2}{dt} = A_1\frac{dy}{dt} \tag{8.19}$$

$$\begin{cases} V_1 = V_{10} + A_1 y \\ V_2 = V_{20} - A_2 y \end{cases} \tag{8.20}$$

By replacing the above relationships and assumptions in Eqs. (8.17) and (8.18) and subtracting these two equations, we can obtain the following relationship:

$$Q_1 + Q_2 = 2A_1\frac{dy}{dt} + \left(\frac{V_{10}}{\beta}\frac{dp_1}{dt} - \frac{V_{20}}{\beta}\frac{dp_2}{dt}\right) + \left(\frac{A_1 y}{\beta}\frac{dp_1}{dt} + \frac{A_2 y}{\beta}\frac{dp_2}{dt}\right) + 2C_{il}(p_1 - p_2) \tag{8.21}$$

By assuming the initial volumes of both chambers are equal to one-half of the total volume of the cylinder (and defining it as the load volume of the cylinder, i.e., $V_{10} = V_{20} = V_L/2$); the piston areas are the same on both sides of a double-rob cylinder $(A_1 = A_2 = A_L)$; the volume change caused by the piston motion is much smaller than the total volume $(|A_L y| \ll 2V_L)$; and the supply pressure from the fluid source is constant during the operation $(d(p_1 + p_2)/dt = 0)$, we can then simplify Eq. (8.21) as follows:

$$Q_L = \frac{Q_1 + Q_2}{2} = A_L\frac{dy}{dt} + \frac{V_L}{4\beta}\frac{dp_L}{dt} + C_{il}p_L \tag{8.22}$$

The system must supply an adequate amount of flow via the control valve to the cylinder load-side chamber. With the assumptions of a consistent pump-discharge pressure, zero back pressure in the tank, and no line pressure drop in this system to simplify the modeling process, a flow control equation for the spool valve can be derived using the control orifice building block in terms of load flow rate and load pressure:

$$Q_L = C_d w_v x_v \sqrt{\frac{1}{\rho}(p_s - p_L)} \tag{8.23}$$

where $C_d$ is the orifice coefficient, $w_v$ is the area gradient (in case of spool valve it is the perimeter of the spool), and $x_v$ is the displacement of the spool.

Assuming the fluid supply pressure maintains a constant value during the operation, we can linearize Eq. (8.23) using the Taylor series (by ignoring the high-order terms), with the control valve spool set at the neutral position (valve opening of $x_{v0}$, and load pressure of $p_{L0}$), and obtain:

$$Q_L = C_d w_v \sqrt{\frac{1}{\rho}(p_s - p_{L0})}\, x_v - \frac{C_d w_v x_{v0}}{2\sqrt{\rho(p_s - p_{L0})}} p_L = k_x x_v - k_p p_L \tag{8.24}$$

where $k_x = C_d w_v \sqrt{\frac{1}{\rho}(p_s - p_{L0})}$ and $k_p = \frac{C_d w_v x_{v0}}{2\sqrt{\rho(p_s - p_{L0})}}$ are defined as the flow amplification coefficient and pressure-flow coefficient of the linearized orifice equation.

Combining Eqs. (8.22) and (8.24), we can obtain a new equation capable of modeling the piston speed and fluid pressure inside the load chamber of a cylinder responding to a valve opening as follows:

$$k_x x_v = A_L \frac{dy}{dt} + \frac{V_L}{4\beta}\frac{dp_L}{dt} + \left(C_{il} + k_p\right)p_L \tag{8.25}$$

Equations (8.15) and (8.25) are the fundamental equations of load dynamics, flow delivery, and flow rate control in modeling a valve-controlled hydraulic cylinder system. In the cylinder-extending operation, the load flow, pressure, load area, and load volume are related to one side of the piston. Similarly, in the cylinder-retracting motion, the corresponding parameters are related to the other side of the piston. As those parameters are normally the same in a double-rod cylinder system, they allow use of the same equation to model the flow control element for both extending and retracting.

**Example 8.1: Spool Valve Power Output and Efficiency**

A spool valve is often used as a power amplification element in a hydraulic system. As the source flow through a four-way valve varies with a load pressure, the system may not always work at its highest possible efficiency. Assume a hydraulic system uses a variable-displacement pump to provide source flow $Q_s$ under pressure $p_s$, and the load flow and pressure are $Q_L$ and $p_L$. Try to determine under what condition the system will reach the highest efficiency.

a. To get the highest possible system efficiency, we need to have the system working under its highest output power. The output hydraulic power $P_h$ from a valve can be determined using the following equation:

$$\begin{aligned} P_h &= p_L Q_L \\ &= p_L C_d w_v x_v \sqrt{\frac{1}{\rho}\left(p_s - p_L\right)} \\ &= C_d w_v x_v \sqrt{\frac{p_s}{\rho}\left(p_L^2 - \frac{p_L^3}{p_s}\right)} \end{aligned}$$

b. From the above equation, we know that $P_h = 0$ when $p_L = 0$. By setting $\partial P_h / \partial p_L = 0$, we can get $p_L = \frac{2}{3}p_s$. This implies that the valve outputs its maximum output power and reaches its highest efficiency under the condition of the spool widely open and $p_L = \frac{2}{3}p_s$:

$$\begin{aligned} P_{h,\max} &= p_L Q_L \\ &= \frac{2}{3}p_s C_d w_v x_{v,\max}\sqrt{\frac{1}{\rho}\left(p_s - \frac{2}{3}p_s\right)} \\ &= \frac{2}{3\sqrt{3}} C_d w_v x_{v,\max}\sqrt{\frac{p_s^3}{\rho}} \end{aligned}$$

c. As the source flow is automatically adjusted to match the load flow in a variable-displacement pump system, the maximum efficiency of the system can be determined:

$$\eta_{\max} = \frac{p_L Q_L}{p_s Q_s} = \frac{\frac{2}{3} p_s Q_s}{p_s Q_s} = 0.667$$

**DISCUSSION 8.1:** The preceding analysis is based on the assumption that the line losses and leaks are negligible. When a system utilizes a fixed-displacement pump, the supplied source flow is fixed regardless of the required load flow, and therefore the maximum efficiency for such a system is lower than that of a system when a variable displacement pump is used.

## 8.2 System Analysis

### 8.2.1 System Block Diagram and Transfer Function

In engineering practice, a **system block diagram** is often used to graphically represent all the pertinent elements of a hydraulic system in modeling analysis. The basic component in such a block diagram is a series of blocks, often called the **black box**, to abstractly represent a device or a system. Figure 8.8 shows a few basic building elements of a system block diagram. It includes a **unit box**, a **summing junction**, and a **takeoff point**. A unit box (Figure 8.8(a)) shows only its **stimuli input** and **output reaction** without any detailed information on how the device or system works to react to the stimuli. Instead it uses a mathematical **transfer function** to represent the relation between the output response to the input stimuli, and therefore it is often called a **black box**. A summing junction (Figure 8.8(b)) adds two or more inputs to form only one output and is equal to the algebraic sum of the inputs, and a takeoff point (Figure 8.8(c)) is used to allow a signal to be used by more than one block or summing point in a block diagram.

In mathematical analysis, a transfer function can often be derived from a **linear differential equation** using the **Laplace transform**. For a valve-controlled hydraulic cylinder system introduced in the previous section, we may get three transfer functions from the three

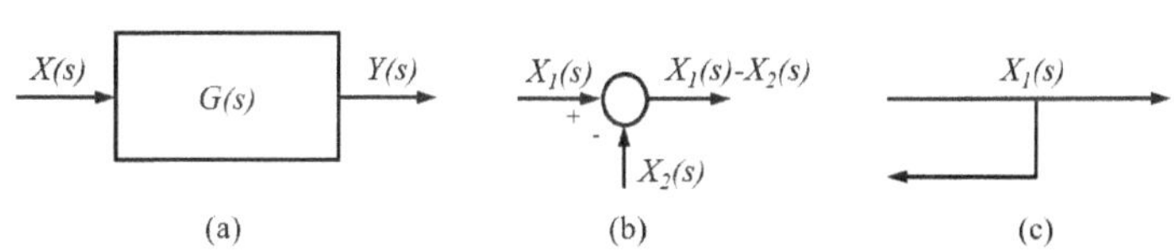


**FIGURE 8.8**
A few basic building elements of system block diagram: (a) a unit black box containing a transfer function along with its stimuli input and output reaction; (b) a summing junction for picking up an additional signal at a particular location; and (c) a takeoff point for allowing the signal at a particular location be used at somewhere else.

fundamental equations for its modeling. Applying a Laplace transform to Eqs. (8.15) and (8.25) and rearranging the new equation obtained, we can get the following two Laplace equations:

$$Y = \frac{1}{ms^2 + cs + K}(A_L P_L - F) \tag{8.26}$$

$$P_L = \frac{1}{k_p + C_{il} + \frac{V_L}{4\beta}s}(k_x X_v - A_L sY) \tag{8.27}$$

where $A_L$ is the piston load area; $V_L$ is the cylinder load volume; $p_L$ is the load pressure in the cylinder; $k_x$ and $k_p$ are the flow amplification coefficient and pressure-flow coefficient of the valve; $C_{il}$ is the fluid-leaking coefficient in the cylinder, $m$ is the mass moved by the piston; $c$ is the piston-damping constant; $K$ is the piston spring constant; $F$ is the total external (nonhydraulic) force acting on the piston; $X_v$ is the spool valve displacement; and $Y$ is the piston-moving distance.

Based on the parametric logic flow presented by Eqs. (8.26) and (8.27), we can create a system block diagram for this valve-controlled hydraulic cylinder system, as shown in Figure 8.9.

In many cases, we would like to know the relationship only between the input stimuli and output response, which can be done by simplifying the system block diagram with fewer blocks to make the system block diagram more graphically illustratable and to show the relevant dynamic relationship(s) between the input stimuli and the output response. **Block diagram transformation** techniques (also called block diagram transformation rules) commonly used to do system block diagram reduction systematically are as follows:

Rule 1: Combining blocks connected in series (cascaded blocks);

Rule 2: Combining blocks connected in parallel (eliminating a feedforward loop);

Rule 3: Combining blocks connected in a feedback loop (eliminating a feedback loop);

Rule 4: Shifting summing junction(s) if there is difficulty with summing point; and

Rule 5: Shifting takeoff point(s) if there is difficulty with the takeoff point.

**Table 8.1** provides a few commonly used self-illustrative block diagram transformation techniques. The five block diagram transformation rules can be repeatedly applied to a

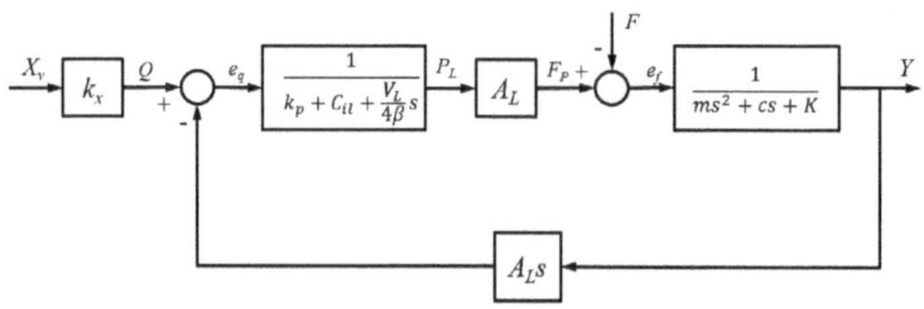


**FIGURE 8.9**
System blocks of a valve-controlled hydraulic cylinder system model (piston motion transformed from load pressure).

**TABLE 8.1**

Lock Diagram Transformation.

| Transformation | Equation | Block Diagram | Equivalent Block Diagram |
|---|---|---|---|
| Combining cascaded blocks | $Y = G_1G_2X$ | 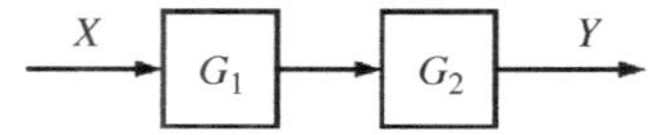 | 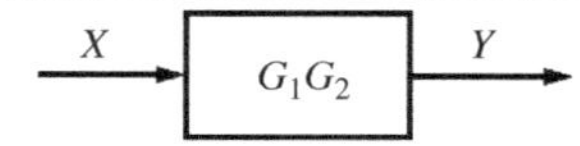 |
| Combining parallel blocks | $Y = (G_1 \pm G_2)X$ | 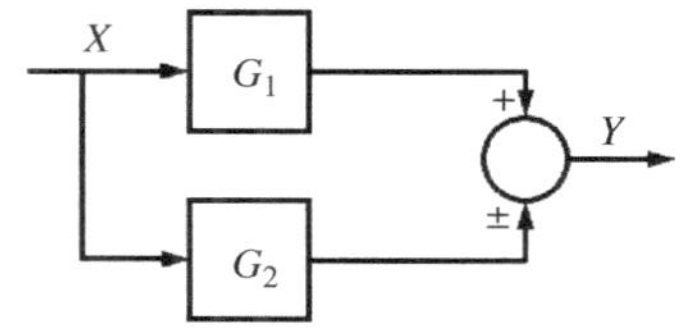 | 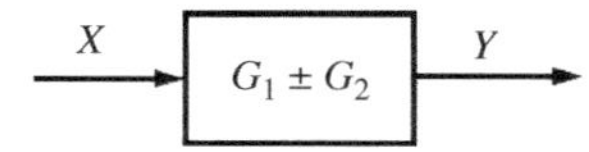 |
| Moving summing point ahead of a block | $Y = X_1G \pm X_2$ | 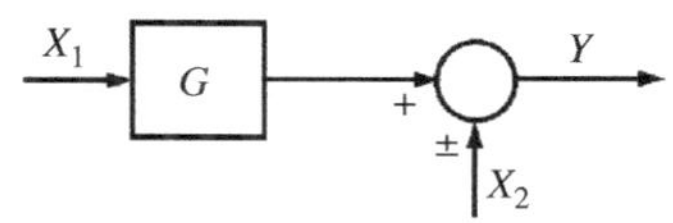 | 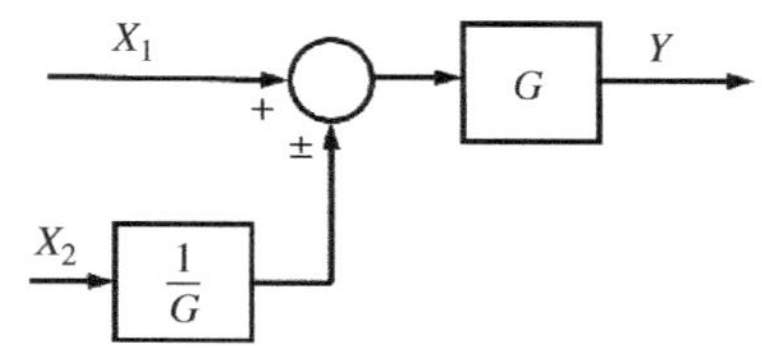 |
| Moving summing point behind a block | $Y = (X_1 \pm X_2)G$ | 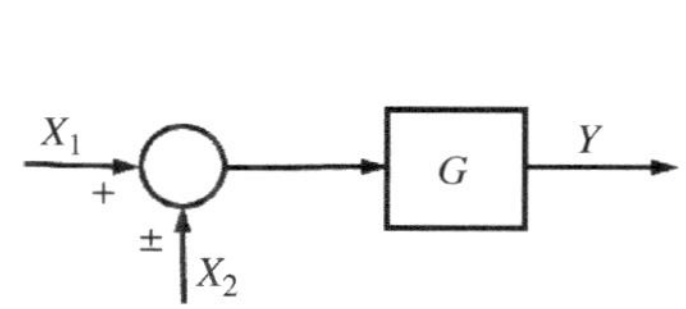 | 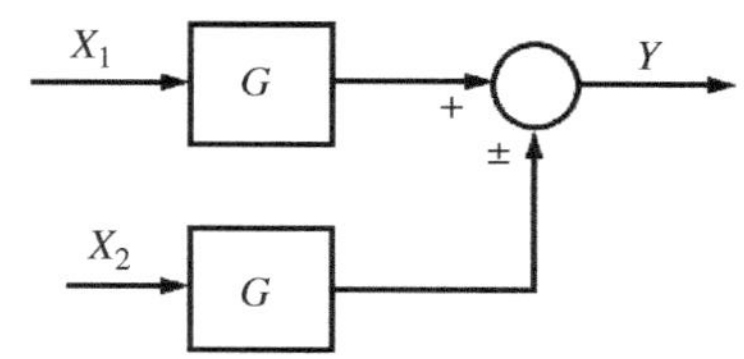 |
| Combing Summing points | $Y = \sum_i X$ | 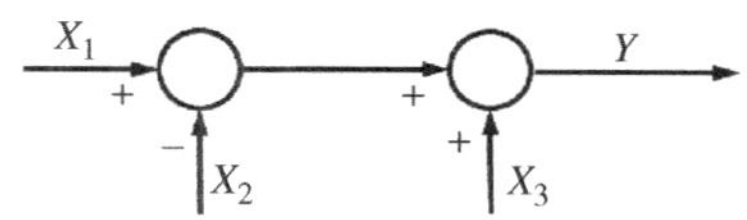 | 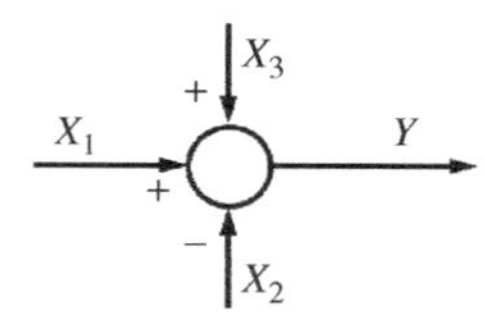 |

(*Continued*)

**TABLE 8.1** ***(Continued)***

Lock Diagram Transformation.

| Transformation | Equation | Block Diagram | Equivalent Block Diagram |
|---|---|---|---|
| Moving takeoff point ahead of a block | $Y_2 = XG$ | 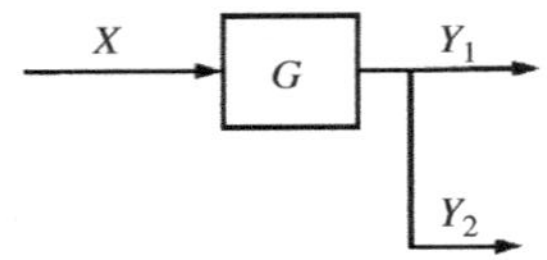 | 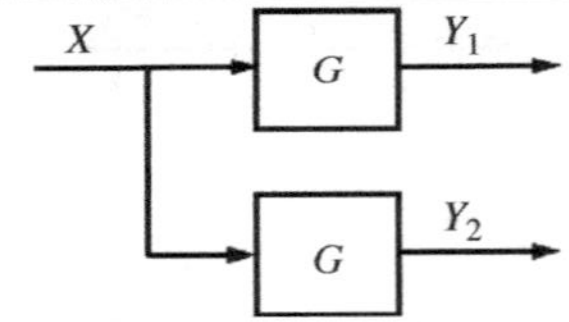 |
| Moving takeoff point behind a block | $Y_2 = X$ | 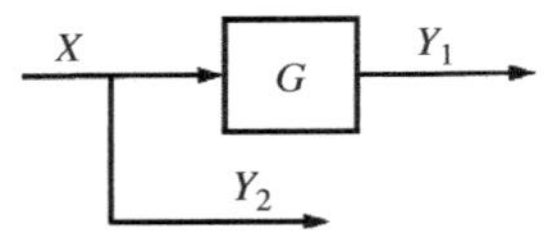 | 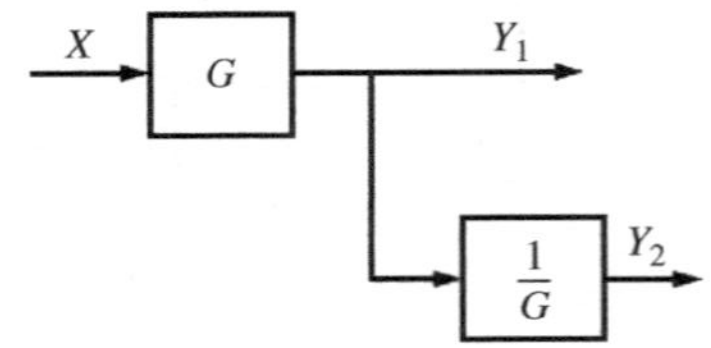 |
| Eliminating Feedback Loop | $Y = \frac{G}{1+GH}X$ | 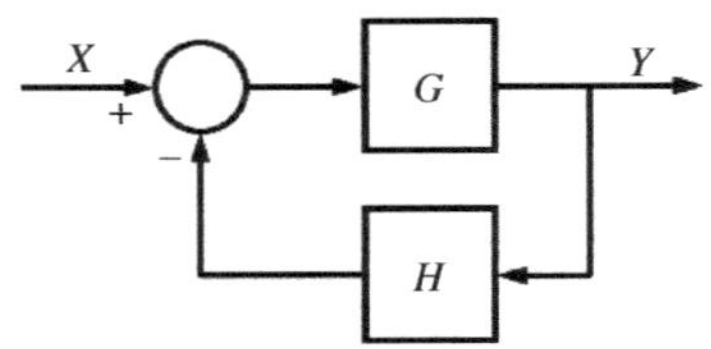 | 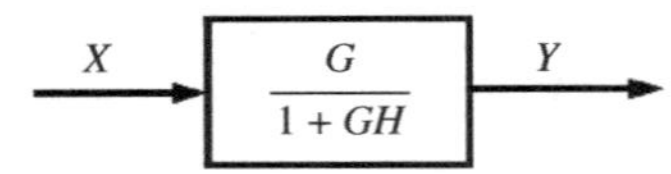 |

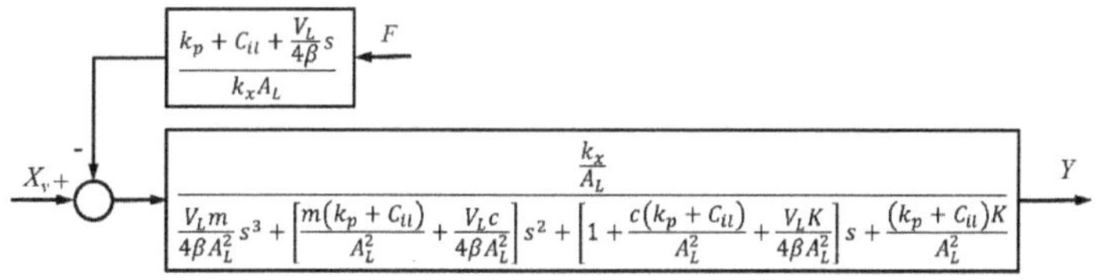


**FIGURE 8.10**
A simplified system block diagram of the valve-controlled hydraulic cylinder system model showing the relationships between the spool valve displacement stimuli, the external force disturbances and the responding piston displacement.

system block diagram until it is simplified to a desirable level. Figure 8.10 shows a simplified block diagram of the valve-controller hydraulic cylinder system by applying these rules. This diagram presents the relationships between the input stimuli of valve spool displacement and the output response of piston displacement under an external force disturbance.

For system analysis in response to a certain valve opening in terms of spool displacement, we can first assume that there is no external force disturbance to the system by setting $F = 0$. The transfer function between the input valve spool displacement and the responding piston displacement can therefore be easily obtained from the simplified system block diagram, as follows:

$$\frac{Y}{X_v} = \frac{\dfrac{k_x}{A_L}}{\dfrac{mV_L}{4\beta A_L^2}s^3 + \left(\dfrac{m(k_p + C_{il})}{A_L^2} + \dfrac{cV_L}{4\beta A_L^2}\right)s^2 + \left(1 + \dfrac{c(k_p + C_{il})}{A_L^2} + \dfrac{KV_L}{4\beta A_L^2}\right)s + \dfrac{K(k_p + C_{il})}{A_L^2}} \tag{8.28}$$

Similarly, we can analyze the influences of the force disturbance to the system response assumed by separating the input stimuli via setting $X_v = 0$. The corresponding transfer function can also be obtained from the simplified system block diagram:

$$\frac{Y}{F} = \frac{-\dfrac{1}{A_L^2}\left(k_p + C_{il} + \dfrac{V_L}{4\beta}s\right)}{\dfrac{mV_L}{4\beta A_L^2}s^3 + \left(\dfrac{m(k_p + C_{il})}{A_L^2} + \dfrac{cV_L}{4\beta A_L^2}\right)s^2 + \left(1 + \dfrac{c(k_p + C_{il})}{A_L^2} + \dfrac{KV_L}{4\beta A_L^2}\right)s + \dfrac{K(k_p + C_{il})}{A_L^2}} \tag{8.29}$$

After the separated responses to input stimuli and external disturbances are determined, we can simply combine Eqs. (8.28) and (8.29) in analyzing the comprehensive responses of piston displacement to both input stimuli ($x \neq 0$) and external disturbances ($F \neq 0$), as follows:

$$Y = \frac{\dfrac{k_x}{A_L}X_v - \dfrac{1}{A_L^2}\left(k_p + C_{il} + \dfrac{V_L}{4\beta}s\right)F}{\dfrac{mV_L}{4\beta A_L^2}s^3 + \left(\dfrac{m(k_p + C_{il})}{A_L^2} + \dfrac{cV_L}{4\beta A_L^2}\right)s^2 + \left(1 + \dfrac{c(k_p + C_{il})}{A_L^2} + \dfrac{KV_L}{4\beta A_L^2}\right)s + \dfrac{K(k_p + C_{il})}{A_L^2}} \tag{8.30}$$

Those interested in analyzing the response of system load pressure to valve opening can simplify the original system block diagram by inputting the spool displacement and

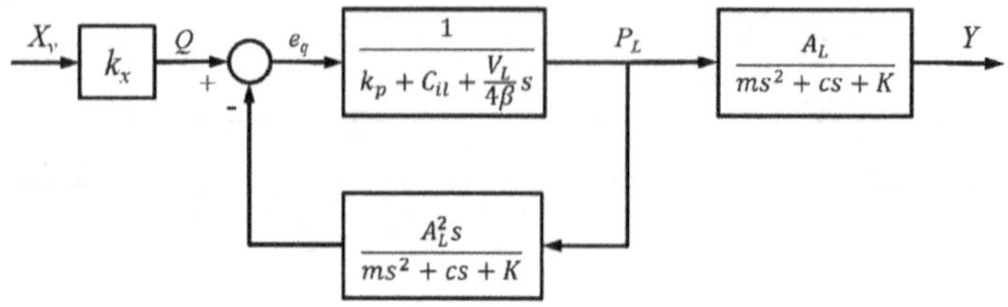


**FIGURE 8.11**
A simplified system block diagram of the valve-controlled hydraulic cylinder system model showing the relationships between the spool valve displacement stimuli and the responding load pressure and piston displacement.

outputting the load pressure (setting $F = 0$). Figure 8.11 shows the resulting simplified block diagram, and the following equation is its transfer function.

$$\frac{P_L}{X_V} = \frac{\frac{k_x}{A_L^2}\left(ms^2 + cs + K\right)}{\frac{mV_L}{4\beta A_L^2}s^3 + \left(\frac{m\left(k_p + C_{il}\right)}{A_L^2} + \frac{cV_L}{4\beta A_L^2}\right)s^2 + \left(1 + \frac{c\left(k_p + C_{il}\right)}{A_L^2} + \frac{KV_L}{4\beta A_L^2}\right)s + \frac{K\left(k_p + C_{il}\right)}{A_L^2}} \tag{8.31}$$

### 8.2.2 Transfer Function Simplification

The transfer functions analytically obtained above are rather complicated. To make the modeling analysis easier in engineering analysis, those transfer functions should be simplified, often using the natural frequency, damping ratio, and hydraulic stiffness to replace some complicated combinations of system parameters.

Hydraulic stiffness is a measure for hydraulic fluid compressibility. When the piston of an ideal cylinder with ports completely blocked is pushed by an external load and moved to the left with a small distance $\Delta y$ as shown in Figure 8.12, it will change the pressure of the fluid in both sides of the piston due to a small change in the volume resulting from the limited compressibility of typical hydraulic fluid, as follows:

$$p_1 = \frac{\beta}{V_1}A_L\Delta y \text{ and } p_2 = -\frac{\beta}{V_2}A_L\Delta y \tag{8.32}$$

This pressure difference will create a potential hydraulic force as expressed by Eq. (8.33) intending to bring the piston back to its original position, just as a spring will do in a mechanical system.

$$A_L\left(p_1 - p_2\right) = \left(\frac{\beta A_L^2}{V_1} + \frac{\beta A_L^2}{V_2}\right)\Delta y = K_h\Delta y \tag{8.33}$$

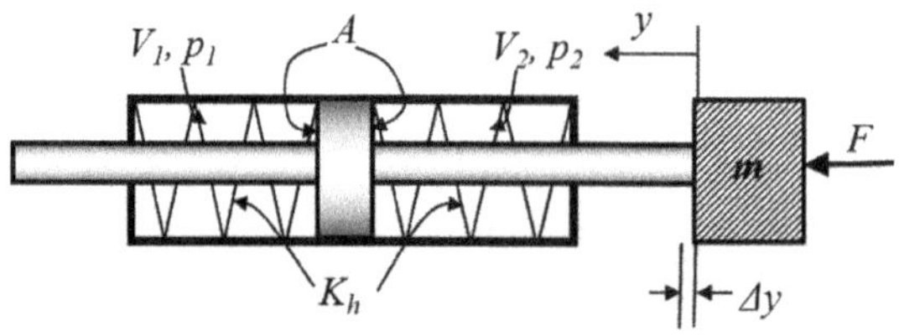


**FIGURE 8.12**
An ideal ports blocked hydraulic cylinder for illustrating hydraulic spring concept.

where $K_h = \frac{\beta A_L^2}{V_1} + \frac{\beta A_L^2}{V_2}$ is defined as the hydraulic stiffness of the cylinder. Corresponding to the assumptions for developing the system model, as well as the transfer function of a valve-controlled hydraulic cylinder system in the previous section $(V_1 = V_2 = V_L/2)$, we can obtain the **hydraulic stiffness** for that model as

$$K_h = \frac{4\beta A_L^2}{V_L} \tag{8.34}$$

When the total mass being moved by the piston is $m$, the potential force created by the hydraulic stiffness will induce an oscillating motion as an overshoot that could induce pressure increase on the piston's other side chamber, just like a mechanical spring-mass system. We can therefore define the **natural frequency** of a hydraulic spring-mass system similar to a mechanical system as follows:

$$\omega_h = \sqrt{\frac{K_h}{m}} = \sqrt{\frac{4\beta A_L^2}{V_L m}} \tag{8.35}$$

In many applications of hydraulic system actuation, the elastic effect of the external load acting on the cylinder is very small, and we could assume $K = 0$ to simplify the analysis. In addition, the term $A_L^2/(k_p + C_{il})$ in transfer functions derived in the previous section represents the damping effect caused by the flow-passing capability under certain pressure in both the control valve and the cylinder, which is much high than damping effects caused by the external load. It leads to $c(k_p + C_{il})/A_L^2 \ll 1$ and can be ignored to simplify the transfer function. Based on the above assumptions and defining $k_{pl} = k_p + C_{il}$ as the **total pressure-flow coefficient** to provide a measure of the efficiency allowing fluid flow under a pressure drop in both the valve and the cylinder, the general transfer function (8.30) could be simplified as:

$$\begin{aligned} Y &= \frac{\frac{k_x}{A_L} X_v - \frac{1}{A_L^2}\left(k_{pl} + \frac{V_L}{4\beta} s\right) F}{\frac{m V_L}{4\beta A_L^2} s^3 + 2\sqrt{\frac{m V_L}{4\beta A_L^2}}\left(\frac{k_{pl}}{A_L}\sqrt{\frac{\beta m}{V_L}} + \frac{c}{4A_L}\sqrt{\frac{V_L}{\beta m}}\right) s^2 + s} \\ &= \frac{\frac{k_x}{A_L} X_v - \frac{1}{A_L^2}\left(k_{pl} + \frac{V_L}{4\beta} s\right) F}{s\left(\frac{s^2}{\omega_h^2} + 2\frac{\xi_h}{\omega_h} s + 1\right)} \end{aligned} \tag{8.36}$$

where $\xi_h = \frac{k_{pl}}{A_L}\sqrt{\frac{\beta m}{V_L}} + \frac{c}{4A_L}\sqrt{\frac{V_L}{\beta m}}$ is defined as the hydraulic damping ratio of the valve-controlled hydraulic cylinder system. The damping effect of the external load is often very small compared to that caused by the hydraulic system and can be ignored. Therefore, the hydraulic damping ratio is often simplified as $\xi_h = \frac{k_{pl}}{A_L}\sqrt{\frac{\beta m}{V_L}}$.

Applying this simplification to transfer functions for piston displacement in response to valve opening (spool position) and applied external load can be obtained as follows:

$$\frac{Y}{X_v} = \frac{\dfrac{k_x}{A_L}}{s\left(\dfrac{s^2}{\omega_h^2} + 2\dfrac{\xi_h}{\omega_h}s + 1\right)} \tag{8.37}$$

and

$$\frac{Y}{F} = -\frac{\dfrac{k_{pl}}{A_L^2}\left(1 + \dfrac{V_L}{4\beta k_{pl}}s\right)}{s\left(\dfrac{s^2}{\omega_h^2} + 2\dfrac{\xi_h}{\omega_h}s + 1\right)} \tag{8.38}$$

As a valve-controlled hydraulic cylinder system is a speed control system by nature, we can easily obtain a speed response transfer function from the simplified displacement function:

$$\frac{\dot{Y}}{X_v} = \frac{\dfrac{k_x}{A_L}}{\dfrac{s^2}{\omega_h^2} + 2\dfrac{\xi_h}{\omega_h}s + 1} \tag{8.39}$$

where $\dfrac{k_x}{A_L}$ can be defined as the speed amplification coefficient of the system.

### 8.2.3 System State-Space Equations

After a system block diagram is created with corresponding determined transfer functions, we can derive the state-space equations of the system from the block diagram. From Figure 8.9, we can get the transfer function of each major block:

$$\frac{F_p}{e_q} = \frac{A_L}{k_p + C_{il} + \dfrac{V_L}{4\beta}s} \tag{8.40}$$

$$\frac{Y}{e_f} = \frac{1}{ms^2 + cs + K} \tag{8.41}$$

where $F_p$ is the hydraulic actuating force generated on the piston by the pressurized hydraulic fluid.

The transfer functions of each summing point are obtained as follows:

$$e_q = k_x X_v - A_L s Y \tag{8.42}$$

$$e_f = F_p - F \tag{8.43}$$

Define the state variables of the valve-controlled hydraulic cylinder system as $x_1 = Y, x_2 = \dot{Y}$, and $x_3 = F_p$. Substituting those state variables to Eqs. (8.40) and (8.41), replacing $x_1 s$ and $x_1 s^2$, using $\dot{x}_1$ and $\dot{x}_2$, and rearranging them, we can get the following set of equations to describe the state for the system:

$$\begin{cases} \dot{x}_1 = x_2 \\ \dot{x}_2 = \dfrac{1}{m}\left(e_f - cx_2 - Kx_1\right) \\ \dot{x}_3 = \dfrac{4\beta}{V_L}\left[A_L e_q - \left(k_p + C_{il}\right)x_3\right] \end{cases} \tag{8.44}$$

The general state-space representation of a hydraulic system can then be written as follows:

$$\dot{x}(t) = Ax(t) + Bu = A\begin{bmatrix} x_1 \\ x_2 \\ x_3 \end{bmatrix} + B\begin{bmatrix} X_V \\ F \end{bmatrix} \tag{8.45}$$

$$Y = Cx(t) \tag{8.46}$$

Matrices A, B, and C can be calculated based on Eqs. 8.40–8.44.

$$A - \begin{bmatrix} 0 & 1 & 0 \\ -\dfrac{K}{m} & -\dfrac{c}{m} & \dfrac{1}{m} \\ 0 & -\dfrac{4\beta A_L^2}{V_L} & -\dfrac{4\beta\left(k_p + C_{il}\right)}{V_L} \end{bmatrix} \tag{8.47}$$

$$B = \begin{bmatrix} 0 & 0 \\ 0 & 1 \\ \dfrac{4\beta A_L K_x}{V_L} & 0 \end{bmatrix} \tag{8.48}$$

$$C = \begin{bmatrix} 1 & 0 & 0 \end{bmatrix} \tag{8.49}$$

The state equations of a system could be different using different ways of analyzing the system. However, they should be able to present the same responding characteristics of the system under the same stimuli input.

### 8.2.4 System Performance Characteristics

From the system transfer functions or state equations we have obtained, we can find that the major parameters determining the performance of a valve-controlled hydraulic system include the flow gain, hydraulic natural frequency, and damping ratio.

The system speed amplification coefficient is often defined as $k_x/A_L$, where $A_L$ is a constant for a system with a finalized design that has been physically built and the valve gain $k_x$ is often changing with valve opening. It is also important to note that all the transfer functions or state equations are obtained through linearization of the system model, based on an assumption of being controlled using a zero-lapped valve. The impact of system nonlinearity, resulting mainly from the physical design of spool valves as discussed in Section 3.1.2, is always an important factor to consider in the design and control of a hydraulic system. Figure 8.13 shows how the nonlinear factors of **deadband**, **saturation**, and inconsistent **modulating gain** affect the flow gain in a typical spool valve-controlled system. The dash-dot line in the figure is the ideal characteristic for controlling the system, which is a straight line between the spool displacement and outputting flow rate, intersecting the origin of the coordinates XOQ. The solid curve is the actual flow-modulating gain of the valve, illustrating major nonlinear characteristics of deadband, saturation, and nonlinearity in modulating gain. The deadband results from the overlap design, which is the standard design for the normal-closed valve, ensuring that the valve can be securely closed by paying the price of having no flow able to pass through the valve at a small spool displacement. Saturation results from the maximum flow passage area of the valve, which limits the maximum amount of flow able to pass through a valve. The valve-modulating gain is defined as how much flow rate increase could result from increasing a unit of valve opening $\Delta q/\Delta x$. As illustrated in Figure 8.13, this gain is normally nonlinear, and engineers often use a linearized gain to simplify the analysis. To obtain a more accurate analysis result, it is also common practice to eliminate the influence of deadband and saturation to linearize the gain, which will increase the modulating gain from $\Delta q/\Delta x$ to $\Delta q'/\Delta x$, as shown by the long dash line in Figure 8.13.

Another important characteristic of a valve-controlled hydraulic system is its hysteresis in flow rate control, as shown in Figure 8.14. Caused by delay in energy dissipation of the fluid passing through the valve, the actual flow rate is either smaller than the theoretical

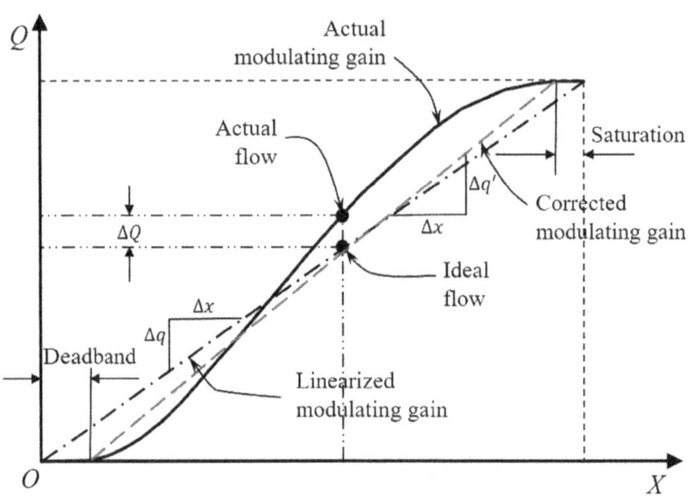


**FIGURE 8.13**
Nonlinear characteristics of deadband, saturation and inconsistent modulation gain in a typical valve-controlled hydraulic system. The nonlinear solid curve represents the flow control modulating curve, the dash-dot line is the theoretical modulating gain and the long-dash line is the corrected theoretical gain by removing the effect of deadband and saturation, and the piecewise linear curve shows an optimized linearization of the nonlinear gain.

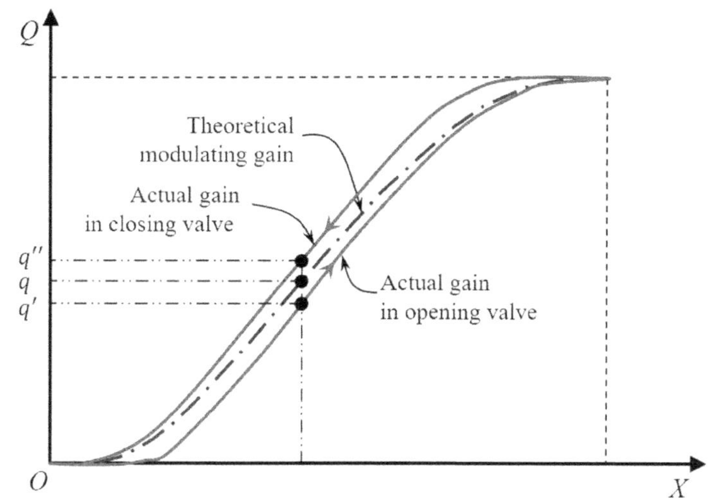


**FIGURE 8.14**
Hysteresis in flow rate control from a valve-controlled hydraulic system.

rate at a certain valve opening or greater than the theoretical rate while closing the valve. Such a difference in the responding flow rate to the same valve opening during opening and closing is defined as the **hysteresis** of the system.

All the introduced nonlinear characteristics of hydraulic systems can result in some difficulty in achieving accurate flow rate regulation, an important measure of system static performance. In engineering practice, some adequately designed control systems could effectively compensate for those nonlinearities.

One of the critical requirements in design of a hydraulic system is to ensure that the system can respond to maneuvering commands promptly and accurately under normal operating conditions. Defined in Section 8.2.2, both hydraulic natural frequency and damping ratio are important characteristics of a hydraulic system and can directly impact the dynamic response of a hydraulic system. Hydraulic natural frequency is collectively determined by the hydraulic system stiffness and the external load mass, and is normally the lowest frequency of a valve-controlled hydraulic cylinder system and determines the responding speed of the cylinder to a control input from the valve. Higher natural frequency results in faster response of cylinder motion to valve-opening changes. Because hydraulic stiffness is affected by the effective fluid mass and the dynamic fluid volume in piston chambers, the system's natural frequency is therefore normally determined when the piston is located at the middle position as the hydraulic stiffness, and therefore the natural frequency, reaches the minimal value at this location.

From Section 8.2.2, we know that many factors can affect the hydraulic damping ratio, with perhaps the efficiency of allowing fluid flow within the system (measured using the system pressure-flow coefficient $k_{pl}$) having the highest impact on it. This coefficient has the lowest value when the spool is at the neutral position, which results in the lowest damping ratio in a hydraulic system. This lowest damping ratio is often used as the system damping ratio, as it will rapidly rise after the valve is opened. Because the hydraulic damping ratio can directly affect system stability, it is a common engineering practice to improve a hydraulic system's stability by increasing system damping.

The hydraulic natural frequency of a hydraulic system is jointly determined by the stiffness of the hydraulic spring and load mass as defined in Eq. (8.35), which decides how fast the system can respond to a control input, and is often the lowest frequency of the entire system. Hydraulic natural frequency is therefore the limiting factor to the response speed in a hydraulic system, and one of the effective ways to increase the hydraulic natural frequency is to reduce the load volume of the system. This can be accomplished mainly by minimizing the noneffective volume in a cylinder and reducing the volume of connecting hoses between valve and cylinder by placing the valve as close to the cylinder as possible.

**Example 8.2: System Load Determination on a Double-Rod Symmetric Cylinder**

Based on the parametric logic flow presented by Eqs. (8.15), (8.22), and (8.24), create a system block diagram for the valve-controlled hydraulic cylinder system shown in Figure 8.9.

a. Taking the Laplace transform of Eqs. (8.24), (8.22), and (8.15), and then rearranging them, we obtain the following three transfer functions:

$$
\begin{aligned}
Q_L &= k_x X_v - k_p P_L \\
\dot{Y} &= \frac{1}{A_L}\left[Q_L - \left(\frac{V_L}{4\beta}s + C_{il}\right)P_L\right] \\
P_L &= \frac{1}{A_L}\left[(ms+c)\dot{Y} + KY + F\right]
\end{aligned}
$$

b. Applying the block diagram transformation rules, we can obtain a system block diagram described by these three transfer functions, as shown in Figure 8.15:

DISCUSSION 8.2: This block diagram and the one in Figure 8.9 represent the same valve-controlled symmetric hydraulic cylinder system. The difference is that this system (Figure 8.15) determines piston displacement in terms of load flow rate ($Q_L$), whereas the one in Figure 8.9 determines it in terms of load pressure ($P_L$). Because of this difference, the approach presented in Figure 8.15 is more suitable for analyzing systems with light load and fast response, and the other is more suitable for systems with large load inertia and system leakage. Simplifying both diagrams by implanting either the load flow rate or load pressure into the system, we can obtain the same overall transfer function between the valve spool displacement, external force, and piston motion for the system.

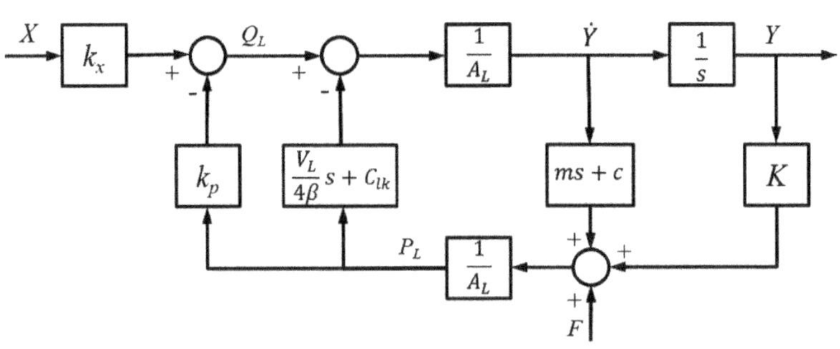


**FIGURE 8.15**
System blocks of a valve-controlled hydraulic cylinder system model (piston motion transformed from load flow rate).

## References


1. Akers, A., Gassman, M., Smith, R. *Hydraulic Power System Analysis*. CRC Press, Boca Raton, FL (2006).
2. Batchelor, G.K. *An Introduction to Fluid Dynamics*. Cambridge University Press, Cambridge, UK (2000).
3. Dort, R.C., Bishop, R.H. *Modern Control Systems (12th Ed.)*. Prentice-Hall, Upper Saddle River, NJ (2011).
4. Cundiff, J.S. *Fluid Power Circuits and Controls: Fundamentals and Applications*. CRC Press, Boca Raton, FL (2002).
5. Gao, Y., Huang, R., Zhang, Q. A comparison of three steering controllers for off-road vehicles. *Proceedings of the Institute of Mechanical Engineers, Part D: Journal of Automobile Engineering*, 222: 2321–2336 (2008).
6. Johnson, J.L. *Questions Answered on Electrohydraulic Control*. http://www.hydraulicspneumatics.com/controls-instrumentation/questions-answered-electrohydraulic-control. Accessed on November 7 (2017).
7. Keesman, K.J. *System Identification: An Introduction*. Springer (2011)
8. Keller, G.R. *Hydraulic System Analysis*. Penton Media Inc., Cleveland, OH (1985).
9. McClay, D., Martin, H.R. *The Control of Fluid Power*. John Wiley & Sons, New York (1973).
10. Manring, N.D. *Hydraulic Control Systems*. John Wiley & Sons, New York (2005).
11. Merrit, H.E. *Hydraulic Control Systems*. John Wiley & Sons, New York (1967).
12. Qiu, H., Zhang, Q., Reid, J.F. Wu, D. System identification of an electrohydraulic steering system. *Journal of Commercial Vehicles*, 108: 361–367 (1999).
13. Qiu, H., Zhang, Q., Reid, J.F., Wu, D. Modeling and simulation of an electrohydraulic steering system. *International Journal of Vehicle Design*, 17: 259–265 (2001).
14. Rovira-Más, F., Zhang, Q., Hansen, A.C. Dynamic behavior of an electrohydraulic valve: Typology of characteristic curves. *Mechatronics*, 17: 551–561 (2007).
15. Stounbaugh, T. *Automatic Guidance of Agricultural Vehicles at Higher Speeds*. Ph.D. Dissertation, University of Illinois at Urbana-Champaign (1997).
16. Stringer, J. *Hydraulic Systems Analysis: An Introduction*. John Wiley & Sons, New York (1976).
17. Ulrich, H.J. Some factors influencing the natural frequency of linear hydraulic actuators. *International Journal of Machine Tool Design and Research*, 11(2): 199–207 (1971).
18. Watton, J. *Fluid Power Systems, Modeling, Simulation, Analog and Microcomputer Control*. Prentice-Hall, New York (1989).
19. Wu, K., Zhang, Q., Hansen, A.C. Modeling and identification of a hardware-in-the-loop hydrostatic transmission simulator. *International Journal of Vehicle Design*, 34: 63–75 (2004).
20. Wu, Z. *Hydraulic Control Systems*, Higher Education Press, Beijing, China (in Chinese) (2008).
21. Zhang, Q., Goering, C.E. *Fluid Power System*. In: Bishop, R. (ed.), *The Mechatronics Handbook*. CRC Press, Boca Raton, FL, pp. 10–11~10–14 (2001).
22. Zhang, Q., Reid, J.F., Wu, D. Hardware-in-the-loop simulator of an off-road vehicle electrohydraulic steering system. *Transactions of the American Society of Agricultural Engineers*, 43: 1323–1330 (2000).


## Exercises

**8.1** What is the stiffness of a hydraulic spring?

**8.2** Does the natural frequency of a hydraulic cylinder system depend on the piston position, and why?

**8.3** In defining the load flow rate, why do we assume that the initial volumes of both chambers of a cylinder are equal to one-half of the total volume of the cylinder ($V_{10} = V_{20} = V_L/2$)?

**8.4** Define the load pressure and load flow in a single-rod asymmetric cylinder.

**8.5** How do hydraulic natural frequency and damping ratio changes affect the system characteristics of a valve-controlled hydraulic system?

**8.6** Identify a few factors that will cause an increase on its damping ratio in a hydraulic system, and explain why.

**8.7** Linearize the orifice equation of a typical spool valve (Eq. 8.23) using the Taylor series expansion technique (assume the fluid supply pressure maintains a constant value).

**8.8** When the zero-lapped four-way control valve depicted in Figure 8.16(a) is used to control the motion of a single-rod asymmetric cylinder, try to develop a model for determining the load flow–pressure relationship controlled by the valve. (Assume that this spool valve has the same orifice area and orifice coefficient for all ports under a certain spool displacement, with the supply pressure being a constant and tank pressure being zero, and ignore the valve internal leakage and fluid compressibility).

**8.9** For a hydraulic system similar to the one as described in Example 8.1, but using a fixed-displacement pump to supply a constant source flow $Q_s$ under pressure $p_s$, assume that the demanding load flow is $Q_L$ and the load pressure is $p_L$, and try to determine under what condition the system can reach the highest efficiency.

**8.10** Assume the motion of a double-rod symmetric cylinder is controlled using a zero-lapped four-way control valve depicted in Figure 8.16(a). Try to develop a model for presenting how the load flow is affected under different load pressures when controlled by the valve (assume all the assumptions specified in the text are valid).

**8.11** Assume that the motion of a single-rod asymmetric cylinder is controlled using a zero-lapped four-way control valve similar to the system depicted in Figure 8.16(a). Try to develop a general equation for modeling the load flow for extending the cylinder when controlled by the valve (assume all assumptions specified in the text).

**8.12** Derive the load flow model of cylinder retraction for the system described in Exercise 8.11.

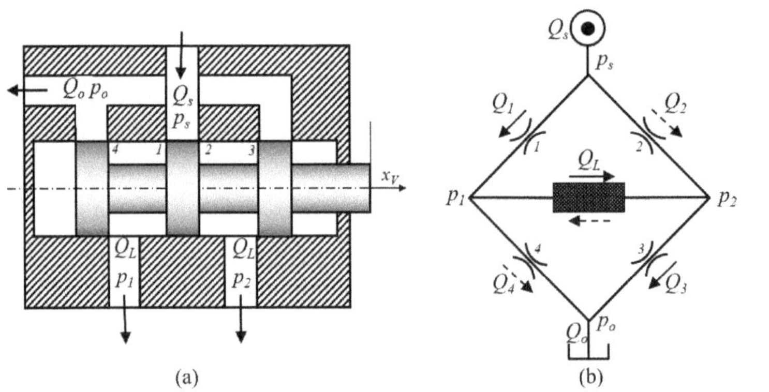


**FIGURE 8.16**
(a) A zero-lapped four-way spool valve and (b) its equivalent bridge circuit.

**8.13** For the same system as described in Exercise 8.11, derive its flow continuity equation when the cylinder is extending (assume all assumptions are specified in the text).

**8.14** Derive the flow continuity model when the cylinder is retracting for the system described in Exercise 8.13.

**8.15** Assume that the hydraulic system explained in Example 8.1 uses a fixed-displacement pump to provide source flow $Q_s$ under pressure $p_s$. What then will be the system efficiency if the needed load flow $Q_L$ under a load pressure of $p_L$ is only one-half of the source flow? (Assume that the line losses and leaks are negligible.)?

**8.16** Assume that the hydraulic system explained in Example 8.1 uses a fixed-displacement pump to provide source flow $Q_s$ under pressure $p_s$. What then will be the system efficiency if the needed load flow $Q_L$ under a load pressure of $p_L$ is 100% of the source flow? (Assume that the line losses and leaks are negligible.)

**8.17** What is the transfer function for a system where the input $y$ is related to the output by the differential equation $a_1 \frac{dx}{dt} + a_0 x = b_0 y$ ?

**8.18** A system has an output $y$ that varies with time $t$ when subject to a step input of $x$ and is described by $\frac{d^2x}{dt^2} + 10\frac{dx}{dt} + 25x = 50y$. What is (a) the undamped frequency, (b) the damping ratio, and (c) the solution to the equation if $x = 0$ and $dx/dt = -2$ when $t = 0$ and there is a step input of size 3 unit?

**8.19** Assume that the forward path transfer function in a control system with a unit negative feedback gain is $G(s) = \frac{K(2s+1)(s+1)}{s^2(Ts+1)}$. What condition should $K$ and $T$ satisfy to ensure that the system will be stable after the loop is closed $(K > 0, T > 0)$?

**8.20** Simplify the system block diagram (Figure 8.17) into a single input–single output block.

**8.21** Determine the stability of a system as presented by the block diagram presented in Figure 8.18.

**8.22** Derive governing equations describing the control principle of a valve-controlled linear hydraulic actuator under the condition illustrated in Figure 8.19. What will be the cylinder pressure at a static condition? (Assume the orifice-opening area of a spool valve to be: $A_v = k_v x$.)

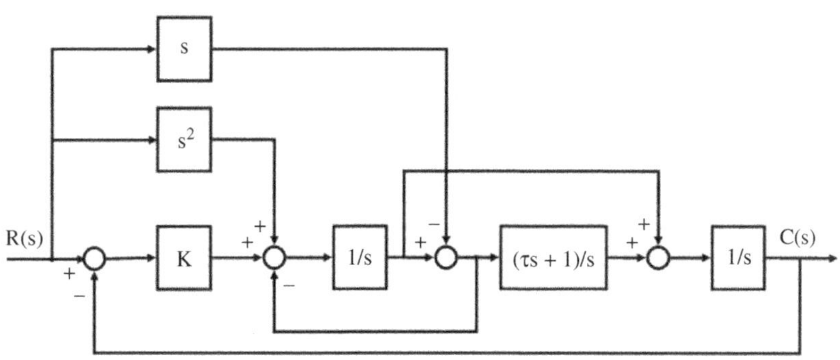


**FIGURE 8.17**
A system block diagram.

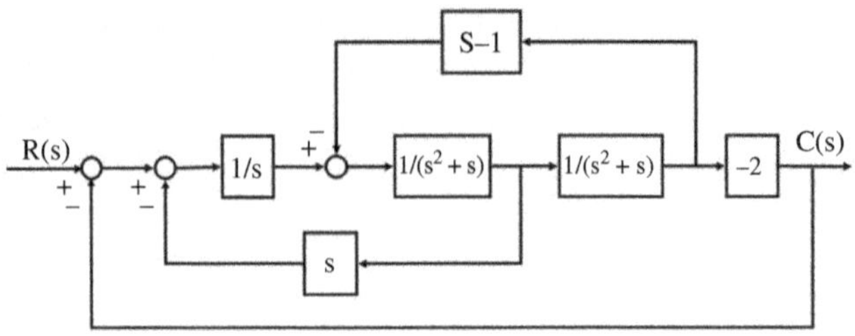


**FIGURE 8.18**
A system block diagram.

**8.23** With a set of appropriate assumptions, the differential equation describing the relationship between the valve spool stroke $x$ and the double-rod cylinder displacement $y$ as shown in Figure 8.19 is:

$$\frac{k_x}{A_{RE}}x = \frac{V_{RE}m}{\beta A_{RE}^2}\frac{d^3y}{dt^3} + \frac{V_{RE}f}{\beta A_{RE}^2}\frac{d^2y}{dt^2} + \frac{dy}{dt}$$

Try to derive the transfer function of the system with zero initial condition, and determine the natural frequency and the damping ratio of the system.

**8.24** Assume the transfer function of a linear E/H system plant is:

$$s^3Y(s) + 2\zeta_h\omega_h s^2Y(s) + \omega_h^2 sY(s) = K\omega_h^2 X(s)$$

Please derive the state-space model of the system, and determine the **A**, **B**, and **C** matrices.

**8.25** Construct a block diagram for a system represented by the following transfer function:

$$\ddot{y}(t) + 2\zeta_h\omega_h\dot{y}(t) + \omega_h^2 y(t) = u(t)$$

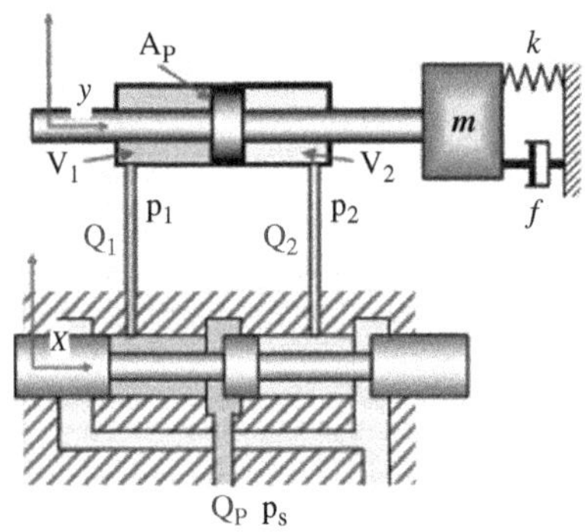


**FIGURE 8.19**
Operation condition illustration of a valve-controlled linear hydraulic actuator.

# 9

# *Electrohydraulic Systems Control*

## 9.1 Concepts of Electrohydraulic System Control

### 9.1.1 Basic Concept of Automatic Controls

One major advantage of an electrohydraulic system over a conventional hydraulic system is its ability to implement automatic controls. To accomplish automatic control of a hydraulic system, taking a valve-controlled hydraulic cylinder actuating system (Figure 9.1) as an example without loss of generality, three fundamental actions are required: (1) obtaining the cylinder-actuating information; (2) comparing the obtained information with the commanded operation to make a decision for further action; and (3) performing the action. Instead of relying on a human operator to observe the actuation to maneuver the system, an automatic control system could use sensors to obtain the desired information, a controller to compare the actual actuation with the commanded operations to determine adequate control actions, and an actuator to implement the required control actions.

An automatic control system needs to have some information to create appropriate control signals before it can take action. The information required by many automatic control systems includes set points and system variables. The set points are often the desired system outputs that are essential, and a control system will be unable to work without them. For example, in a valve-controlled hydraulic cylinder-actuating system, the set points could be the position or velocity of the cylinder motion or the force provided by the cylinder. The value of set points can be entered into an electrohydraulic control system either by the human through analog or digital interfaces in human-operated systems, or by an electronic device receiving signals from a higher layer control system in autonomous or automatically operated systems. The system variables are the variables that can be measured using sensors to indicate how the system reacts to a control action, which can either be the directly measurable actual system outputs or some intermediate variables that may need to be computed to obtain the values of the variables that are not directly measurable. Unlike the set points, system variables are a class of "could-to-have" information, which means that a control system could work without it, but the performance could be improved using these variables.

After gathering the necessary information, the control system will then need to create adequate control signals to drive the implement actuators to take proper actions in order to obtain the desired system outputs. The device used to create control signals is often called the controller. Controllers can be classified by the states of their output: the more states of the output, the more complicated and costly the controller will be. While this is not a textbook on control systems design, the rest of this chapter will introduce some basic concepts useful in designing electrohydraulic systems and their controls.

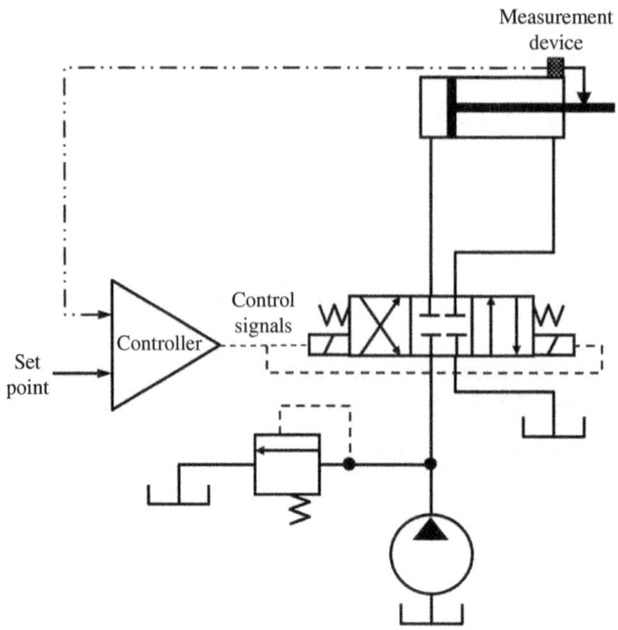


**FIGURE 9.1**
System schematic of an electronically controlled hydraulic cylinder actuating system.

### 9.1.2 Open- and Closed-Loop Controls

Electrohydraulic systems are commonly controlled using either open-loop or closed-loop controllers. An open-loop controller is completely controlled by the input; the output has no responsive effect on the control action. As shown in Figure 9.2(a), the operation of an open-loop control system is very simple: when the controller receives a control input, it creates a control action, independent of the system variables, to drive the actuator producing an expected output. More precisely, the open-loop system uses no responsive feedback to correct the control action, even though sensors may be used to measure the actual output in some of the systems, and it assumes that a desired control output will always be achieved. Open-loop controls could be a better choice when (1) low cost is a priority; (2) output is simply either on or off; (3) output can be easily predictable; and/or (4) output is fairly consistent. Many mobile electrohydraulic motion control systems often use open-loop control as output is usually predictable and the changes in output in those systems are often very small.

The key difference between an open-loop and a closed-loop controller is that the open-loop does not use a feedback but the closed-loop does. As illustrated in Figure 9.2(b), a closed-loop control system measures the current system output, compares it with the set point, and alters the control action if there is an error keeping the output as close to the **set point** (the input) as possible. In other words, a closed-loop control system is a self-adjusting system as the output data flows back to the starting point of the control system, often through a specific amplifier to convert the measured output data to the same form as the control system input, making the controller adjust accordingly to minimize the error between expected and actual outputs. Closed-loop controls could be a better choice when (1) measurement of the output is feasible; (2) the process has a certain degree of predictability; (3) the system may become unstable; and (4) the output is sensitive to external disturbances. Systems requiring accurate control of their output normally use closed-loop control.

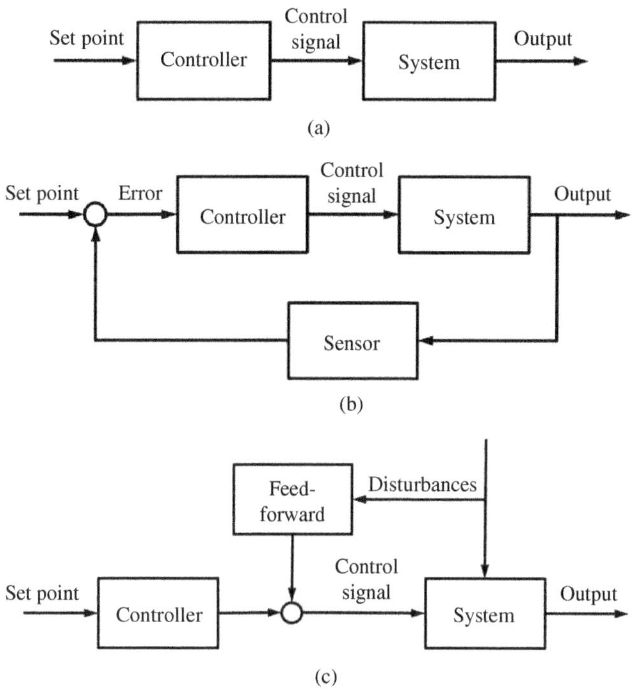


**FIGURE 9.2**
System block diagrams of (a) open-loop control system; (b) closed-loop control system; and (c) feedforward control system.

The major disadvantage of open-loop control is the possibility of losing accuracy. Without feedback, there is no guarantee that the control system will always maintain the output at the expected level when there is a disturbance acting on the system. One solution to this problem is the use of **feedforward control** (Figure 9.2(c)) which uses a model of the system to make a control move to correct output variation caused by disturbances such as an experienced human operator will do. Unlike feedback control, the control variable adjustment in a feedforward system is not error-based. Instead, it is based on knowledge about the system in the form of a theoretical or empirical model of the input–output relationship and knowledge about or measurements of the disturbances. Feedforward control is often used in many mobile electrohydraulic systems.

### 9.1.3 Transient Response and Steady-State Error

The purpose of control is to obtain a desired response from a certain command input. The **transient response** of a control system is therefore one of the most important characteristics, and often requires adjustment until the system provides a satisfactory response. The standard performance measure of the transient response is often defined in terms of the step response of a system. A valve-controlled hydraulic system is a speed control system by nature, and we can use a second-order Laplace transfer function to express the dynamic behaviors of a typical valve-controlled hydraulic cylinder system, as described in Section 8.2. Figure 9.3 shows a typical transient response of a second-order control system to a step input. The speed of response is measured by the **rise time** ($T_r$), **peak time** ($T_p$), **overshoot** ($P_o$), **settling time** ($T_s$), and **steady-state error** ($E_s$), and indicates how fast a control system reacts to the step input.

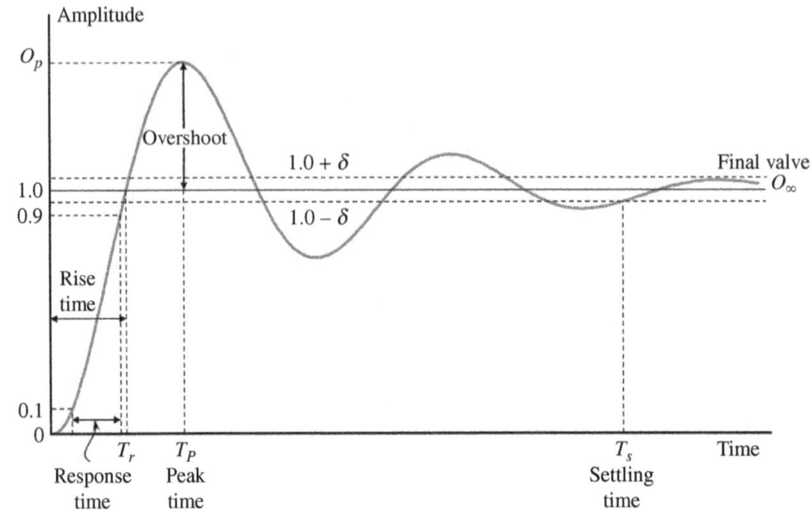


**FIGURE 9.3**
Transient response of an underdamping second-order control system to a step input.

Rise time ($T_r$) is defined as the time that elapses after a step input is applied to the control system to the time the control system output reaches the step high (the set point of the step input) the first time. In engineering practice, a control system could be either underdamped or overdamped, and the output from an overdamped control system may never reach the step high in a finite time. To solve this problem, engineers often use a **response time** ($T_{r1}$), defined as "the time required for the response to rise from x% to y% of its final value," to represent the concept of rise time. While rigorous definition of the response time is still lacking, it is commonly understood that a response time is the time that elapses for the response to rise from 0 to 100% step high for an **underdamped** second-order system, 5 to 95% for a **critically damped** system, and 10 to 90% for an **overdamped** system. Sometimes, people use **delay time** ($T_d$), defined as the time required for the response to reach half the step high value the very first time, to assess how fast a control system can respond to a step input.

In an underdamped second-order control system, the output will exceed the step high and cause some oscillations after receiving a step input. The peak value of the oscillation (often the first peak of the oscillation) in system output is defined as the overshoot, and the time for the response to reach the overshoot value is defined as the peak time ($T_p$). The overshoot ($P_o$) is measured by the percentage of the peak value over the set value, defined as:

$$P_o = \frac{O_p - 1.0}{1.0} \times 100\% \tag{9.1}$$

It takes time to allow the oscillation to settle down. The time taken for the output to be entered and remain within a specified error band of its step high value is defined as the settling time ($T_s$), and the system can be treated as having reached its steady state after this point. The steady-state error ($E_s$) of a control system is defined as the difference between the actual output and the step high value after the system reaches a steady state. For a second-order system with closed-loop damping constant of $\xi\omega_n$, the output response could remain within a 2% error margin after four time constants ($\tau$):

$$T_s = \frac{4}{\xi\omega_n} = 4\tau \tag{9.2}$$

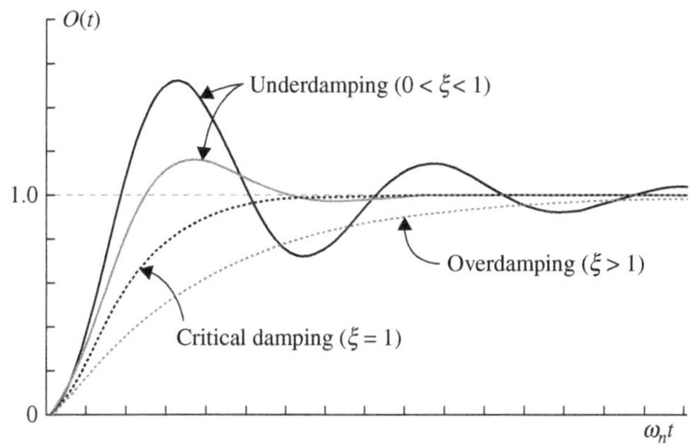


**FIGURE 9.4**
Transient responses of underdamping, critical damping and overdamping second-order control systems to a step input.

As mentioned earlier, a control system can be either underdamped, critically damped, or overdamped. This implies that adjusting the damping ratio of an electrohydraulic system, which is relatively easier to adjust through system design, could result in a big difference in its control system behaviors. Figure 9.4 shows the transient responses of underdamped; critical damped and overdamped second-order control systems to a step input. From this figure one can see that when the system is critically damped ($\xi = 1.0$), there is no overshoot, nor oscillation, and the output reaches its steady-state very quickly. For underdamped systems (such as $\xi = 0.2, 0.5$), their outputs present a convergent oscillation under corresponding damped natural frequency ($\omega_d = \omega_n\sqrt{1-\xi^2}$). The transient response of overdamped systems (such as $\xi = 2.0$) is similar to critically damped systems in pattern, but with a longer converging period.

### 9.1.4 Gain and Feedback

One of the most important concepts of feedback control is its ability to monitor its output and make corrections as needed to ensure that the output remains at the commanded level with an acceptable steady-state error. As the set point to a control system and the output from the controlled system are not always the same parameter, it is essential to understand system gain of a feedback control system to learn how it works.

A system gain is a proportional value showing the relationship between the magnitude of the input set point to the magnitude of the output variable of a control system. It is a product of the steady-state gains of all composing elements of the control system. A common practice in obtaining adequate performance from a control system is to use some methods to alter the gain to provide the system more or less "power" for different situations. Figure 9.5 shows the system block diagram of a typical feedback control system. Being denoted in the system diagram, the system gain ($G_s$) is defined as the ratio of the magnitude of system output ($O$) and input set point (commonly called the "reference point," $R$), as follows:

$$G_s = \frac{O}{R} \tag{9.3}$$

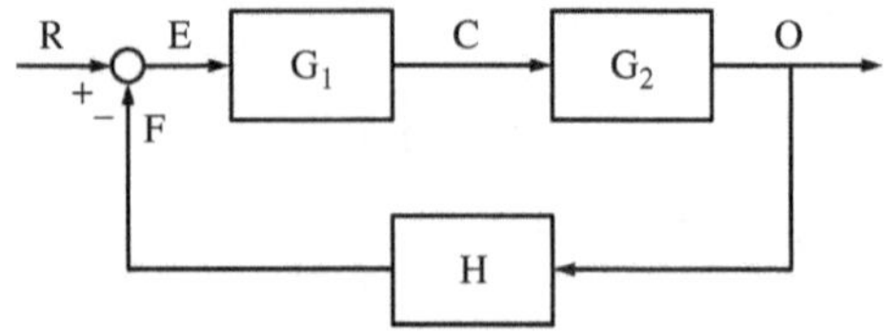


**FIGURE 9.5**
System block diagram of a basic feedback control system.

As the system operates, the feedback signal ($F$) is continuously generated according to the feedback gain ($H$), which is commonly determined by the sensing and monitoring elements of the system:

$$F = OH \tag{9.4}$$

The feedback signal ($F$) is fed to a summing junction of the system to be compared with the reference point ($R$) to see if there is an error ($E$):

$$E = R - F = R - OH \tag{9.5}$$

If there is an error, this detected error ($E$) will then be used to correct control input ($C$) to the system being controlled:

$$C = EG_1 \tag{9.6}$$

This corrected control input ($C$) will then try to bring the system output ($O$) to the desired level:

$$O = CG_2 = EG_1G_2 \tag{9.7}$$

The product of $G_1G_2$ is often called the **feedforward loop gain** of the system. If we substitute the error using EQ. (9.5), the result is:

$$O = EG_1G_2 = (R - OH)G_1G_2 \tag{9.8}$$

When we rearrange Eq. (9.8), we have:

$$O = R\frac{G_1G_2}{1 + HG_1G_2} \tag{9.9}$$

The system gain (also called the **feedback gain**) can therefore be defined as follows:

$$\frac{O}{R} = \frac{G_1G_2}{1 + HG_1G_2} \tag{9.10}$$

In general, increasing the system gain would decrease rise time and increase overshoot and settling time. Increasing the gain beyond a safe zone could cause the system to become unstable. However, responses may not always be the same for all systems; for example increasing gain for a critically damped system will decrease rise time but have no effect on overshoot or settling time.

### 9.1.5 Frequency Response and the Bode Diagram

One basic requirement for an electrohydraulic control system is that the controlled motion of the hydraulic actuator can respond to input control commands promptly and accurately under normal operation conditions. However, the usually low natural frequency and high damping ratio of a hydraulic system often experience a slow response in cylinder motion to input commands, which could stop the system from reaching the required level of speed. Figure 9.6 gives an example obtained from a simple laboratory setup with a valve-controlled double-rod cylinder with no external load. From the results we find that when the rate change of commanded speed is under a certain limit, the system could respond to the input commands promptly and reach a fairly accurate tracking of the commanded cylinder speed (the first cycle). As the changing rate of commanded speed (namely, input frequency) goes up, the cylinder response to the input command frequency will gradually become slower and not be able to reach the demanded maximum speed (the next two cycles), and eventually it will stop responding.

One important system characteristic is the frequency response, which reveals how a hydraulic control system can respond to changes in its input commands, including the changes of magnitude gain and phase delay to the changing rate of input commands. The **Bode diagram**, a set of logarithmic plots generated from the frequency domain transfer function of a system, is commonly used in control engineering to visualize magnitude gain and phase response of the system to different stimuli frequencies. Figure 9.7 is a Bode diagram (often call **Bode plot**) showing the dynamic responses of cylinder displacement $Y$ to input valve displacement $X_v$ under different frequencies in terms of magnitude gain and phase delay. This Bode diagram is based on a transfer function of cylinder motion control (Eq. 8.37), which consists of a proportional flow gain $k_x/A_L$, an integral term $1/s$, and a second-order oscillation term $1\Big/\left(\frac{s^2}{\omega_h^2}+2\frac{\xi_h}{\omega_h}s+1\right)$.

The flow gain $k_x$ is a measure of valve-flow increment, which is highest at spool zero (neutral) position and will be decreased as the spool moves away from that position. System stability is often analyzed using the highest speed amplification coefficient to ensure that the system can be stably operated under the entire range of flow gain if it is able to pass the worst-case scenario test. On the other hand, the flow gain $k_x$ will decrease as system load pressure increases, which will move the crossover frequency $\omega_c$ to the left in the Bode plot and result in a reduced responding speed and control accuracy. Analysis of system

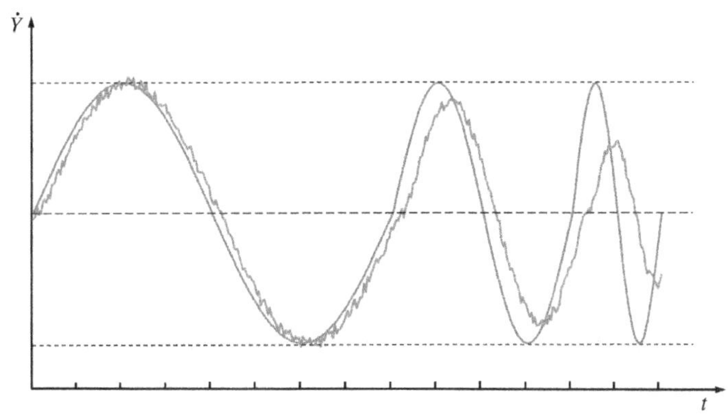


**FIGURE 9.6**
Speed response of a valve-controlled hydraulic cylinder (double-rod with no load) to different commanding rates.

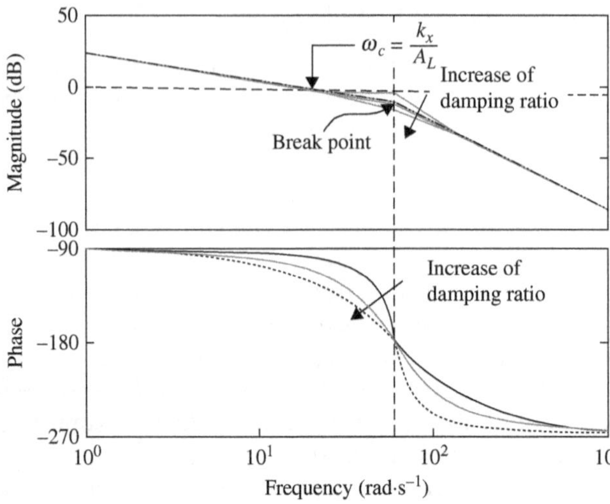


**FIGURE 9.7**
Frequency response (Bode Plot) of a valve-controlled hydraulic cylinder system.

response speed and control accuracy is recommended to use the flow gain at $p_L = \frac{2}{3}p_s$ in order to ensure that the system can respond promptly to a control input and achieve the required control accuracy if passing the test for the worst case scenario.

The damping ratio of a system plays a major role in its phase delay. The higher the damping ratio, the slower the system will respond. As defined in Eq. (8.36), the hydraulic damping ratio can be affected by many factors, but mostly by the pressure-flow coefficient $k_{pl}$ for a specific system. As $k_{pl}$ has the smallest value when the spool is at its zero position, a hydraulic system has the smallest damping ratio when the valve is closed, which is often in the 0.1 to 0.2 range for many valve-controlled hydraulic cylinder systems. Opposite to the flow gain, $k_{pl}$ will quickly be increased as the spool moves away from the zero position or as the system load pressure increases. Under high-load pressure, the damping ratio could be greater than 1.0 or even higher. The large variation range ($\xi_{h,\max}/\xi_{h,\min} \geq 20$ or larger) is one of the most challenging features in analyzing a hydraulic system. As a measure of the relative stability of a system, it is often desirable that the system have an adequate damping ratio under the condition of not sacrificing the response speed.

**Example 9.1: (Closed-Loop System Analysis)**

For the closed-loop system defined in Figure 9.8, try to determine the natural frequency and damping ratio of this system.

(a) From Eq. (9.10) we can get:

$$\frac{O(s)}{R(s)} = \frac{\frac{1}{s}\cdot\frac{100}{50s+4}}{1+\frac{1}{s}\cdot\frac{100}{50s+4}\cdot 0.02} = \frac{100}{s(50s+4)+2} = \frac{50}{25s^2+2s+1}$$

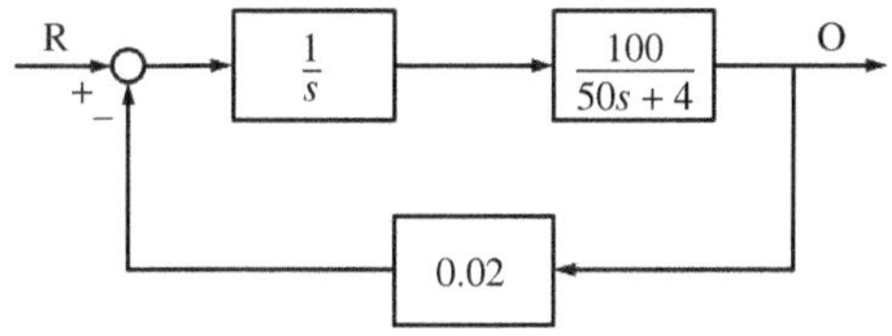


**FIGURE 9.8**
System block diagram of a closed-loop control system.

Compare it with the standard second-order transfer function:

$$\frac{O(s)}{R(s)} = \frac{K\omega_n^{\ 2}}{s^2 + 2\xi\omega_n s + \omega_n^{\ 2}}$$

We can obtain:

$$\omega_n = \sqrt{\frac{1}{25}} = \frac{1}{5} = 0.2\ (rad/sec)$$

$$\xi = \frac{\frac{2}{25}}{2\omega_n} = \frac{1}{25 \times 0.2} = 0.2$$

DISCUSSION 9.1: As the system has a $\xi = 0.2 < 1$, it is an underdamping system. Decreasing the open-loop gain could help to increase the system damping ratio and therefore help to improve the system transient response to be stabilized quickly.

## 9.2 Hydraulic Velocity, Position, and Force Controls

### 9.2.1 Proportional and Servo Actuation in Electrohydraulic Control Valves

One of the major differences between traditional hydraulic systems and electrohydraulic systems is the formerly exclusively used manual-operated directional control valves, often called **bang-bang *control*** valves, and the later use of electrohydraulic **servo** or **proportional** control valves to control flow direction, flow volume, and flow pressure. Electrohydraulic systems are integration technologies of hydraulic power transmission and automated controls. In terms of their control valves (the core component for implementing electrohydraulic control) being used, the electrohydraulic systems are normally classified as servo and proportional systems. With the increasing use of electronics in controlling hydraulic systems, the directional control valves of today can also be controlled using an electromechanical solenoid or servo driver to implement automated controls. (The operational principles of directional, proportional, and servo valves were introduced in Chapter 3.)

**Directional control** basically controls a valve that is to be "on" or "off" to control the flow that is either supplied into or blocked from getting into a specific branch circuit to perform a designated function. With limited capacity, directional control valves can also be used to control the flow volume, attained by selecting an orifice to allow only a specified volume of flow to pass the valve. Supported by an electronic control system and an electronic pressure sensor, it can also achieve programmable pressure control for applications requiring multiple pressure settings. Changing direction, flow, or pressure in a system using only directional control valves would require a separate individual valve for each direction, flow, or pressure desired. It would make the system very complicated; thus, it is applicable only in very simple systems. Direction control in electrohydraulic systems is often performed using a solenoid driver (Section 3.1.6), which is a type of **solenoid control valve**. Solenoid controls are widely applied in many mobile hydraulic systems, including many types of agricultural, construction, and mining equipment.

In engineering practice, many more complicated systems commonly use **proportional control** valves, which allow infinite positioning of the spool to continuously vary the orifice size in proportion to the electronic control inputs and achieve both flow and direction controls via either open-loop or closed-loop controls. Such a feature allows an electrohydraulic system to achieve continuous speed control in both directions by simply adjusting the electrical control signal to maneuver the valve. By carefully tuning the patterns of the input control signal, a proportional control valve can achieve desirable speed control characteristics in motion control without using additional hydraulic components. It gives the system the needed flexibility of implementing a variety of machine cycles at different speeds, smoothly, safely, and productively. As electrohydraulic proportional control valves are mostly actuated using solenoid drivers (Section 3.1.6) to control the spool positions, they are also a type of **solenoid control valve**.

Another type of electrohydraulic control actuation is the servo valve, which is also called the **servo control valve**. Servo control is not a new technology; actually, servo valves were first used in the 1940s. By nature, servo valves are also a type of proportional control valve and can achieve continuous control of the pressure or the flow from zero up to their maximum levels. Such valves commonly use a torque motor (as introduced in Section 3.1.6) in conjunction with sophisticated electronics and closed-loop systems to control the valve position. The feedback controller of the valve can correct the control signal in terms of the feedback signals of the position or force on the actuators (hydraulic cylinders or motors) being controlled, and the servo motor can respond to such corrected control signals promptly. A servo control system can therefore operate with high accuracy, repeatability, and frequency response with very low hysteresis and achieve precise control of the actuator. This type of valve is normally much more expensive to use than solenoid valves but can achieve superior performance. It is commonly used in applications requiring high-precision control over the valve position, such as machine tools.

While both solenoid and servo valves offer proportional control capability on the valve position, a few characteristics of their performances separate their usage. The most distinguishable ones are the frequency response and therefore the control accuracy attributed mainly to actuating methods.

Actuated using a torque motor, a servo valve normally responds to a control signal more quickly and accurately than a solenoid driver can, and as a result there will be a a distinguishable difference in spool positions under a certain input control signal in terms of either increasing or decreasing. As illustrated in Figure 9.9, while an ideal proportional valve should hold a perfect linear relationship between the input control signal and the controlled flow passing through the valve, a servo valve can approximately achieve the desired performance, with a small error attributed to its fairly linear modulating gain and the low hysteresis. In comparison, a solenoid valve often presents a large deadband, highly nonlinear modulating gain, and high hysteresis, which would normally result in a larger control error. Such superior performance of a servo valve derives from a control system that measures its own output to ensure that it can quickly and accurately follow the input

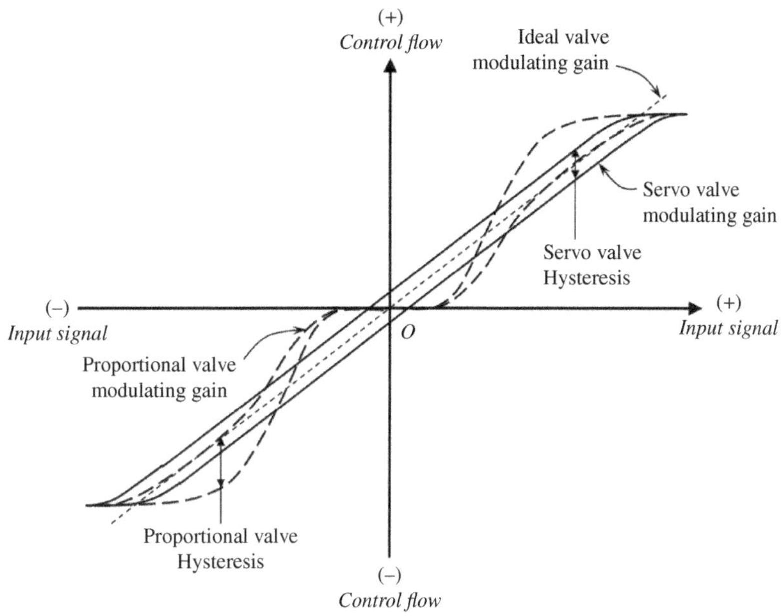


**FIGURE 9.9**
Modulating performance comparison of servo and proportional control valves.

command signal. A servomechanism can be designed to control either motion or pressure in a hydraulic system. Because of its high performance, closed-loop servo systems are gaining acceptance in machine automation where there is a demand for higher precision, faster operation, and simpler adjustment, such as for machine tools, material-handling equipment, and steel-rolling mills.

Like the hydraulic actuators, the dynamic responses of electrohydraulic control valves will also affect the performance of the entire system. As discussed in Section 8.2, the corresponding output value of an electrohydraulic control valve to a specific control input can be described using a transfer function. However, the order of a valve dynamic system is strongly influenced by the hydraulic natural frequency of the fluid power system. When the bandwidth of a valve is close to the hydraulic natural frequency of the system, the dynamic response of this valve can be approximately described using a second-order transfer function. When a valve can respond 5 to 10 times faster than the system natural frequency, it can reasonably be treated as a first-order element. Valves that can respond more than 10 times faster are often treated simply as a proportional gain in the system. In engineering practices, the response time of many commonly used solenoid valves are in the range of 50 to 200 ms, and many servo valves are in the range of 5 to 25 ms. As the natural frequency of many hydraulic systems is lower than 10 Hz, we may reasonably assume that many solenoid valves perform like a second-order element, but servo valves are commonly acting like a proportional gain in an electrohydraulic control system.

### 9.2.2 Velocity Controls

The velocity control on a hydraulic actuator is a common application in electrohydraulic control. Due to the nature of hydraulic power transmission, velocity controls are normally accomplished through flow control, using either a control valve or a variable displacement pump (or motor). The use of a variable-displacement pump to control the actuator velocity works well for circuits where one pump drives one actuator, or in cases where there are multiple actuators in a circuit but only one actuator moves at a time. However, that is not the case in many circuits, and valve-controlled actuator velocity (flow) control is commonly used in many applications.

As described in Chapters 3 and 8, a valve-controlled hydraulic system is commonly maneuvered by regulating an adequate amount of flow to the hydraulic actuator to be controlled using a spool valve, actuated by either a solenoid driver or a servo driver. Upon receiving a control signal, the solenoid or servo drive will push the spool away from its neutral position to open the flow passages in proportion to the input control signal; this will allow a certain flow rate to pass through the valve in terms of the opened flow passage area, as well as the up- and downstream pressures. This controlled flow rate can be quantified using a flow control equation of the spool valve (derived from an orifice equation as described in Section 8.1.2), as follows:

$$Q_L = C_d w_v x_v \sqrt{\frac{1}{\rho}(p_u - p_d)} \tag{9.11}$$

In the equation, $C_d$ is the orifice coefficient (often set $C_d = 0.7$ for spool valves in engineering practice); $w_v$ is the spool wet perimeter; $x_v$ is the spool displacement; and $p_u$ and $p_d$ are up- and downstream pressures across the valve.

Equation (9.11) reveals the principle of a valve-controlled hydraulic power transmission; namely, it controls the power transmission via flow control. When converting the flow control at the valve to motion control at the actuator, which receives the controlled flow from the valve, the directly controlled parameter in such a system is the valve opening that controls the flow rate supplied to the actuator for obtaining the desired velocity. It means that we could modulate the velocity control of the actuator in corresponding to the valve spool displacement as determined by Eq. (9.11). Section 8.2.2 has explained how to derive a velocity control response transfer function for a valve-controlled cylinder. This function can serve as the system dynamic model, as well as the system transfer function of velocity control of a valve-controlled hydraulic cylinder system:

$$\frac{\dot{Y}}{X_v} = \frac{\dfrac{k_x}{A_L}}{\dfrac{s^2}{\omega_h^2} + 2\dfrac{\xi_h}{\omega_h}s + 1} \tag{9.12}$$

This transfer function reveals that the velocity control of a valve-controlled hydraulic cylinder is a second-order system, and its speed transient can reach a new steady state after a step change in less than four cycles of oscillation if the system is properly damped. The integral term in the transfer function makes the velocity gain, that is, the piston motion velocity to valve spool displacement, be proportional to its velocity amplification coefficient $k_x/A_L$ under steady state. The magnitude represents the sensitivity of controlling the cylinder displacement using the selected valve and will directly affect system stability, responding speed, and control accuracy. Increasing this gain may improve the responding speed and control accuracy but may also lead to deterioration of system stability.

### 9.2.3 Position Controls

Position control is probably the most common application of hydraulic system controls. As explained in Section 9.2.2, hydraulic power transmission is by nature a flow (and therefore a velocity) control system, and the position control in such a system is often realized by the accumulation of fluid supplied into an actuator. It means that a position control can be treated as being an integral operation in the dynamics of the hydraulic actuating operation. As discussed in Section 8.2.2, the position control dynamic response for a valve-controlled cylinder system can be expressed using a third-order transfer function:

$$\frac{Y}{X_v} = \frac{\dfrac{k_x}{A_L}}{s\left(\dfrac{s^2}{\omega_h^2} + 2\dfrac{\xi_h}{\omega_h}s + 1\right)} \tag{9.13}$$

The additional integral term in the transfer function makes the position gain that determines the time needed for the piston to move to the position set point from the current position, corresponding to the valve's velocity amplification coefficient $k_x/A_L$ under steady state.

This transfer function reveals that system dynamics will be strongly influenced by three key parameters ($\omega_h$, $\xi_h$, and $k_x$) of the hydraulic control system. Among them, the hydraulic damping ratio can often affect the control performance more directly than the other two as its changing value corresponds to the position change of the actuator as defined by the previous chapter (refer to Eq. 8.36).

As mentioned earlier, a hydraulic system is actuated by supplying pressurized hydraulic fluid to the actuator and is a velocity control system by nature. Thus, accomplishing accurate position control in such a velocity control system requires either a very precise total flow control supplied to the actuator or having some position-tracking capacity using different sensing technologies.

### 9.2.4 Force Controls

Force control is also used widely in hydraulic control, probably because of its unique ability to implement both passive and active force control. In passive force control, the input signal is often a function of the motion of the load being driven, whereas in active force control, its control input is independent of the motion of the load. To distinguish those two types of force control, we can call passive control the driven force control, as the system behavior is influenced by the dynamics of the load motion, and we can call active control the load force control, as its dynamic behavior is independent of the dynamics of the load motion.

Because the force control in a hydraulic system is accomplished by controlling the pressure of the supplied fluid, the force control transfer function for a valve-controlled cylinder system can therefore be expressed in general as follows:

$$\frac{P_L}{X_v} = \frac{\dfrac{k_x}{A_L}\left(\dfrac{s^2}{\omega_m^2} + 2\dfrac{\xi_m}{\omega_m}s + 1\right)}{\left(\dfrac{1}{\omega_h s} + 1\right)\left(\dfrac{s^2}{\omega_c^2} + 2\dfrac{\xi_c}{\omega_c}s + 1\right)} \tag{9.14}$$

where $\omega_m$ and $\xi_m$ are the natural frequency and damping ratio of the load system being driven by the hydraulic system; $\omega_c$ and $\xi_c$ are the comprehensive natural frequency and damping ratio of system, including both mechanical and hydraulic components forming the system; and $\omega_h$ is the hydraulic natural frequency at the operating point.

As active force control can be accomplished independent of the load motion (and actually can often be accomplished without a load motion), the transfer function of an active force control for a valve-controlled cylinder system can therefore be simplified as follows:

$$\frac{P_L}{X_v} = \frac{\dfrac{k_x}{A_L}}{\left(\dfrac{1}{\omega_h s} + 1\right)} \tag{9.15}$$

Therefore, to simplify the analysis in control system design in many applications, we can reasonably treat an active force control system as a first-order control system.

## 9.3 Basic Methods for Electrohydraulic System Controls

### 9.3.1 Bang-Bang Control

**Bang-bang control** (Figure 9.10) is a type of control system that simply turns the hydraulic valve on or off when a predetermined set point has been reached, with or without feedback. Therefore, it is also called an on–off control and is commonly used in hydraulic system controls, especially in many position control applications.

One common use of bang-bang control has no feedback, which is often augmented by human operators who close the loop using their eye–hand coordination to control the actuator. While the actuating velocity can vary under changing system loads, it is often very difficult to achieve accurate control at an arbitrary position, except when the cylinder is fully extended or retracted using this method. However, it is still an adequate method for many applications in hydraulic system control, such as bucket control on a hydraulic excavator.

To furnish more reliable control to electrohydraulic systems when using bang-bang control, **discrete feedback**, using either limit switches or other position detectors, is sometimes used to improve control performance. However, major limitations of this approach are its inflexibility in changing control set points and the fact that it can normally mount only a limited number of detectors where they are required. Therefore, use of some types of programmable controllers is often necessary to provide the needed flexibility in achieving satisfactory control performance.

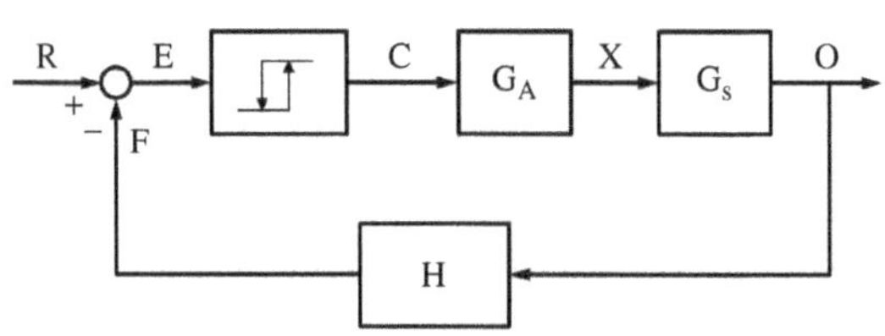


**FIGURE 9.10**
System block diagram of a feedback bang-bang control system.

The most flexible form of feedback is continuous position or velocity feedback. **Continuous feedback** can come from an analog sensor, such as an encoder and a linear displacement transducer (LDT). Continuous feedback can help bang-bang control to achieve good control performance in applications, especially for those applications where motion control must be accurate and repeatable.

### 9.3.2 Modulated Feedforward Control

Spool valves are commonly used in many valve-controlled electrohydraulic systems, and the flow gain from the valve corresponding to the spool stroke controlled by electrically driven actuators is often highly nonlinear, including the deadzone, modulating zones, and saturation zones, resulting from physical configurations of those valves. As explained in Chapter 3, such a high nonlinearity can be described using a **valve transform** curve, presented in the form of flow passing through the valve at different **spool strokes** (Figure 3.8). Such a valve transform curve reveals a few key flow control characteristics of a spool valve controlled system, such as **response delay, modulation gains**, and **flow capacity** in both directions of the spool stroke. One commonly applied control method to compensate for such valve-induced high nonlinearity is the use of **modulated control.**

In practice, modulated control often uses an **inverse valve transform** to determine the appropriate control input to the spool valve in terms of the desired flow rate to compensate for the nonlinearity, and it is a type of feedforward control. The inverse valve transform ($u$) shown in Figure 9.11 is normally converted from an empirical valve transform ($v$) of the electrically controlled spool valve used in the system and can be mathematically expressed using the following equation:

$$u = f^{-1}(v) \tag{9.16}$$

The resulting inverse valve transform is used as the gain being scheduled for achieving the best possible control performance from the feedforward control and therefore may also be called a gain-scheduling control. A typical inverse valve transform is often

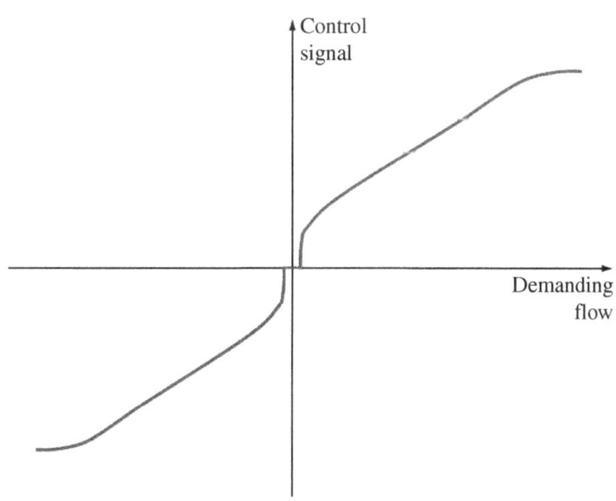


**FIGURE 9.11**
A typical inverse valve transform often used as the scheduled gains for electrohydraulic systems modulated feedforward control.

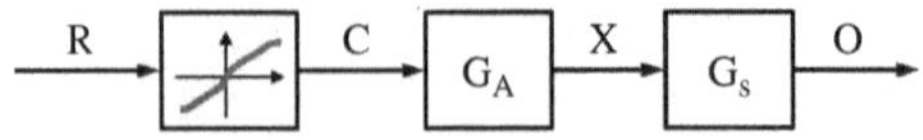


**FIGURE 9.12**
System block diagram of a basic modulated feedforward control system.

attained by modifying an experimentally obtained valve transform of the particular valve by setting the demanding flow rate supplied as the input to the controller, and the required spool position (or the required control signal) to get the required flow as the control signal to drive the actuator. To ensure that the valve can be securely closed, a very narrow no-control-output zone in the inverse valve transform at the neutral position is often set. A sizeable jump in the control signal is often designed from zero to an appropriate level of control input for overcoming the valve deadzone to get a prompt flow control response. The gains are so scheduled in the inverse valve transform that it helps to compensate for the nonlinear and asymmetric flow gains inherited from the structural configurations of typical hydraulic spool valves. From a control system design point of view, a **gain-scheduling** control treats a nonlinear system as a linear one near a specific operating point and uses a family of linear controllers for different operating points to achieve satisfactory control performance from a highly nonlinear hydraulic system.

While many of the modulated feedforward controls for electrohydraulic systems do not require using sensors to track performance (as shown in the example provided in Figure 9.12), there are applications using sensors that provide feedback for achieving more accurate motion control. In some applications, performance of modulated feedforward control can be enhanced by reducing system deadband and improving modulation quality. System deadband is defined as the command level corresponding to the first motion of the cylinder actuator, and modulation quality includes gain variation and linearity on velocity control under varying loads.

Modulated feedforward controls are used in many velocity control applications, either with or without feedback information.

### 9.3.3 PID Control

PID controller stands for **proportional-integral-derivative controller** and is a classical control method with well-developed controller design and tuning methodologies. As one of the most commonly applied control methods, PID controllers have been applied in many fields of automation, including electrohydraulic system controls. Figure 9.13 shows the system block diagram of a typical PID controller that utilizes a feedback signal reflecting the actual operational state of the hydraulic system to improve accuracy in motion control. When a PID controller receives control command (the set point), the controller will first compare with the feedback signal to detect the difference (the error) between the set point and the feedback, and then make a correction on the outputting control signal in order to minimize the error. PID control is the complete form of three-mode feedback controls and can automatically make accurate and responsive corrections to a control function in response to the detected error (P), the error-changing rate (D), and the accumulated total error (I) at any particular moment. Because of its ability to make control adjustments in response to errors in actual system outputs, PID control can effectively reduce the undesirable behaviors induced by external disturbances.

In practice, PID controllers may be applied in a series of different forms, such as P, PI, PD, and PID, to meet different control needs. The proportional (P) control creates a control

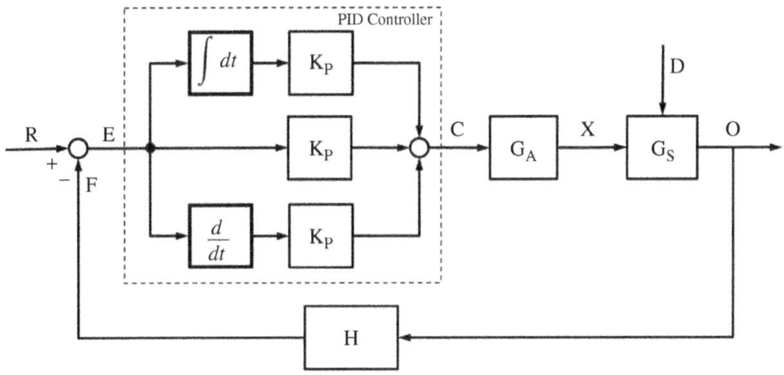


**FIGURE 9.13**
System block diagram of a basic proportional-integral-derivative (PID) control system.

output in proportion to the error with a constant gain, $K_P$. Proportional control is usually simple to tune and implement, and is frequently used in many electrohydraulic system control applications.

Proportional-integral (PI) is another control method that often finds its way into use in electrohydraulic system control. As its integral gain, $K_I$, helps to reduce the error to zero, this type of controller is expected to result in a higher tracking accuracy than P-type controllers, but a 90° phase lag of the gain may also result in a stability issue in the system. The proportional gain in a PI controller is therefore used to increase system stability and responsiveness.

Proportional-derivative (PD) control provides a derivative gain, $K_D$, in a way similar to that of a PI controller, but it offers an improvement in system stability, which may sometimes be important in controlling hydraulic systems. As the derivative gain has a 90° phase lead, it may cause the system to be sensitive to noises; therefore, a PD controller is less commonly used in hydraulic system controls.

### 9.3.4 A Few Other Control Methods

A few other methods have also found a place in electrohydraulic system controls to satisfy performance requirements in some conditions. One such method is feedforward-plus-PID (FPID) control, which is actually an integration of a conventional PID controller with an open-loop feedforward controller (Figure 9.14). In this integrated controller, the

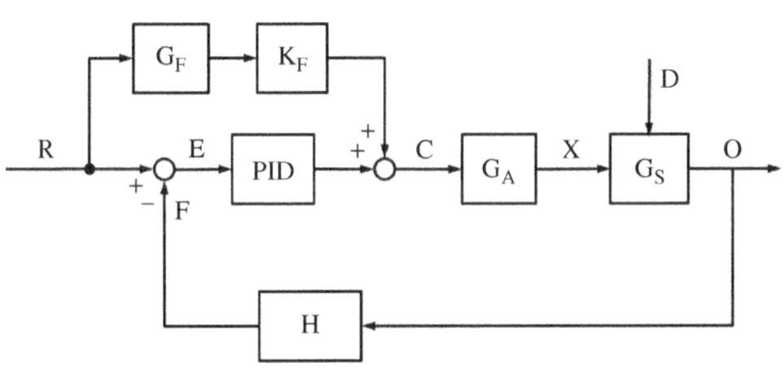


**FIGURE 9.14**
System block diagram of a feedforward-PID (FPID) control system.

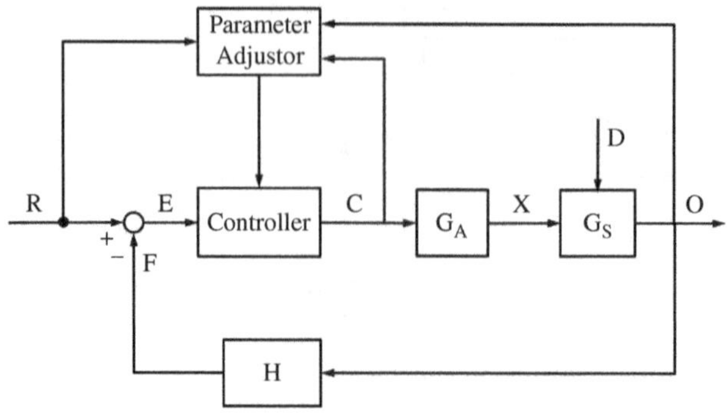


**FIGURE 9.15**
A control scheme of adaptive control.

feedforward loop uses an inverse valve transform function to support forming a responsive control signal to the control set point, which makes it possible to use an initial bias to the control signal for compensating the system deadband and other nonlinearity features inherent to a hydraulic control system. The PID loop formulates a reactive correction signal in terms of the detected error, which performs as a fine adjuster to correct the control errors induced by any cause. Because of those features, FPID offers a practical means to achieve high accurate trajectory tracking from a highly nonlinear hydraulic system.

Capable of automatically adjusting control parameters in reaction to variations in the system dynamics and/or to external disturbances to achieve predetermined design specifications, adaptive control has gained momentum in electrohydraulic system control applications. In order to adapt the aforementioned variations, this control method needs to estimate the value of the system parameters online and then adjust the control parameters based on the outcomes from the estimation. This operation requires having a second feedback loop to perform controller parameters adaptation separate from the standard control feedback loop to implement normal control functions (Figure 9.15). Adaptive control has even been merged with intelligent techniques such as fuzzy and neural networks.

## References


1. Dort, R.C., Bishop, R.H. *Modern Control Systems (12th Ed.).* Prentice-Hall, Upper Saddle River, NJ (2011).
2. Gao, Y., Huang, R., Zhang, Q. A comparison of three steering controllers for off-road vehicles. *Proceedings of the Institute of Mechanical Engineers, Part D: Journal of Automobile Engineering,* 222: 2321–2336 (2008).
3. Gao, Y., Jin, Y., Zhang, Q. Motion planning based coordinated control for hydraulic excavators. *Chinese Journal of Mechanical Engineering,* 22: 97–101 (2009).
4. Hu, H., Zhang, Q. Realization of programmable control using a set of individually controlled electrohydraulic valves. *International Journal of Fluid Power,* 3: 29–34 (2002).
5. Hu, H., Zhang, Q. Development of a programmable E/H valve with a hybrid control algorithm. *Society of Automotive Engineers Transactions: Journal of Commercial Vehicles,* 111: 413–419 (2002).
6. Lee, H.-W., Cho, B.-H., Lee, W.-H. A study on response improvement of proportional control solenoid valve for automatic transmission. In: Proceedings of Seoul 2000 FISITA World Automotive Congress, June 12–15. Seoul, Korea (2000).

7. Manring, N.D. *Hydraulic Control Systems.* John Wiley & Sons, New York (2005).
8. Norvelle, F.D. *Electrohydraulic Control Systems.* Prentice-Hall, New York (2000).
9. Pinsopon U., Hwang, T., Cetinkunt, S., Ingram, R., Zhang, Q., Cobo, M., Koehler, D., Ottman, R. Hydraulic actuator control with open-center electrohydraulic valve using a cerebellar model articulation controller neural network algorithm. *Journal of Systems and Control Engineering,* 213: 33–48 (1999).
10. Qiu, H., Zhang, Q. Feedforward-Plus-PID controller for an off-road vehicle electrohydraulic steering systems. *Proceedings of the Institute of Mechanical Engineers, Part D: Journal of Automobile Engineering,* 217: 375–382 (2003).
11. Qiu, H., Zhang, Q., Reid, J.F. Fuzzy control of electrohydraulic steering systems for agricultural vehicles. *Transactions of the American Society of Agricultural Engineers,* 44: 1397–1402 (2001).
12. Rovira-Más, F., Zhang, Q. Fuzzy logic control of an electrohydraulic valve for auto-steering off-road vehicles. *Proceedings of the Institute of Mechanical Engineers, Part D: Journal of Automobile Engineering,* 222: 917–934 (2008).
13. Tanaka, H. Fluid power control technology-present and near future. *JSME Int. J., Series C,* 37: 629–637 (1994).
14. Walters, R.B. *Hydraulic and Electric-Hydraulic Control Systems (2nd Ed.).* Kluwer Academic Publishers, Norwell, MA (2000).
15. Watton, J. *Fluid Power Systems, Modeling, Simulation, Analog and Microcomputer Control.* Prentice-Hall, New York (1989).
16. Zhang, Q. A generic fuzzy electrohydraulic steering controller for off-road vehicles. *Proceedings of the Institute of Mechanical Engineers, Part D: Journal of Automobile Engineering,* 217: 791–799 (2003).
17. Zhang, Q., Cetinkunt, S., Hwang, T., Pinsopon, U., Cobo, M.A., Ingram, R.G. Use of adaptive algorithms for automatic calibration of electrohydraulic actuator control. *Applied Engineering in Agriculture,* 17: 259–265 (2001).
18. Zhang, Q., Goering, C.E. *Fluid power system.* In: Bishop, R. (ed.), *The Mechatronics Handbook,* CRC Press, Boca Raton, FL, pp. 10–11 ~ 10–14 (2001).
19. Zhang, Q., Meinhold, D.R., Krone, J.J. Valve transform fuzzy tuning algorithm for open-center electrohydraulic systems. *Journal of Agricultural Engineering Research,* 73: 331–339 (1999).
20. Zhang, Q., Wu, D., Reid, J.F., Benson, E.R. Using model recognition to design an electrohydraulic steering controller for off-road vehicles. *Mechatronics,* 12: 845–858 (2002).

## Exercises

**9.1** How does an automatic electrohydraulic control system work?

**9.2** What are the key elements of an automatic electrohydraulic control system?

**9.3** What is the most critical characteristic that separates an open-loop and a closed-loop control system?

**9.4** What are the main features of open-loop control?

**9.5** What are the main features of closed-loop control?

**9.6** How does feedforward control work to control the valve-controlled electrohydraulic system?

**9.7** How does feedback control work in controlling a valve-controlled electrohydraulic system?

**9.8** Try to draw a system block diagram for a valve-controlled hydraulic cylinder velocity control system, and use the block diagram to explain how the system works.

**9.9** Try to draw a system block diagram for a valve-controlled hydraulic cylinder position control system, and use the block diagram to explain how the system works.

**9.10** What is the state of damping for systems having the following transfer functions?

(a) $G(s)=\dfrac{5}{s^2-6s+16}$

(b) $G(s)=\dfrac{10}{s^2+s+100}$

(c) $G(s)=\dfrac{2s+1}{s^2+2s+1}$

(d) $G(s)=\dfrac{3s+20}{s^2+2s+20}$

**9.11** What are the magnitudes and phases of the system having the following transfer functions?

(a) $G(s)=\dfrac{5}{s+2}$

(b) $G(s)=\dfrac{2}{s(s+1)}$

**9.12** What will be the steady-state response of an E/H system with a transfer function $G(s)=\frac{1}{s+2}$ when subject to a sinusoidal input $3\sin(5t+30°)$?

**9.13** Try to draw the asymptotes of the Bode plots for an electrohydraulic system having a transfer function $G(s)=\frac{10}{s(0.1s+1)}$.

**9.14** Try to draw the asymptotes of the Bode plots for an electrohydraulic system having a transfer function $G(s)=\frac{1}{(2s+1)(0.5s+1)}$.

**9.15** A control system is closed-loop with a forward-path transfer function of $1/(s+1)$ and a negative feedback path transfer function of 4s. What is the overall transfer function of the system?

**9.16** A control system is closed-loop with a forward-path transfer function of $1/(s^2+s)$ and a negative feedback path transfer function of 3s. What is the overall transfer function of the system?

**9.17** Try to determine the steady-state step error for a feedback control system with $G(s)=\frac{1}{s+1}$ and $H(s)=\frac{s+2}{s+10}$.

**9.18** Find the amplitude response in decibels when the spool in a spool valve is being actuated at a sinusoidal motion of 6 mm under a 9 mm sinusoidal commanding input.

**9.19** Assume a feedback control system (Figure 9.16) is used to control a valve-controlled hydraulic motor circuit, if the feedforward loop gain of this controller G is

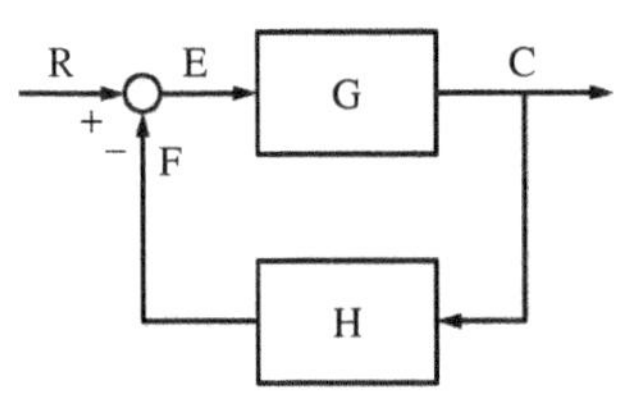


**FIGURE 9.16**
A feedback control system for a valve-controlled hydraulic motor circuit.

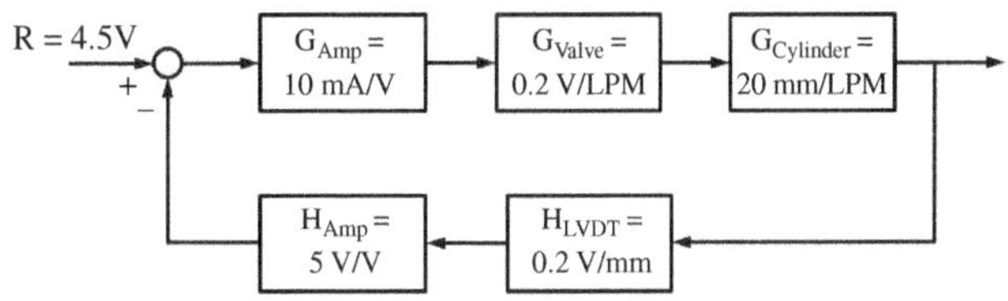


**FIGURE 9.17**
A feedback control system for a valve-controlled hydraulic cylinder circuit.

300 $\text{rmp} \cdot \text{V}^{-1}$ and its feedback loop gain H is 0.2 $\text{V} \cdot \text{rpm}^{-1}$. Try to determine the controlled motor speed when the commanding input is 3.5V.

**9.20** Assume a feedback control system (Figure 9.17) is used to control a valve-controlled hydraulic cylinder circuit. If the gains of each element in the controller are as shown in the system block diagram, try to determine the controlled cylinder position under the inputting command.

# *Appendix A: Hydraulic Power Formulas*

## Newton's Law of Motion

$$F = ma$$

When SI units are used: $F$ = force (N); $m$ = mass (kg); $a$ = acceleration ($m \cdot s^{-2}$).

When English units are used: $F$ = force ($lb_f$); $m$ = mass ($lb_m$); $a$ = acceleration ($in/s^2$).

## Pascal's Law of Pressure

$$p = \frac{F}{A}$$

When SI units are used: $p$ = pressure (Pa); $F$ = force (N); $A$ = area ($m^2$).

When English units are used: $p$ = pressure (psi); $F$ = force ($lb_f$); $A$ = area ($in^2$).

## Hydraulic Cylinder Speed (*An Application of Fluid Continuity Theory*):

$$v = \frac{Q}{A}$$

When SI units are used: $v$ = linear velocity ($m \cdot s^{-1}$); $Q$ = flow rate ($m^3 \cdot s^{-1}$); $A$ = actuator pressure-bearing area ($m^2$).

When English units are used: $v$ = linear velocity (in/s); $Q$ = flow rate ($in^3/s$); $A$ = actuator pressure-bearing area ($in^2$).

## Hydraulic Pump/Motor Speed (*An Application of Fluid Continuity Theory*)

$$n = 60 \times \frac{Q}{D_v}$$

When SI units are used: $n$ = rotating velocity (rpm); $Q$ = flow rate ($m^3 \cdot s^{-1}$); $D_v$ = hydraulic pump or motor displacement ($m^3$).

When English units are used: $n$ = rotating velocity (rpm); $Q$ = flow rate ($in^3/s$); $D_v$ = hydraulic pump or motor displacement ($in^3$).

## Hydraulic Force in Cylinder

$$F = (p_1 A_1 - p_2 A_2)$$

When SI units are used: $F$ = force (N); $p_1, p_2$ = pressure (Pa); $A_1, A_2$ = piston cap-end or rod-end area ($m^2$).

When English units are used: $F$ = force ($lb_f$); $p_1, p_2$ = pressure (psi); $A_1, A_2$ = piston cap-end or rod-end area ($in^2$).

## Hydraulic Torque in Pump and Motor

$$T = pD_v$$

When SI units are used: $T$ = torque ($N \cdot m$); $p$ = pressure (Pa); $D_v$ = pump or motor displacement ($m^3$).

When English units are used: $T$ = torque ($lb_f \cdot in$); $p$ = pressure (psi); $D_v$ = pump or motor displacement ($in^3$).

## Orifice Equation

(An Application of Bernoulli's Equation of Energy Conservation)

In SI units: $Q = C_d A \sqrt{\frac{2}{\rho}(p_1 - p_2)}$

where $Q$ = flow rate ($m^3 \cdot s^{-1}$); $A$ = orifice area ($m^2$); $\rho$ = fluid density ($kg \cdot m^{-3}$); $p_1, p_2$ = pressure (Pa); $C_d$ = orifice coefficient ($C_d = 0.6 \sim 0.8$).

In English units: $Q = C_d A \sqrt{\frac{2g}{\gamma}(p_1 - p_2)}$

where $Q$ = flow rate ($in^3/s$); $A$ = orifice area ($in^2$); $g$ = gravitational constant ($g = 386\ in/s^2$); $\gamma$ = fluid specific weight ($lb_f/in^3$); $p_1, p_2$ = pressure (psi); $C_d$ = orifice coefficient.

## Hydraulic Power

In SI units: $P_h = pQ$
where $P_h$ = hydraulic power (W); $p$ = pressure (Pa); $Q$ = flow rate ($m^3 \cdot s^{-1}$).
In SI units: $P_h = \frac{pQ}{1714}$
where $P_h$ = hydraulic power (hp); $p$ = pressure (psi); $Q$ = flow rate (gpm).

## Mechanical Power in Pump or Motor

In SI units: $P_m = \omega T = \frac{nT}{9554}$
where $P_m$ = mechanical power (kW); $\omega$ = angular velocity ($s^{-1}$); $T$ = torque ($N \cdot m$); $n$ = rotating velocity (rpm).
In English units: $P_m = \frac{nT}{5252}$
where $P_m$ = mechanical power (hp); $T$ = torque ($lb_f \cdot ft$); $n$ = rotating velocity (rpm).

## Mechanical Power in Cylinders

In SI units: $P_m = (p_1 A_1 - p_2 A_2) v$
where $P_m$ = mechanical power (W); $p_1, p_2$ = pressure (Pa); $A_1, A_2$ = piston cap-end or rod-end area ($m^2$); $v$ = piston velocity ($m \cdot s^{-1}$).
In English units: $P_m = \frac{(p_1 A_1 - p_2 A_2) v}{550}$
where $P_m$ = mechanical power (hp); $p_1, p_2$ = pressure (psi); $A_1, A_2$ = piston cap-end or rod-end area ($in^2$); $v$ = piston velocity (ft/s).

# Appendix B: Orifice Area Formulas of a Few Typical Shaped Orifices

1. *Orifice area of round holes*

$$A = \frac{\pi}{4} d^2$$

where $d$ = diameter of the orifice hole.

2. *Orifice area of a disk valve*

$$A = \begin{cases} \pi d x, & x < \dfrac{d}{4} \\ \dfrac{\pi}{4} d^2, & x \geq \dfrac{d}{4} \end{cases}$$

where $d$ = diameter of the flow passage in front of the valve, $x$ = valve opening.

3. *Orifice area of a needle valve*

$$A = \pi x \tan\frac{\alpha}{2}\left(d - x\tan\frac{\alpha}{2}\right)$$

where $d$ = diameter of the flow passage in front of the valve, $x$ = needle valve opening, $\alpha$ = flow angle.

4. *Orifice area of a simple ball valve*

$$\text{when } D \geq 1.3d$$

where $d$ = diameter of the flow passage in front of the valve, $D$ = diameter of the ball, $x$ = valve opening.

5. *Orifice area of a simple spool valve*

$$A = \pi d(x - x_0) \text{ when } x \geq x_0$$

where $d$ = diameter of the spool, $x$ = spool displacement, $x_0$ = spool overlap.

# *Appendix C: Some Useful Conversion Factors*

## Energy

$$1(\text{Btu}) = 1.055\ (\text{kJ})$$
$$1(\text{Btu}) = 9{,}331\ \left(\text{lb}_f \cdot \text{in}\right)$$
$$1(\text{J}) = 1\ (\text{N} \cdot \text{m})$$

## Power

$$1(\text{hp}) = 745.6\ (\text{W})$$
$$1(\text{hp}) = 550\ \left(\text{ft} \cdot \text{lb}_f/\text{s}\right)$$
$$1(\text{W}) = 1\ \left(\text{J} \cdot \text{s}^{-1}\right)$$

## Volume

$$1(\text{gal}) = 3.785\ (\text{L})$$
$$1(\text{gal}) - 231\ \left(\text{in}^3\right)$$
$$1(\text{L}) = 1 \times 10^{-3}\ \left(\text{m}^3\right)$$

## Force

$$1\left(\text{lb}_f\right) = 4.448\ (\text{N})$$
$$1(\text{N}) = 1\ \left(\text{kg} \cdot \text{m} \cdot \text{s}^{-1}\right)$$

## Torque

$$1(\text{lb}_\text{f} \cdot \text{in}) = 0.113\ (\text{N} \cdot \text{m})$$

## Mass

$$1(\text{lb}_\text{m}) = 0.454\ (\text{kg})$$
$$1(\text{lb}_\text{m}) = 1/386\ (\text{lb}_\text{f} \cdot \text{s}^2/\text{in})$$

## Pressure

$$1(\text{psi}) = 6{,}895\ (\text{pa})$$
$$1(\text{psi}) = 1(\text{lb}_\text{f}\ /\ \text{in}^2)$$
$$1(\text{Pa}) = 1(\text{N} \cdot \text{m}^{-2})$$

## Velocity

$$1(\text{ft/s}) = 0.305(\text{m} \cdot \text{s}^{-1})$$

## Gravity

$$1(\text{g}) = 386(\text{in/s}^2)$$
$$1(\text{g}) = 9.8(\text{m} \cdot \text{s}^{-2})$$

# *Appendix D: Solutions to Selected Exercise Problems*

## Chapter 1

| | | | | |
|---|---|---|---|---|
| **1.9** | a) 10 MPa | b) 5 kN | c) 1 cm | |
| **1.10** | a) 5.0 MPa | b) 2.5 kN | c) 2 strokes | |
| **1.11** | a) 2.5 kN | b) 2.0 $mm\cdot s^{-1}$ | | |
| **1.12** | a) 1256 N | b) 844 N | c) 412 N | |
| **1.13** | a) 1178 N | b) 1678 N | c) 855 kPa | |
| **1.14** | a) 39.2 $kN\cdot m$ | b) 625 kPa | c) 4.0 kW | d) 377 $L\cdot min^{-1}$ |
| **1.15** | 20.6 $L\cdot min^{-1}$ | | | |
| **1.16** | 6.1 $mm^2$ | | | |
| **1.17** | a) 5.0 kW | b) 5.0 kW | c) 5.9 kW | |
| **1.18** | a) 15.0 kW | b) 12.75 kW | | |
| **1.19** | a) 8.0 kW | b) 86 $N\cdot m$ | | |
| **1.20** | a) 7.5 $L\cdot min^{-1}$ | | | |

## Chapter 2

| | | | | |
|---|---|---|---|---|
| **2.8** | 10.6 $L\cdot min^{-1}$ | | | |
| **2.9** | a) 112.5 $L\cdot min^{-1}$ | b) 12.5 $L\cdot min^{-1}$ | c) 88.9% | |
| **2.10** | a) 80.0% | b) 93.3% | c) 82.9% | |
| **2.11** | a) 199 $N\cdot m$ | b) 41 $N\cdot m$ | | |
| **2.12** | a) 90.5% | b) 76.5% | | |
| **2.13** | a) 83.8% | b) 89.3% | c) 93.9% | d) 83.8% |
| **2.14** | 93.8% | | | |
| **2.15** | 84.9% | | | |
| **2.16** | a) 8.1 $L\cdot min^{-1}$ | b) 1.1 kW | | |
| **2.17** | 63.6 $L\cdot min^{-1}$ | | | |
| **2.18** | a) 83 cc | b) 100 $L\cdot min^{-1}$ | c) 396 $N\cdot m$ | |
| **2.19** | a) 83.3% | b) 90.9% | c) 91.6% | |
| **2.20** | a) 4.75 kW | b) 749 rpm | | |

## Chapter 3

| | | | |
|---|---|---|---|
| **3.10** | a) 4 MPa | b) 9 MPa | |
| **3.11** | $71.4\ L \cdot min^{-1}$ | | |
| **3.12** | 0.41 | | |
| **3.13** | 1.2 kW | | |
| **3.14** | a) 7.5 MPa | b) 10 MPa | |
| **3.15** | a) 16.7 MPa | b) 2.0 cm | |
| **3.16** | $23.7\ mm^2$ | | |
| **3.17** | a) $100\ L \cdot min^{-1}$ | b) $9.5\ mm^2$ | |
| **3.18** | a) 15 cm | b) 4 cm | c) 5 cm |
| **3.19** | a) $8.5\ kJ \cdot s^{-1}$ | b) 72.6 °C | |
| **3.20** | a) 87.9°C, cannot work properly | | b) 77.9°C, on the boundary, can work |

## Chapter 4

| | | | |
|---|---|---|---|
| **4.10** | a) 1.27 MPa; 2.15 MPa | b) $0.22\ m \cdot s^{-1}$; $0.24\ m \cdot s^{-1}$ | |
| **4.11** | a) $0.105\ m \cdot s^{-1}$; 50.9 kN | b) $0.059\ m \cdot s^{-1}$; 85.5 kN | |
| **4.12** | a) $0.083\ m \cdot s^{-1}$ | b) 47.7 kN | |
| **4.13** | a) $14\ L \cdot min^{-1}$ | b) 35.6 MPa | |
| **4.14** | $1.56\ m \cdot s^{-2}$; 360 kPa | | |
| **4.15** | $318.5\ N \cdot m$ | | |
| **4.16** | a) $300\ N \cdot m$ | b) 90% | |
| **4.17** | a) 7.54 L | b) 4.0 MPa | |
| **4.18** | a) 83.3%; 93.3%; 77.7% | b) 27.2 kW | |
| **4.19** | 64.2 kW | | |
| **4.20** | a) 6.0 kW | b) 10.0 kW | c) 0.38 L; $20.3\ L \cdot min^{-1}$ |
| | d) $22.5\ L \cdot min^{-1}$; 12 cc | | |

## Chapter 5

| | | |
|---|---|---|
| **5.11** | 0.16 m | |
| **5.12** | a) 3.75 kN | b) 1050 kg |
| **5.13** | a) 1.2 L | b) 17 L |
| **5.14** | 65 L | |
| **5.15** | a) 3.0 L | b) 0 L |
| **5.16** | a) $1.9\ L \cdot min^{-1}$ | b) 78.4 MPa |
| **5.17** | a) 1.73 | b) 1.41 |
| **5.18** | a) $0.3\ m \cdot s^{-1}$; 40 kN | b) $0.6\ m \cdot s^{-1}$; 20 kN |
| **5.19** | $\sqrt{2}$ | |
| **5.20** | a) 266 L | b) 434 L |

## Chapter 6

**6.14** 300 L (or 200~400 L)
**6.15** 99.98%; 99.00%; 98.67%; 99.00%
**6.16** 500, 89, 16, 3 and 1 counts of particles 5~15 μm, 15~25 μm, 25~50, 50~100 and >100 μm
**6.18** a) 23.8 kW b) 17.7°C
**6.19** a) 200 W b) 2.5°C
**6.20** 30.3 kW

## Chapter 7

**7.13** a) 2.0 MPa b) $0.39\ \mathrm{m \cdot s^{-1}}$
**7.14** a) $1.29\ \mathrm{cm^2}$ b) $0.82\ \mathrm{cm^2}$
**7.15** a) 4.4 MPa b) 5.2 MPa
**7.16** a) $0.15\ \mathrm{m \cdot s^{-1}}$; 113 kN b) $0.26\ \mathrm{m \cdot s^{-1}}$; 64 kN
**7.17** 75%
**7.18** 100%
**7.19** $645\ \mathrm{N \cdot m}$
**7.20** a) 210 cc; 79 cc b) 100%

## Chapter 8

**8.17** $\dfrac{b_0}{a_1 s + a_0}$

**8.18** a) 5 b) 1.0 c) $x = -(32t + 6)e^{-5t} + 6$

**8.19** $3(2K + 1) > T$

**8.20** 1

**8.21** $T(s) = \dfrac{-2}{s^5 + 2s^4 - s - 2}$

**8.22** a) $A_P p_1 = m\dfrac{d^2y}{dt^2} + f\dfrac{dy}{dt} + Ky$ b) $Q_1 = C_d k_v x\sqrt{\dfrac{2}{\rho}(p_s - p_1)}$

c) $Q_1 = A_P\dfrac{dy}{dt} + \dfrac{V_1}{\beta}\dfrac{dp_1}{dt}$

**8.23** a) $G(s) = \dfrac{K}{s\left(\dfrac{s^2}{\omega_n^2} + \dfrac{2\zeta}{\omega_n}s + 1\right)}$ b) $\omega_n = \sqrt{\dfrac{\beta A_{RE}^2}{V_{RE}m}}$ c) $\zeta = \dfrac{1}{2}\sqrt{\dfrac{V_{CE}f^2}{\beta m}}$

**8.24** a) $\dfrac{d}{dt}\ddot{x} + 2\zeta_h\omega_h\ddot{x} + \omega_h^2\dot{x} = K_q\omega_h^2 u$

## Chapter 9

**9.11** a) $\tan\phi = -\frac{\omega}{2}$ b) $\tan\phi = -\frac{1}{\omega}$

**9.12** $0.56\sin(5t - 38°)$

**9.15** $\frac{1}{5s+1}$

**9.16** $T(s) = \frac{1}{s(s+3)}$

**9.17** $e_{est} = \frac{1}{6}$

**9.18** $-3.52dB$

**9.19** 17.2 rpm

**9.20** 4.39 mm

# Index

Note: *Italicized* page numbers refer to figures, **bold** page numbers refer to tables.

## T



## U



## V